AMERICAN Plastic

AMERICAN  Plastic

A CULTURAL HISTORY

JEFFREY L. MEIKLE

RUTGERS UNIVERSITY PRESS

NEW BRUNSWICK, NEW JERSEY

LIBRARY OF CONGRESS CATALOGING-IN-PUBLICATION DATA

MEIKLE, JEFFREY L., 1949–

    AMERICAN PLASTIC : A CULTURAL HISTORY / JEFFREY L. MEIKLE.

      P.   CM.

    INCLUDES BIBLIOGRAPHICAL REFERENCES AND INDEX.

    ISBN 0-8135-2234-X (CLOTH : ALK. PAPER). — ISBN 0-8135-2235-8

(PBK. : ALK. PAPER)

    1. PLASTICS—HISTORY.   2. PLASTICS INDUSTRY AND TRADE—UNITED STATES—HISTORY.   I. TITLE.

TP1117.M45   1995

303.48′3—DC20   95-15187

                                    CIP

BRITISH CATALOGING-IN-PUBLICATION INFORMATION AVAILABLE

PUBLISHED BY RUTGERS UNIVERSITY PRESS, NEW BRUNSWICK, NEW JERSEY

DESIGNED BY KAROLINA HARRIS

MANUFACTURED IN THE UNITED STATES OF AMERICA

For Alice, Vanessa, and Jason,

and for my parents, Arlene and Wendell

C O N T E N T S

ILLUSTRATIONS

PLATES

FIGURES

Illustrations

*T*ry to imagine a world without plastic. At first it seems easy. Lots of things could be made of other materials. Eyeglasses, for example, could be made of glass and metal just as they were for centuries. Computer cases could be made of sheet metal. Kitchen counters could be surfaced in ceramic tile, and linoleum would do just fine for floors. Automobiles would become heavier and consume more fuel, but we could live with that. Glass milk bottles would enable us to avoid paper cartons with their irritating flakes of wax. Frisbees would not exist, nor would compact discs. The list could go on and on. Without belaboring the obvious, few of the objects we use daily would remain unchanged. Some would be impossible. We would inhabit a different world, parallel perhaps, but certainly not the same place. Commenting on plastic from the opposite perspective, novelist Mark Helprin imagined an inhabitant of the 1890s magically transported a hundred years into the future. As he awakens, unaware of what has happened to him, and becomes more alert (and more disoriented), he notices that "the way things were shaped, and the materials of which they were made, seemed almost otherworldly . . . everything seemed to have grown smooth, to have lost its texture." We are told that "never before had he seen plastic."[1]

It is hard to do justice to plastic because it serves so many functions, assumes so many guises, satisfies so many desires, and so quickly recedes into relative invisibility as long as it does its job well. In little more than a century, plastic has spread through the material world, moving from almost no presence at all to near ubiquity. Owing to our culture's ambivalent relationship with plastics—perceived variously as high-tech miracle materials and as cheap substitutes or imitations—the word plastic has taken on metaphoric significance beyond any direct reference to particular chemical substances used in manufacturing or construction or packaging. From the early twentieth century onward, plastic has signified greater human control over nature. Its innate formlessness has suggested the outlines of a material world ever more malleable in the face of human desire. At the same time, its proliferation in everyday life has echoed the increasing economic abundance of the American middle class. While some observers have celebrated a material transcendence to be achieved through plastic, others have criticized its wastefulness and superficiality. Either way, plastic extends beyond the purely material realm to the central meaning of a culture that is itself ever more malleable and inflationary.

To understand the historical development of plastic's presence in American life and consciousness requires a shifting sequence of perspectives as the narrative moves chronologically from the 1870s, when celluloid was introduced, to the final years of the twentieth century. The interdisciplinary strategies of the American Studies method come into play as the narrative itself takes on a certain plasticity, touching in turn on the histories of technology and invention, of industry and marketing, of industrial design and consumer culture. Inventors of new materials had to define them. They had to persuade manufacturers to use them in products and had to present them convincingly to the public who ultimately consumed them. In the process, a new industry came into being, its promoters perpetually caught between a sense of wonder at plastic's utopian potential as a democratizing agent, and a recurring suspicion that plastic exemplified the cheap, the shoddy, and the meretricious. Ordinary people only complicated matters by interacting with the definition process, sometimes rejecting plastic's laboratory image and demanding that new materials be domesticated or naturalized.

As plastic gradually appeared less a discrete product rationally controlled by its makers and more a catalyst or mirror of unpredictable change, the narrative shifts after 1945 to a broader cultural history with excursions into journalism, literature, art, social criticism, and speculation. When ecological concerns emerged on a popular scale around 1970, plastic's ability to transcend nature often no longer seemed utopian but instead simply disastrous. Even so, everyday experience became ever more malleable, ever more a matter of plasticity, with both place and personality apparently open to creative remolding and reshaping in a postmodern society. As the twentieth century wound to a close, the locus of plasticity shifted again, from polymers to electrons, from material plastic to the immaterial promise of virtual reality. The very plasticity of plastic throughout its history suggests that a variety of perspectives best reveals its significance as material and metaphor—a conclusion that yields the multiplicity of approaches of this study.

AMERICAN **P***lastic*

INTRODUCTION: A MATTER

OF DEFINITION

*N*atural limits of wood, stone, and metals long made it hard to conceive of material desires beyond the traditional. That situation changed in the twentieth century when chemists learned to synthesize substances that had never before existed and to specify their properties. With so many different plastics available to us we no longer recall just how radically such materials as vinyl and polyethylene extended the limits of possibility. The makers of Bakelite, the first synthetic plastic, called it "the material of a thousand uses" and used the mathematical symbol of infinity as a trademark. To an ever-increasing degree, the things of everyday life are molded, extruded, foamed, stamped, vacuum-formed, or otherwise fabricated of plastic. The very stuff of existence in the late twentieth century is synthesized in chemical refineries from petroleum, a universal medium of exchange. We live and work with machines, appliances, and furnishings whose visual appearances and tactile qualities would have seemed unnatural to our ancestors. Objects of plastic have so proliferated that we take them for granted. Plastic has been naturalized.

The postwar generation grew up with plastic. More precisely, baby boomers grew up while the plastic industry was growing and plastic was expanding into their lives. Before the Second World War, when their parents were coming of age, visible uses of plastic in consumer goods remained limited to a few products, to such things as celluloid dresser sets and Bakelite radios. Although the new industry's publicists predicted a utopia molded out of "miracle materials" derived from coal, water, and air, not until the war did synthetic materials gain wide application. Annual production in the United States nearly tripled between 1940 and 1945, a year in which 818 million pounds went for such military uses as aircraft cockpit covers, mortar fuses,

1

bayonet scabbards, helmet liners, and even the atom bomb. Expanding explosively after the war, with a growth curve steeper than that of the GNP, annual production exceeded six billion pounds by 1960.[1] Baby boomers played with Wham-O hula hoops and frisbees, Barbie dolls and Revell airplane models, Lego blocks and Mattel machine guns. They ate breakfast at Formica dinettes, spilled milk from polyethylene tumblers onto vinyl floors, and left for school clutching disposable Bic pens. Their families experienced a flood of new plastic products—Tupperware, garbage pails and laundry baskets, Melamine dishes, appliance housings, Saran Wrap and dry cleaning bags, picnic coolers, scuff-proof luggage, Naugahyde furniture, Mylar recording tape, Corfam shoes, shrink-wrapped meats, Styrofoam egg cartons, artificial Christmas trees, and endlessly on. Even life's greatest status symbol, the automobile, took on more plastic each year—while that desirable new car smell came from plasticizers migrating into the air. Proliferation of plastic coincided with expansion of the postwar economy. The people of plenty consumed an ever-increasing quantity of plastic as they experienced what *Time* referred to as a "flood of new products . . . transforming . . . the American way of life." It was certainly true, as a historian observed, that "more material objects are being produced today, and destroyed, than ever before in history."[2]

From its beginnings plastic embodied considerable hubris. Early imitative uses exhibited pride in the ingenuity of illusion. Celluloid, introduced around 1870, imitated the layering of ivory, the mottling of tortoiseshell, the hard translucency of amber and semiprecious gems, the weave and stitching of linen, the veining of marble. But the attractions of plastic extended beyond surface imitation. Promotion of various types early in the twentieth century emphasized substituting cheap substances for scarce natural materials. Modern science was making former luxury goods available to democratic man and woman. Most important, celluloid and newer plastics often seemed superior to natural materials with their irregularities and flaws. By comparison to ivory or shell, even celluloid was a homogeneous material that lent itself to mass production. Only superficially imitative, plastic was in essence artificial. It seemed nearly miraculous. In the words of a biblical paraphrase, it was "a new substance under the sun" offering evidence of an ability to improve upon nature, even to transcend it.[3] The artificiality of materials and the proliferation of things have changed our perception of reality by making it seem more malleable, less permanent, even ephemeral. Often smooth and unnatural to the touch, most plastic is so lightweight compared with metal that its impact appears a matter of surface alone. Easier to shape and color, plastic has given everyday life a sense of greater possibility or plasticity. Compared with nineteenth-century artifacts of iron and steel, those of the present seem light, impermanent, replaceable, and disposable. As far back as 1927 publicists for the industry proclaimed a Plastic Age equal in historical significance to earlier ages of bronze, iron, gunpowder, or steel. A few years later, in 1942, Williams Haynes, an economist and historian of the chemical industry, declared that synthetic materials would have "more effect on the lives of our great-grandchildren than Hitler or Mussolini." He claimed further that "new materials" could "compel the course of history as greatly as any man." Such statements are impossible to prove. Yet plastic has indeed contributed to shaping twentieth-century culture.[4]

JUST ONE WORD . . .

The word plastic entered the folklore of the baby boom generation just as its oldest members were coming of age. Thousands of moviegoers watched in 1968 as Dustin Hoffman, starring as *The Graduate*, received some advice from a family friend. Offering the secret of life, or at least of worldly success, he told the confused young man, "I just want to say one word to you. Just one word. . . . Plastics. . . . There's a great future in plastics." This odd pronouncement, isolated in the film's opening scene, convulsed audiences and became a line "repeated into classicdom by a whole generation of kids." The scene made a permanent impression—and not just among dismayed plastics executives whose trade journal could not bring itself to refer to that "tired old joke about plastics" until 1986.[5] Most viewers would have had trouble explaining their laughter. Some perceived a comment on the banality of business, others an attack on comfortable middle-class materialism. Still others, recalling plastic's simulation of traditional materials, understood the scene as metaphoric commentary on the rhetoric of the Great Society. A few, catching an ominous note, entertained fleeting thoughts of science fiction nightmares, of technology run amok. And some merely relished the absurd elevation of the commonplace. Whatever the reasons, the scene hit a nerve and entered communal memory.

Plastic itself, by its very nature, complicates efforts to think about it. Able to assume many degrees of shape, texture, hardness, density, resilience, or color, the myriad varieties are united only by a word—plastic—that has defied most attempts to promote specific trade names. What do we mean when we talk about plastic? Is it indeed a material capable of shaping the limits of history, or is it an amorphous substance passively receptive to any psychological or cultural projection? The "mythology" of *le plastique* intrigued Roland Barthes more than ten years before *The Graduate*'s release. When the French critic's short essay on plastic was translated in 1972, it echoed the uneasiness of moviegoers at the invocation of "plastics" in a scene with no obvious context. For Barthes suggested that plastic was nothing without context. At the same time it threatened to dissolve context. It was, in fact, everything and nothing. Written as a meditation on a trade show where Barthes had observed perfectly formed novelties emerging from an injection-molding machine, his essay had plastic fulfilling the promise of medieval alchemy by bringing about "the transmutation of matter." Unlike other materials, plastic was "more than a substance"; it embodied "the very idea of . . . infinite transformation." Possessed of endless possibility, it triggered amazement at "the proliferating forms of matter." With its protean capabilities, plastic evoked "the euphoria of a prestigious free-

wheeling through Nature." Eventually, however, plastic's freedom to become anything would reduce everything to nothing by dissolving all differences. With plastic
as a universal solvent, Barthes observed, "the hierarchy of substances is abolished: a
single one replaces them all: the whole world *can* be plasticized."[6]

The material's inherent formlessness attracts such inflated rhetoric. And the word
itself invites semantic or metaphoric fluidity. To understand plastic's resonance requires some knowledge of how the word evolved in relation to the material's emergence. One might argue that plastic did not exist until it was so named. Although
most of the substances now commonly indicated by the word are recent developments, with only celluloid appearing before 1900, the word has a long history. In
Greek the verb πλασσειν (plassein) meant to mold or shape a soft substance like
clay or wax; the adjective πλαστικοs (plastikos) referred to something capable of
being molded or shaped. Cognates referred to potters, sculptors, and others with
the power to shape (the Creator, for example, in New Testament Greek). On occasion, metaphorically, the verb could mean to fabricate or forge dishonestly—to imitate, in other words—but that connotation disappeared as the adjective moved from
Greek into Latin, then French, and finally into English.[7]

Samuel Johnson's famous dictionary cited a line from an early eighteenth-century
poet ("Benign creator! let thy *plastick* hand / Dispose its own effect") to illustrate
plastick as an adjective meaning "having the power to give form." This use of plastic
to describe an active shaper of people or things extended back to the early 1600s
and forward into the early 1800s, when an American writer described a child's mind
as "like wax beneath the plastic hand of the mother." By then, however, the adjective
was beginning to refer not to active shapers but to passive materials "capable of
being modeled or molded into various forms, as plaster, clay, etc." By extension,
according to *The Century Dictionary* of 1890, the word meant "capable of change or
modification; capable of receiving a new bent or direction: as, the mind is *plastic* in
youth." Most dictionaries, slow to reflect technological innovation, added nothing
to these adjectival meanings until the 1930s. One exception was the first dictionary
to note plastic as a noun roughly corresponding to our current usage. In 1910 *The
Century Dictionary Supplement* described plastic as "the commercial name for any
one of a class of substances, such as celluloid or viscose, which are worked into
shape for use by molding or pressing when in a plastic condition." This meaning did
not reappear in a major American dictionary until 1934 but acknowledged an
emerging class of materials known mostly to industrial chemists and engineers.[8]

During the nineteenth century various plastics of natural origin had gained wide
application. Shellac, a resinous insect secretion gathered in India and southeast Asia,
was used with wood flour to mold daguerreotype cases and other small objects.[9]
Gutta percha, from the milky sap of a Malayan evergreen, was used to insulate tele-
graph cables and to mold small consumer goods. The most profitable natural plastic
was caoutchouc or rubber, derived from the *Hevea brasiliensis* tree of South America.
When vulcanized with sulphur it became vulcanite, ebonite, or hard rubber and was

used for buttons, fancy goods, and electrical insulation. Supplementing these natural molding materials were celluloid, introduced in 1869 by John Wesley Hyatt, an American mechanic, and Bakelite, invented in 1907 by Leo H. Baekeland, an industrial chemist who had emigrated from Belgium. Celluloid came from a doughy mixture of nitrated cellulose (obtained from pulped paper or cotton) and camphor (from the bark of the Formosan camphor tree). Pressed into blocks under heat and pressure, it was sliced into thin sections and then, after being partially resoftened by heat, was cut, stamped, pressed, or molded into inexpensive consumer goods. Bakelite, the product of a condensation reaction between phenol and formaldehyde conducted under heat and pressure, was the first chemically synthetic plastic. When it first appeared as a substitute for shellac and hard rubber, it was used to mold electrical insulation.

For several decades confusion reigned over just what constituted a plastic. Among the inventors, chemists, and businessmen who promoted new materials, the noun referred to the manufacturing process. A plastic, whether natural or synthetic, was something that could be molded or shaped when soft, and then hardened. In 1903 the United States Patent Office created the classification "Plastics" from out of the former "Caoutchouc and Minor Plastics." Although the new category explicitly excluded glass and butter, it admitted such oddities as "scraps of cork, leather, etc." compressed to make "articles of definite shape." Even Baekeland, later referred to as the father of plastic, bragged in 1909 that his Bakelite was "far superior to . . . all plastics." In fact it was "a *non-plastic*" because its chemical reaction rendered it permanently hard, infusible, and insoluble. Eventually a definition emerged that encompassed even Bakelite. The industry's first trade journal, *Plastics*, founded in 1925 by two partners who "didn't even know what plastics were," borrowed a definition from *The Century Dictionary Supplement* of 1910. "Plastics"—changed to plural form to suggest a variety of types—indicated a "commercial . . . class of substances . . . worked into shape for use by molding or pressing when in a plastic condition." As a category plastics was more commercial than scientific, encompassing an array of materials united by similar manufacturing processes, shared markets, and a common name.[10]

As the word plastics attained general recognition, the new industry worried about its public image. Using new materials to imitate or substitute for traditional materials tended to imply inferiority. After all, celluloid often imitated ivory or tortoise-shell and Bakelite substituted for shellac or hard rubber. But most promoters downplayed imitation and substitution. By the late 1930s, after the appearance of urea formaldehyde, cellulose acetate, polymethyl methacrylate, and polystyrene, the noun plastics connoted new artificial materials, products of chemical synthesis distinct from any natural substances. Addressing potential customers, the Du Pont company magazine in 1938 defined plastics as "man-made combinations of basic chemicals and materials." No longer "substitute materials," they were designed "by man to his own specifications." In 1945, when *Webster's New Collegiate Dictionary* finally provided a technical definition, plastics was a noun encompassing a specific industry's commercial products based on organic chemistry.[11]

Confusion reigned, however, as the general public experienced postwar expansion of the varieties of plastic. The soft waxy vinyl of raincoats shared few properties with the dull hard Bakelite of distributor caps, the tough transparent acrylic of aircraft enclosures, or the brittle polystyrene of children's toys. According to a promoter, the "average person" of 1947 had trouble explaining "just what a plastic is." Soon, however, leaders of the industry were wishing the public was *not* so familiar with the word plastic. Urged on by magazine articles about synthetic materials in the postwar home, American consumers accepted them as a class and based opinions of all plastics on experiences with those they had used. But failures of some products began to give plastic a bad name, and a trade editor lamented that "the word plastics means only one material to too many people." The public had "not yet learned that cellulose acetate, polystyrene, phenolics, ureas and other materials are as different from each other as iron, steel, copper or lead." By and large the public never learned that lesson. Like the material, the word succeeded so well that in 1968 a linguist found "other uses of the word . . . losing ground."[12]

AMBIVALENT METAPHORS

By then plastic was undergoing its greatest shift in meaning since the appearance of new materials. Expansion of plastic mirrored a civilization that seemed to be abandoning its ideals in pursuit of material goods. To some young people, especially those of the middle class who enjoyed the benefits of the postwar economy, American values seemed superficial, as phony as woodgrain laminate or vinyl leatherette. These baby boomers had experienced cheap toys and heard their parents curse shoddy plastic products. The material did not ring true; nor did the society that produced it. Whether the opening scene of *The Graduate* catalyzed the transformation or merely reinforced it, plastic became an adjective meaning fake or insincere— referring especially to the older generation, its activities, its accomplishments.

The word's transformation seemed instantaneous, but that was not the case. The process had already begun in 1962 when novelist Ken Kesey described the head nurse of a psychiatric ward as having a "terrible cold face, a calm smile stamped out of red plastic," and later compressed the metaphor as "her fixed plastic smile." Similar transfers of meaning, based on ambivalent feelings about use or value of other materials, were not unknown. In a play written in 1898, for example, one of George Bernard Shaw's characters accused another of being a "fraud," a "humbug," a "miserable little *plaster* saint." More directly appropriate, the cinema's use of celluloid film inspired D. H. Lawrence to complain in 1928 that "something has gone out of the *celluloid* women of today." Plastic as a pejorative adjective thus had precedents. And when use of plastic to mean fake or insincere first became common, the younger generation had no monopoly on it. "Now that so many of the young seem to wear their hearts on their sleeves," Russell Lynes wrote in *Harper's* in 1967, it is "hard to tell which ones are real and which are plastic."[13]

But for the most part the word stood for an adult society its children wanted no part of. A young woman in San Francisco's Haight-Ashbury district complained in 1967 that few "square women" exhibited female qualities "because their nest is all chrome and plastic." A male counterpart insisted that "nobody wants to be plastic"; instead "one wants to be real." The new meaning so quickly attained general currency that in 1970 Yale law professor Charles A. Reich could promote a visionary "Consciousness III" as the source of "a culture that is not plastic or artificial."[14] The word entered glossaries of slang as "phony; unreal; dehumanized; superficial; valueless."[15] A new edition of *Webster's New Collegiate Dictionary* in 1973 added a definition of plastic as "formed by or adapted to an artificial or conventional standard: synthetic"—a definition that paled next to a thesaurus's ragbag of synonyms: "sham, meretricious, counterfeit, substitute, factitious, spurious, synthetic, artificial, specious, bogus, factoidal."[16] A final ironic twist came at mid-decade as journalists began referring to "plastic credit" or "plastic money." Introduced around 1955, credit cards of rigid plastic became so common that by 1980 people referred to them simply as "plastic."[17] Both as material and as concept, plastic encompassed the era's inflationary tendencies—including a fear that what had been created was ephemeral or unreal.

Given the word's evolution, it is not surprising that some manufacturers of synthetic materials spurned plastic as a general name, choosing instead to promote distinctive materials with an array of generic and trade names. Some of them caught on. Nylon retained its identity and positive reputation. Vinyl and polyester gained independence but acquired their own negative connotations. Plexiglas, Teflon, and Kevlar were among the few trade names to become widely known. Urban professionals of the 1980s who cringed at the touch of synthetic fabrics thought nothing of using Kevlar tennis rackets. As innovation focused on carbon or boron reinforcing fibers embedded in a matrix of synthetic resin, the industry spoke of composites just as innovators of the 1930s had referred to plastics. This shift owed as much to the word's low esteem as to the makeup of the new materials. When the University of Delaware established a research center for synthetic materials in the 1980s, the debate over a name revolved around reputation. Eventually it became the Center for Composite Materials to avoid association with plastic.[18]

Even so, plastic's connotations were already improving. Wolfgang Schivelbusch has noted that "only during a transitional period" do new technologies evoke "a sense of loss" in those who experience them. Soon people develop "a new set of perceptions" and accept as "natural," even desirable, events or objects formerly experienced as violations of their expectations of the world.[19] Familiarity breeds acceptance, as acknowledged by a journalist who suggested that "children whose first experience of the world comes from Toys 'R' Us may be developing a different set of material values than adults who grew to maturity surrounded by wood, stone, and metal."[20] Computer housings, electronics components, automobile interiors, and high-tech sports equipment—not brittle polystyrene toys—pervaded the experience of the 1980s generation. Bright artificial colors and textures of many plastic artifacts indicated acceptance of high-tech styling. The variety and whimsy of their forms celebrated an unprecedented degree of material possibility. Despite an improving reputation, however, most people ignored the industry's claim that the United States entered the "Plastics Age" in 1979 when the annual volume of plastic exceeded that of steel.[21] They identified the era not with plastic but with information, with devices for recording, storing, reproducing, and manipulating sounds, words, and images. Though physically sheltered by the built environment, people found their emotions and thoughts ever more stimulated by immaterial experiences. Even the plastics that facilitated these synthetic experiences by means of film, tapes, discs, and coatings receded from view and from consciousness.[22]

The 1980s and 1990s saw the arrival to positions of cultural influence of people who never knew a world without plastic and who grew up with the information age. They were bombarded by images of all times and places, the fantastic or exaggerated as well as the authentic. The storehouse of materials held an embarrassment of riches—not only infinitely moldable plastics but also databases of digitized images with potentially infinite yield through recycling and recombination. Experience promised to become more malleable, more synthetic, through computer animation and holographic projection. Immateriality triumphed as the medium of artifice shifted from plastics to electronics, from long-chained molecules to flowing electrons. This apparent dissolving of limits made anything and everything seem possible—and also, as Roland Barthes had already hinted, of lessened value.

On the other hand, plastic's potential recalled a traditional faith in the openness of experience to intentional shaping. Optimistic Americans had often considered personal character, the land itself, even the cosmos, as receptive to reforming impulses. The nation's most characteristic philosophers, Ralph Waldo Emerson and William James, had asserted the plasticity of experience in the face of human desires. In an 1837 essay on "The American Scholar," Emerson attacked a "mischievous notion" that "the world was finished a long time ago." Instead it was "plastic and fluid in the hands of God" and remained so to godly men. "In proportion as a man has any thing in him divine," Emerson insisted, "the firmament flows before him and takes his signet and form." This idea was echoed by James, for whom the very cosmos was fluid and unfinished, beyond comprehension yet open to human influence. In 1907, coincidentally the year of Bakelite's invention, James argued in *Pragmatism* that recent transformations in science had proven "how plastic even the oldest truths . . . really are." Truth, he suggested, is "not a stagnant property"; rather "truth *happens* to an idea." Arguing that "the world stands really malleable, waiting to receive its final touches at our hands," James considered it "wholly plastic" until proven otherwise.[23]

When James wrote about plasticity, he was not thinking about plastic, but the connection extended beyond words of similar meaning. Pragmatism's concept of a malleable universe open to human influence appealed to desires similar to those motivating plastic's development. To shape the stuff of existence at a fundamental chemical level, to imbue it with properties, textures, and colors unknown to earlier generations, to mold from it objects and environments unknown to prior civilizations—all these marked a degree of human control over nature that James would have applauded at first, though eventually he might have feared its hubris. Whether used to imitate traditional materials or to create seamless artificial surfaces, plastic established unprecedented control over the material environment. Taken to extreme, such control implied the possibility of stifling humanity in a rigidly ordered artificial cocoon—or, in the event of loss of control, the possibility, as a retired Du Pont chemist predicted in 1988, that humanity would "perish by being smothered in plastic."[24] There was something about plastic's chemical artificiality, as Barthes implied, that evoked the permanence of death—a finality conveyed by the hard *k* sound of the word plastic. On the other hand, plastic promised material freedom, a malleable environment whose openness was evoked by the flowing *c* sound of plasticity. This duality ran throughout the American experience of plastic, sometimes as undercurrent, sometimes as conscious statement. An advertisement played to this ambiguity by saying that "plastic is forever . . . and a lot cheaper than diamonds."[25] The history of plastic—its invention, industrial promotion, consumer uses, popular reactions, images, and metaphors—reveals much about American culture of the twentieth century, especially its tendency, backed by new technologies, of moving toward an inflationary proliferation of experiences.

CELLULOID: FROM IMITATION

TO INNOVATION

*T*he very word celluloid sounds archaic. It recalls an era when ladies secured their tresses in place with decorative combs and gentlemen chafed at the discomfort of stiff collars. Such relics of celluloid as combs and hand mirrors, powder boxes and hair receivers, that are now mellowed to an uneven golden hue, suggest a lost world of comforting simplicity. But despite celluloid's power to conjure up a mythical past, the clumsiness of the word exposes that era's makeshift confrontation with ongoing material transformations. The sound of celluloid suggests confusion about the significance of new materials; so do other trade names for the same basic stuff, such odd, wonderfully resonant words as Xylonite, Viscoloid, and Ivorine; Fiberloid, Coraline, and Pulveroid; Pasbosene, Cellonite, and Pyralin. The mixed allusions of these vaguely chemical tongue-twisters indicate both an exuberant pride in technological manipulation and a lack of foresight regarding its potential. Several of them, and such others as Eburite and Crystalloid, also indicate fascination with imitation and substitution. The story of plastic begins, in theme as well as in time, with a material that often pretended to be something it was not.[1]

The idea of substituting one material for another motivated celluloid's invention. About 1863 a printer in Albany, New York, came across an offer of $10,000 from the firm of Phelan & Collender for anyone who could devise an acceptable substitute for ivory in billiard balls. John Wesley Hyatt's first attempt involved collodion, a solution of nitrocellulose used to coat photographic plates and by physicians to protect minor cuts with a hard, thin film.[2] After noticing some spilled collodion that had hardened into a horny mass, Hyatt thought he had found a material that might be molded into an ivorylike billiard ball. He soon discovered, however, that collodion evaporated slowly, that its solvent (a mixture of alcohol and ether) was expensive, and that it shrank and cracked as it dried. Possibly aware that Alexander Parkes of Great Britain had recently patented a molding compound based on an inefficient process of evaporating "wet" solutions of nitrocellulose, Hyatt in 1869 hit on the idea of a "dry" process using camphor as a solid solvent. Disregarding professional chemists who warned of violent explosion, Hyatt mixed ground camphor with nitrocellulose, heated the mixture to dissolve and soften it, compressed it in a mold, and upon cooling obtained a hard, shiny, durable substance unlike any of its constituents.[3]

Seeking to exploit the new material, he applied for a number of patents and convinced his brother Isaiah, a newspaper editor, to serve as business manager of the Celluloid Manufacturing Company, founded in 1871. It was Isaiah who contributed the name, whose meaning ("like cellulose") illogically suggested an imitation even though it was made from real cellulose—pulped cotton nitrated in acid to form nitrocellulose. In collaboration with others, especially John H. Stevens and Charles F. Burroughs, the Hyatts developed most of the chemical processes, fabricating techniques, and processing machines that comprised the celluloid industry. While continuing to experiment with billiard balls, they also manufactured denture plates, combs, knife and brush handles, harness fittings, piano keys, and small novelties, most of which directly substituted for ivory, tortoiseshell, or horn. Despite the material's eventual success, there is no evidence that Hyatt ever received the $10,000 prize.[4]

Awareness of substitution, imitation, and innovation as three distinct motives for using plastic emerged almost simultaneously with Hyatt's invention of celluloid. These three motives appeared in well-defined form as early as 1878 in an advertising circular describing the material's properties and advantages.[5] At least in the abstract, it seemed, celluloid was a material of innovation. Described in the opening sentence as "a new material in the arts," celluloid was "a hard, durable substance, almost entirely unaffected by acids or alkalies; unchangeable under ordinary atmospheric conditions." Unlike natural materials subject to chance irregularities, celluloid possessed "a uniform and perfect consistency." Its potential manifestations exceeded those of other materials. After being "rendered plastic by heat," it could be "molded into any desired form." Colorless in a pure state, it could assume "any desired shade." The pamphlet summed up these virtues in a classical allusion that soon became a promotional cliché: "like the fabled Proteus, celluloid appears in a thousand forms." It seemed certain, at the very least, that celluloid had "an important future before it." As an innovative material it promised unprecedented, previously inconceivable results.

Even so, the pamphlet moderated its celebration of innovation by referring frequently to both substitution and imitation. While celluloid enjoyed a versatility and uniformity unknown to natural materials, understanding its strengths required comparing it functionally to the materials for which it substituted. Thus celluloid

11

was "tough as whalebone; elastic and dense as ivory." It seemed attractive not because it extended the reach of human desire but because it allowed familiar desires to be fulfilled more easily and cheaply through substitution. "As petroleum came to the relief of the whale," the pamphlet proposed, so too "has celluloid given the elephant, the tortoise, and the coral insect a respite in their native haunts; and it will no longer be necessary to ransack the earth in pursuit of substances which are constantly growing scarcer." While serving as a functional substitute, celluloid also often imitated the surface appearances of materials it replaced. According to the pamphlet, celluloid offered "perfect imitations of tortoise-shell, agate, coral, amber, malachite, and other materials" so convincing they would "defy detection." Industries as varied as clothing and furniture benefited from celluloid's power of mimicry, with collars and cuffs "in imitation of linen" and "veneers representing different colored woods." Celluloid might assume "a thousand forms," but it usually appeared reassuringly familiar to consumers and users of traditional materials. Imitation obscured the fact of substitution and helped neutralize any fears of innovation. In fact, the emphasis on innovation in this early booklet marked it as unusual. Imitation and substitution dominated most nineteenth-century discussions of celluloid. Later, as celluloid and newer plastics gained wider use in the early decades of the twentieth century, the balance gradually shifted from imitation to innovation, and promoters began to celebrate the frankly artificial (Plate 1).

THE PLEASURE OF IMITATION

Our distance from the celluloid era was suggested by Brian Moore's visionary novel *The Great Victorian Collection*, in which a history professor of the 1970s dreams into existence a vast collection of nineteenth-century objects, each an exact replica of an artifact known to him through museums or exhibition catalogues. Under scrutiny by academic experts and media representatives, this phenomenal collection begins to decay. The professor notices, for example, that a few of the blocks of glass composing a replica of a fountain from the Crystal Palace have taken on "the plastic lightness of Lucite, a substance unknown in Victorian times." These stand out as "obvious imitations, blatant shams, marring the perfection of the others." Eventually the entire collection fades in like manner.[6] The very idea of such a collection offends modernists as kitsch and excites postmodernists to meditate on ironies of reproduction and simulation. To a progressive citizen of the nineteenth century, however, a fountain of Lucite imitating glass would have seemed one more glorious example of the ability to cast the material world into any desired shape or surface.

Far from indicating shoddiness or dishonesty, imitation in the nineteenth century expressed a brash exuberance and offered provocative evidence of the extension of human artifice through new technologies. Imitation often occurred for no reason other than its technical feasibility—as if designers were calling attention to the ingenuity of the processes at their command. Building fronts of cast iron assumed outlines and textures of brick and stone. Stamping and molding of papier-mâché and gutta percha liberated ornate effects in furniture from any reliance on wood. Not even the realm of high art escaped, as museums filled galleries with plaster casts of sculpture from around the world—intended as objective surveys of styles, with each artifact an exact replica down to color and texture. By comparison, imitative celluloid was modest, hardly worthy of attention, though it evoked a similar pleasure and reached a larger audience.[7]

Modernist disdain for imitation as dishonest and immoral had its roots in nineteenth-century British critics of industrialization who promoted a romantic aesthetic of craftsmanship. As early as 1841 A.W.N. Pugin attacked "the false notion of *disguising* . . . articles of utility." Such "cheap deceptions" as cast iron painted to resemble wood or marble enabled people "to assume a semblance of decoration far beyond either their means or their station." A few years later John Ruskin criticized the imitation of marble by marbleizing, rare woods by graining, and hand craftsmanship by casting or machine carving. He considered such practices "utterly base and inadmissible" because they implied a history of prior events—whether geologic change, botanic growth, or human labor—that had not actually occurred. Objects produced by imitative processes falsely claimed a natural or cultural richness they did not possess. More directly concerned with mechanical inroads, William Morris in 1881 advised designers of commercial products to "get the most out of your material" by making "something that could not be done with any other." If forced to design for machine production, they should "make it mechanical with a vengeance" and "as simple as possible" without trying "to make stone look like ironwork, or wood like silk, or pottery like stone."[8]

This critique of imitation reached a wide audience through Charles Eastlake's *Hints on Household Taste*, reprinted in the United States in 1872, and was familiar to upper-class American followers of Ruskin and Morris in the Arts and Crafts movement.[9] Even so, imitation retained its fascination. Ruskin himself described imitation as one of the five sources of pleasure in art. "Whenever anything looks like what it is not, the resemblance being so great as *nearly* to deceive," he observed, "we feel a kind of pleasurable surprise, an agreeable excitement of mind." Ruskin insisted that the resemblance had to be "so perfect as to amount to a deception"; secondly, there had to be "some means of proving at the same moment that it *is* a deception."[10] If imitation ranked with truth and beauty as a source of pleasure in the high arts, how much more pertinent might it not be to the enjoyment of everyday objects? No one, for example, would have long mistaken for true marble such a typical celluloid object as a delicately colored and marbleized soap dish, so lightweight and thin that it became hazily transparent when held up to the light (Plate 2). This ordinary object provoked the pleasure of imitation described by Ruskin. Even more, it triggered a sense of delight at the expanding powers of science, that could so easily imitate the natural. By doing so in such an unexpected context, science passed into the realm of the unnatural. Imitation became one more sign of increasingly precise human control and thus, paradoxically, a sign of innovation. Far from indicating shoddiness or dishonesty, imitation offered provocative evidence of the extension of human artifice through new technologies.

The sheer pleasure of imitation did not alone guarantee its success as the most frequent mode of treating the surfaces of celluloid products. Imitation also involved more mundane concerns. Objects of celluloid often imitated other materials to conform to tradition or custom. It would have been possible (and cheaper) to make collars for men's shirts from shiny smooth sheets of pink celluloid (or any other color), but custom dictated that they imitate the weave of linen—at considerable extra cost. And cost-conscious manufacturers of combs and brushes adopted labor-intensive processes for imitating in celluloid the surfaces of ivory and tortoiseshell even though solid colors were cheaper to produce because custom demanded the appearance of traditional materials. Artificial or non-natural design vocabularies did not yet exist. As one material substituted for another, imitation eased the transition. Only after a new material had gained wide acceptance for a particular application did it stop imitating the old material. Imitation became most visible or most obvious when it was no longer necessary, just before it was abandoned.

THE PROMISE OF SUBSTITUTION

William Morris would have approved Du Pont's decision at the end of the 1920s to market celluloid toiletware that imitated no other materials but frankly appeared as artificial products of chemical manipulation. If something had to be mass produced by machine, he had said, then let its appearance directly reflect its origin. Given the popularity of imitation, however, the general public would hardly have agreed with Morris during the first fifty years of celluloid's existence. Whether imitation disguised the fact of innovation or announced the ingenuity of the innovators (and it did both), it also pointed up the economic role of celluloid as a substitute for other materials. By replacing materials that were hard to find or expensive to process, celluloid democratized a host of goods for an expanding consumption-oriented middle class. Robert Friedel has argued that celluloid appeared "not as a cheap, utilitarian stuff but rather as a raw material for artistry and ornament . . . a vehicle for spreading a taste for luxury and the 'finer things.'"[11] Not many of the products newly rendered in celluloid had been luxuries available only to a wealthy elite, as publicists liked to claim, but it would have been hard to increase their volume of manufacture without a substitute material like celluloid. And had celluloid not imitated the surfaces of the materials for which it substituted, then the resulting products, lacking in tradition and devoid of aesthetic reference points, would have seemed cheap and undesirable. The conjunction of imitation and substitution ensured both cultural identity and economic success.

In most promotions of celluloid, imitation and substitution were never far apart. Around 1890, for example, a Celluloid Company brochure touted the material as capable of "counterfeits" of "coral, ivory, malachite, tortoise shell, amber, turquoise, lapis lazuli, agate and carnelian" so exact that they deceived even "the eye of the expert." All the same, their major "superiority over the genuine articles" lay in their "moderate cost." A "few dollars invested in Celluloid," according to the company's optimistic projection, equalled "hundreds expended in the purchase of genuine products of nature."[12] More than thirty years later, in 1921, the *Du Pont Magazine* laid out a similar argument, stating that Pyralin (Du Pont's brand of pyroxylin or nitrocellulose plastic) could be used "to reproduce grained ivory, tortoise-shell, amber, horn, mother of pearl, abalone pearl, coral and such woods as ebony, mahogany and oak." Equally as important, unlike most materials it replaced, Pyralin was "tough and strong; will not crack or splinter; does not rust or corrode; is free from grit; can be cut, sawed, turned on a lathe like wood, and in plastic form can be bent or molded in any way desired."[13] Substitution thus extended beyond surface imitation.

More neutral sources also emphasized the lower cost of celluloid's raw materials, the lower cost of processing and fabricating celluloid, and its functional superiority to the traditional materials it replaced. For example, an article published in 1906 in *Scientific American* defined the industry's goal as "supply[ing] certain objects of perfect make . . . at a cheap price." Imitating other substances, many of them "exceedingly valuable," made possible "production on a large scale." As long as the results of substitution remained "really worthy and in good artistic taste," then the particular material employed to satisfy a given need was "of quite secondary importance." A few years later Edward Chauncey Worden, whose two-volume *Nitrocellulose Industry* provided a massive account of the development and processing of celluloid, characterized it as "essentially an imitative industry" providing "a forgery of many of the necessities and luxuries of civilized life." Its continued success depended on "how closely it can reproduce that which is beautiful and scarce, and hence sought after and costly." Successful imitation merely ratified successful substitution, however. According to Worden the varieties of celluloid, "unlike many forgeries," possessed "properties superior to those of the originals which they are intended to simulate."[14]

The highest level of generalizing about substitution appeared in an article written in 1906 by Robert Kennedy Duncan, a chemical promoter who was about to initiate a series of industrial fellowships at the University of Kansas. Enthusing in *Harper's* over "The Wonders of Cellulose," Duncan referred to it as the organic equivalent of such treasures of the "non-living mineral realm" as gold and silver. His hyperbolic prose described cellulose (and the celluloid made from it) as "too tough a morsel for time to swallow; when pure, it rusts not, neither does it decay, and it can endure throughout all generations." Its near indestructibility rendered it a veritable "organic archetype of conservatism." Most important, unlike such mineral treasures as gold and silver, cellulose was "the commonest of common things." And therefore, though Duncan failed to reach this conclusion, cellulose and its chemical derivatives,

15

however physically conservative they seemed to him, actually functioned in a socially and economically expansive manner by promoting proliferation of cheaply mass-produced goods.[15]

Such was the intellectual construct of celluloid among the people who produced and promoted it. But how accurate was this characterization? Was celluloid indeed cheaper as a raw material than ivory, tortoiseshell, or horn? Did it indeed yield a homogeneous material easier and less costly to fabricate than the irregular natural materials for which it substituted? Was celluloid truly more adaptable to mass production? And finally, was celluloid in fact more durable or otherwise more satisfactory to the consumer? These questions are simply stated but the answers are complex because it is hard to isolate their elements. The best approach to evaluating celluloid's success as a substitute involves keeping these questions in mind while conducting a more general survey of celluloid in relation to the many materials it replaced. Even then, the most accurate conclusion, however unsatisfactory, may have to be simply that celluloid's expanding success as a material proved its superiority.[16]

Celluloid's promoters certainly considered their products superior to those they replaced. In 1878 a Hyatt patent specification described deficiencies of traditional combs. Those of wood warped and became "foul or viscid" when "coated with oil or dampened by water." Horn, the most common comb material, also had a tendency "to sliver or fray on the edges of its teeth." Metal, though not often used, was subject to corrosion. Hard rubber, another recent contender, generated static electricity, became brittle when cold, and exuded "a disagreeable odor" when warm. As for ivory, in the words of the Celluloid Company a dozen years later, it would "crack . . . warp . . . speedily wear . . . [and] quickly and easily become discolored." According to a Du Pont Pyralin catalogue, horn was "readily attacked by certain vermin," and even ivory, "the most valued of all materials for the making of combs," suffered by comparison with celluloid owing to "prohibitive" cost and a tendency to scale and crack. Not that celluloid was perfect. A champion of ivory criticized the substitute's "camphory" odor and its flammability, while even Du Pont privately admitted problems with warping, shrinking, brittleness, and discoloration. On the whole, celluloid functioned as well as the materials it replaced. Survival for decades of numerous celluloid artifacts, often as good as new, revealed it as the tough, durable, flexible material its promoters had claimed. And beauty proved, as always, in the eye of the beholder. In 1937 a plastics executive described a discolored forty-year-old celluloid box as "mellowed with age" like "natural ivory."[17]

More certain than celluloid's functional superiority was its cost advantage as a raw material. Although manufacturers suffered from a Japanese monopoly of Formosan camphor after the turn of the century, celluloid remained a material of easy accessibility.[18] By contrast, throughout celluloid's period of prominence from Hyatt's early years into the 1920s, promoters emphasized the increasing rarity and cost of competing natural materials. Tortoiseshell, a richly mottled yellow-brown material prized for large decorative combs used to restrain Victorian tresses, was perhaps the most scarce. Found in the tropics, the hawksbill turtle yielded six to eight pounds of workable shell, which by 1920 commanded $2 to $25 a pound depending on quality.[19] So valuable was tortoiseshell that methods of imitating it with horn were common by the early 1800s, long before celluloid took over. Even horn, traditionally the cheapest natural equivalent of celluloid, became scarce by 1900 as ranchers adopted the practice of dehorning cattle before shipping them to market and thus packinghouses no longer offered a limitless cheap supply.[20]

Ivory attracted particular attention as an exotic substance whose supply seemed threatened. As the material most often imitated, ivory seemed to define celluloid's raison d'être. To cite only one example of the industry's fixation on ivory, a Du Pont salesmen's handbook from 1919 observed that "in earlier times only the very rich could enjoy the luxury of ivory toiletware." Without celluloid, "ivory" toiletware would have remained a luxury. Because "the great herds of elephants" were "now practically extinct," it had become nearly "impossible to obtain any natural or rough ivory whatever—even for a king's ransom." In fact, as Friedel has argued, "the ivory problem existed more in the mind than in the market." Although fear of an impending shortage stimulated development of nitrocellulose plastic during the mid-nineteenth century, by 1919 comments such as those of Du Pont served primarily to associate celluloid with ideas of luxury and rarity, to suggest that the American housewife enjoyed comforts formerly available only in a sultan's harem. No evidence suggested a scarcity of ivory during the early twentieth century. Between 1884 and 1911, the fluctuating price of ivory was largely unrelated to an increase of nearly 250 percent in annual consumption. During this period the average value per pound of ivory imported into the United States reached a high of $3.76 in 1890 and a low of $1.99 in 1901; in 1911 it stood at $2.51. During these same years of fluctuating prices, annual imports increased steadily from 221,000 to 534,000 pounds. Despite considerable increase in consumption, there seemed little chance of the elephant being killed off.[21]

On the other hand, if ivory showed no signs of becoming unavailable or increasing in price during these years, celluloid production increased phenomenally, and prices, already below those of ivory, continued to fall. In 1879 the Celluloid Manufacturing Company, then virtually the only supplier, produced 213,000 pounds of celluloid with an average price of $1.37 per pound. By 1913, production of celluloid in the United States reached approximately 8.4 million pounds, excluding photographic film, with such standard prices as fifty cents a pound for white or black, sixty cents for amber or shell, seventy-five cents for ivory, and $1.25 for pearl. In other words, in 1913 roughly sixteen times as much celluloid was produced as ivory imported, and that celluloid could be had for less than a third as much per pound.[22]

Some of the increase in celluloid consumption reflected innovative uses for which ivory was inappropriate, such as transparent sheeting for automobile windows. Even so, for applications in which celluloid could and did substitute for ivory, the

17

artificial material definitely served as a democratizing agent, making possible production of a much higher volume of goods at lower prices. Celluloid did deserve credit for relieving pressure on supplies of natural ivory, which could be reserved for expensive toiletware and jewelry and for the billiard balls that ivory-celluloid compositions were never quite able to duplicate. Without celluloid as a substitute, the price of ivory would have risen inexorably, and a shortage of ivory might indeed have developed. Without celluloid, things that had to be made from ivory would have become ever more expensive luxuries. With celluloid, on the other hand, everyone could enjoy combs, brushes, and mirrors with "graining so delicate and true that you would think it could only have come from the gleaming tusks of some fine old elephant." Such formulations partially disguised the fact that lower price—and higher volume—constituted the bottom line of substitution. Rather frank in this regard, the Fiberloid Corporation noted in its catalogue for 1923 that its own brand of celluloid was superior to "animal ivory" primarily because it was "decidedly less expensive."[23]

Celluloid succeeded as a substitute raw material precisely because it *was* less expensive. But celluloid was hardly the alchemical material of Roland Barthes's mid-twentieth-century meditation, shapeless and therefore malleable, magically assuming any desired form, flowing continuously from chemical refineries into injection molding machines from which emerged the pristine artifacts of commerce and consumption. The process for making celluloid was too physical, dirty, and impure to inspire such rhetoric. To the extent that celluloid suggested an alchemical metaphor, it referred more accurately to the alchemy of history rather than legend, to an uneducated but clever craftsman working in isolation, jealously guarding unwritten secrets and operating by rule of thumb to create an uncertain material of variable quality. If the celluloid industry endeavored "to imitate the imperfections of nature," as a promoter once observed, it did so by means of imperfect methods that only gradually tended toward the perfectly controlled artificiality announced by Barthes.[24]

PROCESSES AND PRODUCTS

Plastic succeeded as a material of choice for manufacturing in the twentieth century not only owing to lighter, cheaper raw materials but because one-shot automatic molding operations eliminated the cost of separate fabricating, finishing, and assembling operations. But celluloid introduced no such savings. Manufacturing techniques remained those of the nineteenth century, with celluloid fabricators often directly borrowing techniques from the horn or tortoiseshell industries they were displacing. Despite the industry's claim that "standardization" was "the keynote of progress," it remained a labor-intensive, craft-oriented business.[25] Its competitive edge derived from its offering a somewhat more homogeneous raw material, an artificial substitute lacking some of the imperfections and irregularities of natural materials.

Techniques devised by Hyatt and his associates for manufacturing and fabricating celluloid long remained standard among the Celluloid Manufacturing Company and its competitors. With little regard for patent law and an aptitude for industrial espionage, a number of rivals set up shop in the 1870s and 1880s. During the 1890s, after Hyatt's basic patent expired, three of these proved viable: the Fiberloid Company, the Viscoloid Company (organized to supply comb manufacturers in Leominster, Massachusetts, with a substitute for scarce horn), and the Arlington Company of New Jersey.[26] Each of these firms not only adopted the same basic processes for producing celluloid as a raw material of manufacture but also engaged in fabricating and distributing similar end products—collars and cuffs, combs and brushes, and novelties. Producing the raw material for fabrication entailed a complex process: nitrating the cellulose; mixing nitrocellulose with camphor and other solvents to arrive at celluloid; creating various imitative and decorative effects with the material; and forming it into sheets, rods, and tubes, the raw materials of fabrication. The imprecision of the chemical operations, the predominance of mechanical manipulations, the frequent intervention of workers, the length of time involved, even the lumpy materiality of the process—all these conspired to produce an impression of gross physical rather than subtle chemical change.[27]

The first step, nitrating the cellulose, was also the most critical. Too much nitric acid yielded guncotton, a high explosive, rather than the moderately nitrated mixture appropriate for celluloid. An early rival of Hyatt, the Merchants' Manufacturing Company of Newark, organized in 1881, failed precisely because its nitration technique was faulty. Not until joining forces in 1886 with Joseph R. France, who knew something of nitration but nothing of molding or fabricating, did the group succeed as the Arlington Company. Even then nitration was carried out in small earthenware crocks stirred by hand under the eye of "an old German chemist who had largely forgotten his chemistry."[28] Very quickly, however, Arlington and other companies adopted a nitration process for which Hyatt had received a patent in 1878. At first the industry was careless about its sources of cellulose, but eventually it settled on white tissue paper made from cotton shirt cuttings to avoid impurities. Some companies even acquired paper mills to ensure a constant source of uniform supply made with pure water. After the rolls of tissue paper arrived at the plant, they were shredded into flakes no more than an inch across. The flakes were washed in water, dried, and finally stirred into a bath of nitric and sulfuric acids, the latter used to absorb the water given off during the nitration reaction. At its conclusion, the nitrated flakes were "whizzed" in a spinning wire basket to throw off excess acid, thereby reducing risk of explosion and recovering sulfuric acid for reuse. A series of fresh water baths, alternating with further spinning, completed the acid removal process. Until about 1907, the water was in turn removed by whizzing, blotting, and pressing the nitrated flakes; after that date, an alcohol bath typically served the same purpose, followed by a pressing out of the alcohol. If destined for transparent celluloid, the flakes were then bleached, and finally, still looking very much like shredded tissue paper, they were ground to a pulp.

While nitration proceeded in a subtle, almost invisible manner, as a series of washing and drying operations during which flakes of tissue paper retained their original texture and color, the next phase of manufacture entailed a dramatic change in appearance of the material, which assumed a more heavily physical presence. Mixing camphor as a solvent with nitrocellulose transformed the material into celluloid. Absolutely essential to the process, camphor was its weak link.[29] The supply of cotton tissue paper seemed cheap and endless. Du Pont, for example, touted its Pyralin celluloid as "American through and through," the offspring of "good old King Cotton himself, that benign potentate who rules so merrily over Dixieland."[30] That was not the case with camphor. A crystalline distillate of the wood of the camphor laurel tree, the solvent came from as exotic a source as any of the materials celluloid replaced and evoked equal fear of scarcity. It was originally imported from China, Japan, and Formosa, but by 1900 deforestation had left Formosa the only source. Mitsui & Company, an agent of the Japanese government, monopolized Formosan camphor in an unpredictable manner.[31] A German process for making synthetic camphor was announced in 1901 and promised an alternate source, but American celluloid producers established their own camphor plantations in Florida (Celluloid in 1907 and Arlington in 1914). After planting thousands of acres, both companies found their investments destroyed by pests and by the cost of labor. Various small American chemical companies toyed with making synthetic camphor, most notably during the First World War, but not until 1933 did Du Pont begin serious production.[32]

By whatever means camphor was obtained, it was combined with nitrocellulose in a mechanical mixer similar to those used by commercial bakeries. Additives included pigments, stabilizers, and enough alcohol to enable the camphor to dissolve the nitrocellulose at a relatively low temperature, despite "the statement sometimes made that a so-called dry method is in use."[33] After several hours of mixing, assisted when necessary by jacketed steam heat, the mixture assumed the appearance of heavy, puffy dough, or, in the words of a more technical comparison, it took on "the shape of a plastic mass somewhat resembling very stiff rubber." The resulting raw celluloid was "soft enough to be cut in sheets with a knife, yet hard enough so that the sheets will not lose their shape when being handled." It had become, according to the meaning of the word in 1919, "a pyroxylin plastic—a homogeneous mass ready to be rolled and worked."[34] The batch then entered a steam-heated box of paired rollers or masticators, each about eighteen inches in diameter, that worked it for several hours, the precise length of time depending on how well the mixer had done its job. The operator judged this calendering process to be finished when "a freshly cut edge of the mass" appeared "uniform and free from grains like sand." He then removed it, tightened the rollers, and ran it through again to produce a thin sheet from which "dirt, tacks, pieces of zinc, babbitt and other undesired objects" could be easily plucked.[35] Whatever else celluloid might have been, it was not a material of inhuman precision and purity (Figure 1–1).

From the rollers onward, each batch received different treatment depending on

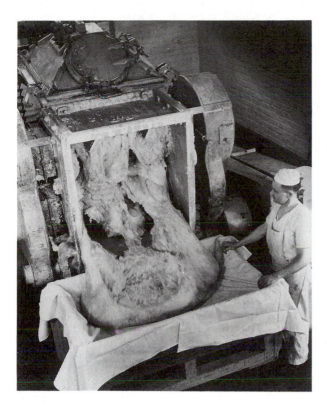

1–1. RAW CELLULOID AT THE DU PONT PYRALIN PLANT, ARLINGTON, NEW JERSEY, FEBRUARY 1932

the desired shape and color of the end product. Material destined to be hydraulically pushed through "stuffing machines" to extrude tubes and rods of varying diameters received extra time on the rollers to ensure complete homogeneity, while material destined to be planed into sheets of varying thicknesses still had several remaining stages of treatment to ensure its refinement. Some of the imitative effects, especially those based on mottling or marbling of colors, occurred at this point. To create imitation tortoiseshell, for example, an operator combined dough from batches of two different colors: a transparent amber and a richly translucent dark brown. While a thin sheet of amber turned through the rollers, the operator scattered onto it strips or blobs of brown in a pattern of contrived randomness. Often the resulting sheet was folded over or cut up and run through several times to obtain the desired effect.[36] Whether mottled, transparent, or solid-color, the continuous sheet formed by a single batch was removed from the calendering rolls, cut into rectangular sheets of uniform dimension, and stacked between the two plates of a hydraulic "cake press." While hot water circulated through the jacket of the press, hydraulic pressure of several thousand pounds further homogenized the material by compressing the layers into a single "cake" or block, which was solidified by circulation of cold water, removed from the press, allowed to cool further for a day, and then sent to the sheeting or planing machine.

In the early days Hyatt took the sheets straight from the rollers, but he soon discovered that uneven evaporation of solvent cracked the edges, and in 1878 he patented the sheeting technique that remained standard into the 1930s. The moving bed of the sheeting machine carried a block of celluloid under a stationary razor-sharp blade, thereby curling off a sheet of desired thickness—anything from three-hundredths of an inch to somewhat more than an inch. After retraction and raising of the bed the required amount, it was ready for the next cut from the block. Sheets of raw celluloid were "flabby, like soft rubber" owing to high alcohol content, and streaked by the knife.[37] During the first several hours after planing, they were strung up inside for immediate evaporation of most of the alcohol. Then they were moved to heated sheds for seasoning, a process that lasted from several weeks to six months depending on thickness. Owing to uneven solvent content and evaporation, the sheets emerged from the seasoning sheds in a wavy or buckled condition and had to be flattened and polished (to remove the knife streaks) by stacking them in a press between highly polished, mirror-smooth nickel-finished plates. Each sheet of celluloid came from the press "as straight as a die and as smooth as glass" and would "stay in this condition practically indefinitely."[38]

Some imitative effects required additional processing of seasoned sheets prior to polishing. Most common was imitation ivory, formed by an extra blocking and sheeting process. To simulate ivory's fine graining, workmen stacked up alternating sheets of rich cream and light sepia. Compressing the sheets in a hot press created a solid block, which was in turn upended and run endwise through a sheeting machine. The result was a sheet of cream with parallel lines of sepia running through it. But imitation ivory looked perhaps too regular to be mistaken for real ivory. In fact, a competitor of "Ripley's Believe It or Not" claimed in 1937 that flaws were "*intentionally made* in imitation ivory, manufactured from pyroxylin," owing to the fact that "the public once *refused to buy it* because it *was too perfect!*"[39] Most ivory celluloid revealed few if any deliberate flaws, but as one expert observed in 1911, a simple folding or crimping of the sepia-colored sheets before pressing yielded nature's arbitrary irregularities.[40] Whatever its surface colors, whatever its imitative characteristics, each sheet emerged from the final press polishing to receive equal treatment, to be trimmed, inspected, stacked between tissue paper, weighed, and packed with identical sheets in boxes—ready for sale to other manufacturers or for in-house fabrication.

The appearance of unnatural precision in a carefully trimmed and packaged stack of celluloid sheets represented a consummate illusion. Though certainly more regular in consistency and quality than a wagon load of cattle horns or a pile of elephant tusks, stacks of celluloid sheets in a warehouse were only as homogeneous as a highly imprecise process could make them. The manufacturing process itself exuded uncertainty and risk. During a single month in 1917, for example, more than a hundred injuries occurred at the Arlington plant, the most severe being a hand crushed in a mixer, another caught and lacerated in a rolling machine, and three fingers cut off under a sheeting knife.[41] More serious to management and investors was the danger of fire and explosion. Minor fires were frequent and major conflagrations not that rare. Arlington experienced a "disastrous explosion" in 1888, for example; triggered in the drying room, it killed two workers and brought about "an almost total destruction of the company's works." Only ten years later, a major fire that began in a seasoning room destroyed several buildings and led the company to purchase additional land on which to construct unconnected seasoning sheds.[42]

If the process was uncertain, so was the product. Jasper Crane, an Arlington chemist who continued on the board of directors after Du Pont's 1915 takeover, remembered "shell" that would "puff up, and give off acid vapors" if exposed to heat, and so-called "transparent" so murky that it could "almost be seen through" when held up to a bright light. And as for uniformity of color, full sets of Arlington's ivory toiletware were assembled by sorting through bins of supposedly identical pieces to create an ensemble in which everything almost matched. Nine pages of instructions for salesmen warned of complaints expected from customers: sheets that warped or shrank, mottled varieties that separated or bubbled at the break between colors, areas of imperfect transparency or opacity, imperfect color matching in a shipment of sheets, alcohol spots, and contamination with foreign matter. Consumers also had to worry about flammability, after hearing reports of an Ohio congressman whose celluloid visor caught fire, engulfing his head in flames, as he tried to light a cigar.[43]

There was no doubt about it. Celluloid was a messy, irregular, grossly variable, sometimes even dangerous material, an improvement on horn or ivory but a far cry from later plastics with their smooth precision and predictability (Figure 1–2). Not really innovative in most of its applications, celluloid substituted for natural materials not only in terms of physical properties and appearance but also in the way it was used. For all its vaunted uniformity, celluloid was only as good as the methods of working it, and most of those came directly from traditional industries. In its basic forms—sheets and tubes, that is—celluloid directly substituted for other materials. Tubes of celluloid, most often worked by being cut into short lengths for bracelets, napkin rings, and the circular walls of hair receivers and makeup jars, replaced

1–2. Celluloid sheets, rods, beading, and tubes manufactured by Du Pont Pyralin, 1919

sections cut from hollow tusks or horns. In similar fashion, celluloid sheets, the most common form of the material, substituted for horn or tortoiseshell that was softened by hot water and flattened into crude but workable sheets. Celluloid shipped from the factory thus served—like ivory, tortoiseshell, or horn—as a raw material of manufacture.

Fabricating celluloid entailed nearly as many operations as making the raw material. These operations required skilled workers using hand tools or small machines. Methods for making celluloid combs, for example, differed little from those for combs of horn or ivory. An operative used a small blanking press to cut comb-shaped blanks from a celluloid sheet. Teeth were formed either by cutting out the waste material between, or, in a "twinning" procedure more economical of material, by cutting a wide zigzag down the middle of a double blank that separated to form two combs. Piano keys, checkers, dice, and cutlery handles were cut from sheet stock and then stamped or molded to create curvature, surface indentations, or surface texture. To make circular hair receivers or other containers, a thin celluloid sheet was fastened to a concave die, softened in hot water, forced into the die by a jet of water, and immersed in cold water to harden. Or a section of tube, forming the wall of the container, was cemented with acetone to a circular base—a cheaper method that required less equipment and more labor. The frame of a hand mirror was blanked from relatively thick sheet stock, then pressed in a heated mold to obtain the desired contours. Alternately, for a cheaper mirror, thin sheets were stamped to fit a wooden core, then wrapped around and glued down. In all cases, fabrication required softening of the material in boiling water, on a steam table, or on a dry heat table covered with felt to prevent combustion. Sawing and turning operations occurred under spraying jets of cooling water, again to prevent fire. Polishing with pumice and buffing with acetone restored and heightened the finish of the material at the end of the fabricating process.[44]

Only ingenuity limited the ways of using celluloid to create complex artifacts, but all of them entailed many discrete cutting, shaping, and joining operations. Most ingenious was the process for making detachable collars of imitation linen, the first in a never-ending procession of plastic goods promoted for the fact that if soiled they could be "renovated with a moistened sponge," as the New York Times enthused in 1880. Detachable collars appealed to an emerging class of urban clerical and service workers who had to look presentable though unable to afford a clean shirt each day. Even the daily cost of a freshly laundered linen collar proved too steep for some people—thus stimulating imitations in celluloid or paper. In 1918 Du Pont pitched its advertising for Challenge Cleanable Collars to commuters, traveling salesmen, and railroad workers; two decades later waiters, ship stewards, and clerics comprised the chief customers. As Worden noted in 1911, the celluloid collar had to be "produced at a sufficiently low price to enable it to compete in quality with a linen collar on the one hand and in cost with a [disposable] paper collar on the other." Unlike some celluloid imitations, which satisfied only the criterion of visual

1–3. Challenge Cleanable shirt
collar of celluloid

appearance, a celluloid collar had to imitate the "thickness, strength, weight, color, resiliency, stiffness and flexibility" of "a starched linen fabric." It succeeded uncannily (Figure 1–3).[45]

To create the illusion, workers began by punching out collar-shaped blanks in two sizes—one slightly larger in outline than the other—from stacks of paper-thin white celluloid softened with castor oil. Raw material for a single collar consisted of a large blank, then a thin inner sheet of muslin for body, then a smaller blank. The edges of the larger blank were folded over the smaller and ironed down by hand to conceal the presence of the muslin and prevent its unraveling. Then the collar was pressed between linen toweling or between electrotyped plates to impart the texture of linen fabric. Exposure to steam heat for a week removed the excess solvent. Final steps included pumicing and polishing each collar between rollers, roweling its edges to simulate hem stitching, and punching in the button holes. To the casual touch of an unpracticed late-twentieth-century hand, the result could not be distinguished from stiff fabric.[46]

More innovative was a method for making cheap lightweight toys by blowing air between softened sheets of celluloid. As perfected by 1920 using a German process, two extremely thin sheets of celluloid were inserted between two heated dies and welded together at the edges by pressure, thus forming an airtight unit. Forced air introduced between the heat-softened sheets pushed them outward to line the cavity of the two-part mold. The process yielded a hollow one-piece three-dimensional toy, an animal or other figure, up to several inches long, with the shape, details, and texture of the mold (Figure 1–4). Depending on the size of the toy, a single mold could incorporate from six to more than a hundred cavities and thus produce that many toys in each cycle. This blow-molding process required less human intervention than most other celluloid fabrication techniques and directly foreshadowed plastic technology of the mid-twentieth century. All the same, manufacturers often combined several thin blow-molded parts to create intricate toys. And the parts frequently went outside the plant to sweatshops and individual families for assembly by hand, or for a detailed, many-hued application of paint.[47]

1–4. Blow-molded
celluloid rattles

In other words, celluloid required an inordinate number of processing steps even in its most innovative applications. All the same, development and use of celluloid suggested a drive to regularize and expand control over the things of everyday life, to limit human dependence on the vagaries of nature. On a geopolitical level this desire fueled a search for artificial substitutes for potentially scarce raw materials that had become indispensable to civilized life. International crises frequently threatened to disrupt supplies of ivory and tortoiseshell, gutta percha and shellac, and it seemed clear that a permanently expanding middle class threatened entirely to consume them. Even horn, a prosaic domestic material once in limitless supply, was no longer available, a victim of the cattle industry's rationalization. Although celluloid replaced some natural materials, it too depended on a dwindling natural resource of exotic origin, camphor, which in its turn had to be replaced by a synthetic substitute.

While celluloid promised to introduce order into the supply of raw materials, it also contributed to a relatively more rationalized manufacturing process. Even though traditional fabricating techniques predominated, celluloid proved a more tractable material than those for which it substituted. Horn had to be clarified, cut, flattened into irregular sheets. Plates of tortoiseshell had to be rasped smooth and compressed with others to form rough sheets. Ivory tusks often had soft or discolored patches, and their tapering, twisted forms contrasted unfavorably with precisely regular tubes of celluloid. Even the value of scraps and trimmings was greater in the case of celluloid. Lesser secondary uses existed for waste from ivory or horn, but celluloid waste went right back into the mixer to be combined with virgin "dough" and reused without loss of value. The inconsistency and impurity of celluloid seemed minimal by comparison with that of natural materials. Sheets of uniform size and precisely calipered thickness fostered routine production in a way that irregular sheets of natural materials, however processed and treated, simply could not. Blow-molded toys, even though economically insignificant and assembled by sweatshop labor, suggested an unprecedented degree of control over the manufacturing process. Gradually, despite the imitative aura of celluloid, its innovative tendency became apparent. New materials yielded artifacts that not only satisfied new desires, but whose artificial surfaces also acknowledged them. Going beyond nature's uncertainties toward heightened control of the material world seemed not only possible but desirable.

A WONDERFUL NEW MATERIAL

Celebration of innovation as celluloid's primary focus—rather than imitation or substitution—became typical by the late 1920s, with Du Pont leading the way. Known as a manufacturer of explosives, the company had been seeking other applications for its nitrating plants since 1907, when adverse "Powder Trust" publicity led the federal government to pursue antitrust prosecution and build its own munitions plants. Du Pont's promotion of peacetime uses of nitrocellulose increased after the First World War, which left the firm with tremendous profits but idle plants. Through a policy of expansion that began in 1910 and continued into the 1920s, Du Pont shifted its focus from explosives to chemicals by acquiring established companies or obtaining exclusive licenses in a number of cellulose-based industries, not only celluloid but also artificial leather (pyroxylin-coated fabric), artificial silk (viscose rayon), cellophane film, and paints and lacquers. In this new corporate strategy celluloid no longer owed its definition to its end use as a substitute for natural materials. Instead it gained its significance from an origin shared by many products with a similar chemistry—a fact that liberated Du Pont to emphasize celluloid's innovative potential as "a chemical product which science has developed and perfected" or, more grandly, a "wonderful material, the product of American cotton fields and chemical plants."[48]

This attitude achieved its apotheosis in a wonderfully ahistorical booklet published in 1928 to promote *Du Pont Pyralin: Its Manufacture and Use*. Designed with stylized *art moderne* graphics depicting chemists, retorts, microscopes, and factories, the booklet wrongly called Pyralin a new material, a "modern industrial plastic." Purportedly answering demands from "forward-looking manufacturers" for a material "readily adaptable to all sorts of mechanical processes," Pyralin took its place among the contributions "modern industry has exacted . . . from science during the twentieth century." Nowhere did the word celluloid appear, though the booklet outlined a manufacturing process little changed from that of Hyatt. An executive casually glancing over photographs of laboratories received an impression of Pyralin as a wholly new product of "the world-famous du Pont method of chemical control."[49]

This rhetoric indicated a conscious effort to overcome celluloid's reputation as an imitative substitute and to establish it as a material appropriate for truly innovative applications, for things impossible to achieve with ivory or horn or tortoiseshell. In that same year Du Pont first introduced celluloid toiletware that did not imitate natural materials. While continuing to market traditional patterns of ivory, shell, and amber under the Pyralin trade name, the company offered a new line supposedly "fashioned from a new material by a new process." As Du Pont informed dealers, colorful translucent Lucite made possible "a complete and basic re-styling of toiletware . . . so radically different, so smart and so modern" that women who already owned dresser sets would replace them. In reality, Lucite (not to be confused with the acrylic plastic of a decade later) was the same old celluloid enhanced by bright colors and "modernistic" styles. By 1934 the company had relegated imitative ivory, shell, and amber to the back of its catalogue and was focusing on garish Lucite artifacts "designed expressly to stress the streamline idea."[50]

With this change the most traditional celluloid objects shared in a modern spirit that marked several less ostentatious uses of celluloid. Into the 1920s, for example, the automotive industry used transparent celluloid sheeting ("clear as glass" and "flexible and tough as canvas") for side curtains, rear windows, and even windshields.[51] Eventually, during the 1920s and 1930s, as the closed car replaced the open touring car, celluloid sheeting literally vanished, placed between two sheets of glass as the plastic interlayer in safety glass (later replaced by other plastics).[52] In the meantime transparent celluloid sheeting filled the side and overhead windows of most airplanes, including Charles Lindbergh's *Spirit of St. Louis*, and appeared in safety goggles, tail light lenses, instrument dials, and drafting tools.[53] Other new uses ran from the mundane to the frivolous, as in thin veneers of pearlized colors applied to toilet seats, laundry hampers, and various smaller items (Plate 3), or transparent green visors like that sported by tennis champion Helen Wills in 1926—which became "a life saver for the sheeting end of the business" before the fad died away, apparently at the hands of bank tellers and professional gamblers.[54] Celluloid's most innovative application, for photographic film, proved more than a fad. According to Friedel, celluloid "entered the realm of 'high technology' " in photography and cinematography, but the industry "was foreign to this realm."[55] Leaving Eastman Kodak to manufacture its own film, other celluloid manufacturers remained stuck in traditional ways and maintained an illusion of innovation through weak expedients like Du Pont's *art moderne* Lucite.

Celluloid's curse, as well as its genius, was its propensity for imitation. But it also acquired a second curse, that it differed little in composition and fabricating techniques from materials it replaced. By 1937 a host of new plastics had emerged not from cotton fields and paper mills but from stills and condensers of chemical synthesis. In that year polymethyl methacrylate, a tough, rigid, lightweight transparent plastic, stole away the very name Lucite and joined the ranks of cellulose acetate, polyvinyl chloride, and polystyrene—truly innovative plastics that could be molded using compression or injection techniques, without the cost of fabrication by hand, to produce artifacts whose properties clearly transcended those of prior materials. In that same year industrial designer Jean Reinecke bemoaned imitation as "one of the greatest injustices to the plastics." The versatility that enabled plastics to imitate other materials also revealed them as substances "so definitely distinctive that they cannot be simulated by others." Rather than masquerading as the very materials they superseded, plastics had to express their uniqueness. According to Reinecke, "the design of a plastic product should be such that it cannot be effectively copied in another material."[56] Ironically these comments suggested nothing not implicit in the Celluloid Manufacturing Company's advertising booklet of 1878. But the balance was shifting from imitation to innovation; from replacing scarce natural materials to creating unprecedented materials; from improving old ways of doing things to creating new ways of conceiving the material world.

As a substance of the nineteenth century, celluloid had brought the romance of the exotic into the twentieth century. It was linked to the "huge pink maps" of empire, to the attractions of African safaris, native traders and carvers, the romance of sailing, and the promise of paradise of the South Seas.[57] As a material imitating vanishing rarities, celluloid looked to the past and promised to universalize luxuries once enjoyed only by adventurers and those who profited from their risks. But celluloid's techniques of manufacture remained little different from those of ivory or tortoiseshell. Celluloid addressed the scarcity of materials but barely brushed the surface of true democratization of manufactured artifacts. Eventually the plastic industry experienced a conceptual quantum leap by consolidating into a single molding operation the making of a solid material and the fabricating of a finished product—using as raw material powders or pellets derived from chemical synthesis.

This transition was already well under way when Du Pont purchased the Arlington Company in 1915 after technical advisers reported there were "no substitutes for celluloid." That need had supposedly long "been the subject of diligent study by able men but without results."[58] Two decades later, however, the Du Pont Plastics Department was injection-molding cellulose acetate combs at its Leominster plant, site of the former Viscoloid Company. A single injection machine with a multiple-cavity mold yielded ten fully formed, highly polished combs every fifty seconds. To publicize the achievement, Du Pont released photographs of two Leominster comb-makers, father and son, each posed with his daily output. While the old man gestured to a small pile of 350 celluloid combs, his son was surrounded by heaps of injection-molded combs, ten thousand in all. Celluloid combs sold in 1930 for about a dollar; those of cellulose acetate went for anywhere from ten to fifty cents at the end of the decade. And while the consumer thus benefited, Du Pont intended the continuity of generations to suggest that workers had not suffered.[59] The town of Leominster, which had easily retained its reputation as "the Comb City" after celluloid replaced horn around 1900, soon became known as "the Plastic City," a place where combs made up only a drop in the plastic bucket.[60] Celluloid itself almost disappeared in the last half of the twentieth century. After Du Pont closed down its celluloid operations in 1958, production in Leominster lingered on until 1972 in the form of pearlized toilet seats and heel protectors for women's shoes. Fifteen years later celluloid Ping-Pong balls remained, for which apparently no substitute existed.[61]

Long before the transition announced by Du Pont's father-and-son photographs of the late 1930s, celluloid had stopped embodying the innovative aspects of the plastic industry. In fact, injection molding of such materials as cellulose acetate and polystyrene challenged the position not only of celluloid but of compression molding materials like the phenolic resin Bakelite. If celluloid gave plastic its longstanding image as a substitute or an imitation, ingeniously clever or hopelessly inferior, it was Bakelite that during a quarter century of dominance established plastic's novelty, its protean versatility, its unique ability to become whatever one wanted. Bakelite challenged inventors to discover its uses and taxed designers to give it form. Its undefined status, so full of potential and so unlike that of celluloid, existed from the very moment of its inception in 1907. How to define a new material, Leo Baekeland found himself asking, and how to present it persuasively to potential users? In exploring these questions he and his associates defined the innovative aspect of plastic.

BAKELITE: DEFINING AN

ARTIFICIAL MATERIAL

*T*wo years into a campaign to make Bakelite a household word, the Bakelite Corporation in 1926 issued a booklet promoting *The Material of a Thousand Uses*. Stylish color illustrations traced a family's daily progress through a Bakelite world (Figure 2–1). Morning's shaving brush and percolator handles yielded to father's pipe, baby's teething ring, and a pen used by mother to check off purchases. Telephones, office machines, and billiard balls of Bakelite all came into play. Grasping the gearshift knob of the family car, mother encountered Bakelite without realizing its presence in the distributor cap under the hood. After listening to a radio with Bakelite parts in the evening, the couple retired to twin beds of Bakelite-lacquered brass. This "average day with an average family" suggested an "infinite number of present and future uses." The text invoked the "reality of modern magic, created by that modern magician—the Chemist," master of "the science of the transformation of matter."[1]

This abstract Chemist assumed heroic proportions in a 1920s scenario promoting Bakelite as a material of innovation rather than substitution. Aware of his goal from the beginning, the Chemist had expended "years of research work" to satisfy the world's desire for a material with "properties found wanting in other materials." The result, a condensation product of the reaction between phenol and formaldehyde, could be molded into any shape but became hard and infusible in its final phase. It resisted heat, electricity, and harsh chemicals. The Chemist himself described it as nearly immortal, possessing "the ultimate degree of chemical stability." This miraculous substance not only substituted for inferior materials like shellac, hard rubber,

AT breakfast, your wife pours you a cup of coffee; the handle she takes hold of on the percolator is of made Bakelite, as well as the button under the table she presses for service, and the twin-light plug from which are carried the wires to the toaster.

The Material *of a* Thousand Uses — BAKELITE

2–1. BAKELITE FOR BREAKFAST, 1926

celluloid, and porcelain. It also extended control over nature by harnessing electricity for lighting, manufacture, transportation, and communication. The Chemist had presented humanity with "a better material than any which Nature unaided has provided."[2]

Although Leo Baekeland, the real person behind the abstract Chemist, still controlled the enterprise he had founded, his name did not figure prominently in this promotional campaign. Maybe it seemed better to give the material an aura of scientific inevitability. Baekeland once rebuffed a journalist who wanted to profile him because, as he put it, "Men of my own class . . . look unfavorably at any such publicity." Even so, the popular press long regarded Baekeland as an inventive genius equal to Goodyear or Edison, a man who had overcome poverty and ignorant taunts to realize his vision. When *Time* put Baekeland on its cover in 1924, his celebrity allowed the magazine to accompany his portrait and name with no more than an allusive incantation—"It will not burn. It will not melt." Fifteen years later *Time* regarded him simply as "the father of plastics." After the Second World War, with a flood of new plastics and the disappearance of invention into corporate labs, Baekeland's stock fell. By 1985 *Newsweek* could refer to him as "a forgotten genius." At about the same time, however, a historian revived the Baekeland myth in a magazine devoted to great discoveries of the twentieth century. After describing him as "father to the family of versatile, exasperating, indispensable materials" that had made "the very feel of the twentieth century unique," Robert Friedel insisted that Baekeland had achieved something "so very different from anyone before him that the very stuff of which the world was made began to change."[3]

These comments introduced a new theme to the story of Baekeland as inventor. Earlier versions had presented him as in full control of a process that delivered Bakelite to a public which appreciated both the material's radical innovation and its logical necessity. Therein lay a paradox. "To invent" implied perceiving the need for something that did not exist, bringing it into being, and envisioning its ultimate potential. "To discover" a substance like Bakelite, on the other hand, however "beyond Nature" its essence, implied somehow its prior existence, or the inevitability of its emergence. Either way, the myth of the inventor allowed no fumbling starts and stops, no hesitation beyond the moment of conceptual illumination supposedly central to every true invention. Although Friedel recycled the notion of Baekeland as the father of plastic, the phrasing suggested not the heroic myth of invention but the uncertain passivity of biological parentage, its offspring subject to simple accident and the subsequent influence of others. Friedel had it right when he implied that Bakelite, not Baekeland, contributed to shaping the material parameters of the twentieth century—and in ways no one, certainly not Baekeland himself, could have foreseen or controlled.

To understand Bakelite's significance requires knowing how the material was interpreted—by Baekeland himself, by manufacturers who adapted it to commercial needs, by promoters who gave it popular definition, and by ordinary people who found it cropping up in the artifacts of everyday life. The material Baekeland "invented" or "discovered" in 1907 was indeed a product of chemical synthesis, a substance distinct from its reactive parts, a substance with no direct analogue in nature. But it did not become "the first synthetic plastic," a cultural as well as technological innovation, until its makers, promoters, processors, and users had incorporated it into the material world through a complex interactive process.[4]

2–2. LEO BAEKELAND, CA. 1900

AN INNOCENT ABROAD

Beneath an appearance of bourgeois stability Leo Baekeland (Figure 2–2) was a man
of margins, straddling uncertain borders, living between Europe and America, between a traditional working class and a new industrial aristocracy, between the purity of academic science and the brashness of industrial chemistry, and, finally, between the nineteenth and twentieth centuries. His career revealed a talent for putting
things in new perspectives. Baekeland was born in 1863 in a village outside Ghent,
Belgium, itself a country of margins.[5] His mother, a domestic servant, supported her
son's desire for education against the opposition of his father, an uneducated shoe-
maker.[6] Although Leo was apprenticed in the same trade at thirteen, his mother
encouraged him to attend night classes at a vocational school from which he graduated in 1880. A city scholarship enabled him to study chemistry at the University of
Ghent. Working with Theodore Swarts, a former student of organic chemist Fried-
rich Kekule, the young man moved quickly through his studies. After completing a
doctorate in 1884, Baekeland simultaneously taught at Bruges and served as his
mentor's research assistant. Although Swarts intended an illustrious academic career
for his most promising student, bringing him back to Ghent as assistant professor in
1888, Baekeland was attracted by the more tangible challenges—and rewards—of

industrial chemistry. A passionate amateur photographer, he became a consultant to a dry-plate photographic business and launched a commercially disastrous venture to market a developing solution. Swarts, who was anxious to save Baekeland for pure chemistry, criticized his applied work. The two often quarreled. To complicate matters, Baekeland had fallen in love with his mentor's daughter Celine. Two days after marrying in 1889, the couple left for the United States using funds from an award for academic study abroad. Officially Baekeland was on leave, but he did not return, either to the University of Ghent or to pure chemistry.

Soon after arriving in New York, he went to work for Richard A. Anthony, whose photographic company was locked in competition with Eastman Kodak.[7] There Baekeland worked on improving photographic printing paper, the greatest remaining obstacle to mass amateur photography after Kodak's introduction of roll film. By 1893 Baekeland had left Anthony and joined investor Leonard Jacobi in forming the Nepera Chemical Company at Yonkers to perfect and market Velox, a photographic printing paper. Most papers had to be printed under natural sunlight, but Velox could be printed under artificial light—enabling photographers to control the process with greater precision. Baekeland soon learned it was not enough to market a superior product, however; he also had to teach people to use it. One dissatisfied customer described Velox as "the greatest photographic swindle of the age." Another, unable to make good prints with Velox, insisted that "you cannot blame my insuccess on faulty manipulation, because I am a professor of chemistry." As Baekeland observed, the complainers "*knew too much* about photography and . . . never gave themselves the trouble of even glancing at the printed directions." He realized the significance of this experience only later when introducing Bakelite to manufacturers who tried to use it like rubber or celluloid or shellac rather than recognizing it as a unique innovation. Except for those on the margins receptive to new perspectives, it was sadly true that "routine holds sway over this world."[8]

The Belgian emigré became independently wealthy in 1899 when George Eastman offered him a cool million for the sale of Velox. Baekeland then established himself as an independent "research chemist and chemical expert" with a small laboratory in a barn on his estate, Snug Rock, overlooking the Hudson north of Yonkers. He installed his growing family in the main house, a rambling shingled structure of rustic stone with a round tower to which Baekeland escaped from the "unnecessary complications and artificialities" of life among the wealthy. Legend portrayed his retreat from active commerce to Snug Rock as a leisurely retirement, but it was nothing of the sort. Boredom with photographic chemistry provoked a search for other areas on the verge of commercial development. In later contemplating his career's radical shifts, his involvement in "opposite corners of chemistry," Baekeland explained how an "element of novelty" frequently revived his creativity and prevented him "from becoming too one-sided."[9]

Hydroelectric projects at Niagara Falls were then stimulating development of new chemical manufacturing processes. After traveling to Europe to survey the state of research in electrochemistry, Baekeland in 1903 became involved in a long-term effort to commercialize the Townsend cell, a device for electrolyzing sodium chloride to yield caustic soda and chlorine. The process was invented by Clinton P. Townsend, a former patent examiner, and Elmer A. Sperry, later inventor of the gyroscope. Abandoning the project to become a patent attorney, Townsend sold his rights to Elon H. Hooker, who engaged Baekeland almost full-time as a consultant supervising the installation of two cells in Brooklyn, next door to an Edison generating station. This pilot plant was so promising that Hooker began full-scale production at Niagara Falls in January 1906. For the next two years Baekeland traveled there frequently to iron out problems at the new plant, the nucleus of the Hooker Chemical Company. These trips provoked rising irritation, despite the income, because his career had again taken a radical turn.[10]

By then Baekeland was an enthusiastic supporter of all things new and different. He named his son for George Washington and struggled to overcome a Flemish accent. "How lucky," he exclaimed on a visit to Europe, that "I escaped these depressing surroundings to develop myself in [a] better more inspiring environment." His attitude toward America was like his wife's toward New York City, a place in which she took a "newcomer's pride and delight" as if "it had been her own creation." Baekeland's enthusiasm for his adopted country merged with a passion for its most visible, transcendent new technologies—those of transportation. He engaged in the sport of motoring before it became routine. At times he expended more energy on tinkering with automobile engines or his yacht's steam engine than he devoted to business. But he reserved his greatest enthusiasm for aviation. One Sunday morning in May 1910, as he and Celine sat on Snug Rock's piazza overlooking the Hudson, their reverie was shattered by a cannon salute from the yacht club. "There flies [Glenn] Curtiss," Celine shouted, and Baekeland got his "first sight of a flying machine in operation . . . gracefully flying ahead in a straight line over the river towards New York, about the height of the Palisades." It was "an admirable sight to behold . . . an intelligent man backing his intelligence by his courage." In a new century marked by such wonders the chemist indulged a mood of expectancy as he investigated the reaction that yielded Bakelite.[11]

A CHEMISTRY OF GUNKS

Baekeland was not the first chemist to investigate the condensation reaction of phenol (derived from coal tar) and formaldehyde (derived from wood alcohol). His own scientific papers revealed the subject had long received inconclusive attention.[12] In 1872 the German organic chemist Adolf Baeyer announced synthesis of a recalcitrant substance from phenols and aldehydes. Looking for a synthetic dye, Baeyer had no use for a product that neither crystallized nor yielded to chemical analysis. Nearly twenty years later, in 1891, Baeyer's student Werner Kleeberg returned to the reaction as part of a wider investigation of formaldehyde, which had just become commercially available. He heated phenol and formaldehyde in the presence of hydrochloric acid and obtained an irregular pasty resinous mass. It slowly became extremely hard, but violent release of gas at the beginning of the reaction rendered it porous, brittle, and useless as a practical material.

Around 1900 other chemists took up the phenol formaldehyde reaction as the expanding electrical industry created greater demand and higher prices for the natural resin shellac, used for varnishes to insulate coils and for molded insulators pressed from layers of resin-impregnated paper.[13] Some chemists began seeking shellac substitutes from among the so-called novolaks, phenol formaldehyde products that did not permanently harden but remained soluble in alcohol and fusible at high temperatures. Other investigators hoped to devise substitutes for hard rubber, which was also used for electrical insulation. Results differed from person to person and from day to day. A host of variables—proportions and sources of reactants, particular acids or bases used as catalysts, presence or absence of water, amount and timing of heat, amount of pressure—all conspired to produce a confusing array of syrups, porous masses, and amber solids, some of them soluble, some of them not, variously resistant to heat and electricity, and all of them resistant to precise chemical analysis. Nothing had changed since Baeyer and Kleeberg except clear economic motive, which convinced researchers to embrace a formerly despised "chemistry of gunks."[14]

Baekeland owed his success in exploiting the phenol formaldehyde reaction to a certain mental plasticity, a capacity for "interpretative flexibility"—that is, an ability to look at a situation from multiple perspectives.[15] As he once observed of Hyatt's invention of celluloid, "Sometimes it is a real advantage for a man of genius not 'to know too much' " because "too much book knowledge" might "petrify the mind" in a "one-sided point of view." Rather than focusing on a synthetic substitute for a single substance, Baekeland remained open to multiple applications. As it turned out, the phenol formaldehyde reaction offered something for everyone, a fact that stimulated Baekeland's eager promotion of the new material but also attracted a throng of potential users who besieged him with contradictory needs. A brief company history confessed that in the beginning Bakelite's "greatest drawback was the multiplicity of applications." This "Protean adaptability to many things" complicated matters but offered a key to Baekeland's success for the very reason that he recognized it as such.[16]

Just as important, he remained flexible while conceptualizing various processes for condensing phenol and formaldehyde. The search revolved around two general approaches, one by analogy from the process for making celluloid, the other by analogy from the process for vulcanizing hard rubber. Like so many other investigators, however, Baekeland began by seeking a shellac substitute for use as a varnish. The project probably began at the end of 1904 when he hired Nathaniel Thurlow, a Niagara Falls chemist who had patented a process for making synthetic camphor from turpentine.[17] They first investigated the novolaks as the phenol formaldehyde products most similar to shellac in susceptibility to heat and solvents—and thus in ease of handling. But that very convenience marked novolak as no improvement over shellac. Baekeland was next attracted to the infusible, insoluble, but porous and brittle product Kleeberg had described in 1891. Its hardness and chemical stability suggested an exceptional improvement over all natural resins, but the reaction had to be brought under control. Adolf Luft, a German chemist, had modified the Kleeberg process in 1902 by adding such solvents as camphor, glycerine, or alcohol toward the end of the reaction. As Baekeland later noted, use of camphor suggested Luft was following a celluloid analogy, trying "to make a plastic similar to celluloid."[18] Baekeland considered the result inferior to celluloid in toughness and flexibility. Furthermore, the solvent had to be evaporated slowly (rather than by heat) to avoid the porousness of a violent reaction, and this slow drying cracked and warped the final product.

All the same, Baekeland's assistant Thurlow continued thinking in terms of a celluloid model because his earlier work on synthetic camphor predisposed him to it. He hoped that some solvent, used with novolak or with the ground-up product of the Kleeberg reaction, would render phenol formaldehyde as valuable as camphor had rendered nitrocellulose. This fixed point of view delayed Baekeland's solution of the phenol formaldehyde puzzle and proved so frustrating that he and his assistant often quarreled. In June 1907, as Baekeland was finally recording success in his diary, he exulted that he had just emerged from "an exceedingly active period which allowed me to clear many mysterious reactions with which Thurlow has been struggling unsuccessfully since over a year." Almost a year later, with commercialization under way, Baekeland continued to complain about time wasted owing to his assistant's fixation on the "soluble products" of the celluloid model.[19]

Baekeland achieved success in 1907 by abandoning the idea of modifying Kleeberg's final product with a solvent and instead devising a means of controlling the violent reaction Kleeberg had observed. Baekeland's so-called "heat and pressure" method involved curing or hardening of a mixture of phenol and formaldehyde at a temperature of 150°C in an enclosed vessel pressurized to about a hundred pounds per square inch—enough to suppress the violent gaseous foaming that otherwise induced porousness. His product, whether applied to objects as a varnish and then cured under pressure, or used as a molding compound, retained the desirable qualities of hardness, stability, and resistance that had proven attractive in Kleeberg's fatally porous mass.

The earliest surviving outline of Baekeland's heat and pressure process is in a lab notebook entry dated June 20, 1907. By then his understanding was so advanced that the description differed little from his public announcement of Bakelite nineteen months later. Initial heating of phenol and formaldehyde using hydrochloric acid or zinc chloride as a catalyst (later replaced by the base ammonia) yielded a "liquid condensation product" referred to as A. Soluble in alcohol, acetone, or additional phenol, A became pasty with further heating and eventually became B, an "elastic rubberlike product" that was only partially soluble and could be softened by heat. Continued heating produced C, an "infusible, insoluble hard gum" that softened in boiling phenol and seemed to be the "last condensation product." When introduced into a "horizontal digester" (soon referred to as a "Bakelizer") and heated under pressure at about 150°C, any of the three (A, B, or C) would be transformed into D, the final product, completely infusible and insoluble under all circumstances. "I call it Bakalite," he declared (using a spelling that soon became standardized as Bakelite). In refining his description of the process, he soon merged products B and C into a single category B (the so-called "intermediate condensation product") and referred to the "Bakalized" material itself as C (the "final condensation product").[20]

Using countervailing pressure to contain gas that otherwise rendered the final product porous, brittle, and useless seems simple, obvious, elegant, and therefore a mark of genius. But Baekeland did not pull the idea out of thin air. The heat and pressure method was analogous to processes used with shellac and rubber, materials Bakelite eventually replaced. Luft had imitated the celluloid process because he hoped to make a celluloid substitute. Thurlow had done so because his experience came from areas related to celluloid production. Baekeland, who had little interest in celluloid, instead turned for inspiration to the electrical industry among whose chemists and engineers he had just spent several years. His heat and pressure method drew on processes for making electrical insulation from shellac and rubber. Shellac, for example, was "hardened" by heating it at 125°C in the presence of hexamethylenetetramine ("hexa"), a base that soon became a standard accelerator of the Bakelite reaction. This hardening process improved shellac's viability for heavy electrical applications by raising its melting point. Vulcanizing of hard rubber, which usually involved both heat and pressure, offered an even more direct parallel to Baekeland's heat and pressure method. Typically a mixture of rubber and sulphur was heated several hours at 250 or 300°C; the process was speeded by subjecting the mixture to a pressure of fifty psi. To make electrical insulators and other molded goods, heat and pressure were applied within a hydraulic press, later also the case with Bakelite; to make sheet goods, vulcanizing occurred in a sealed steam-heated vessel called a vulcanizer. Baekeland admitted the analogy as early as February 1907 in his first phenolic patent application, which described a preliminary version of his eventual process as "similar in some respects to the vulcanization of rubber products." Another chemist, H. Lebach, later bluntly recorded Baekeland's reliance on "a modified form of autoclave or vulcanizer called a *Bakelizer*." Just as Bakelite was superior to shellac, however, which often melted at high temperatures generated by dynamos and electric motors, so too was it superior to hard rubber, which shrank, cracked, and deteriorated with age. Bakelite, as Lebach put it, was "not a substitute for any natural or artificial material" but could be "judged on its own merits."[21]

Baekeland owed his material's commercial success to the fact that he immediately recognized it as superior to many other materials for many different applications. Capable of shifting from Thurlow's celluloid process model to a process model based on shellac or hard rubber, Baekeland was equally flexible in conceiving of practical applications. At the time of his discovery in 1907 he described numerous possible uses. For several months he had been concentrating on a varnish for "hardening wood with condensation products," as he phrased it on June 18, 1907, when he began his first Bakelite lab notebook.[22] Two days later, however, he had switched to heating test tubes of Bakelite A packed with asbestos fibers. The resulting "hard compact stick" suggested profitable use as "hard insulating masses." By varying the fillers used with the resin he would be able to "make a substitute for celluloid and for hard rubber"; using rubber itself as a filler would yield "an excellent rubber compound." Despite the reference to celluloid, he recognized that his material's strength came from the fact that unlike Hyatt's permanently plastic substance, Bakelite C could not be melted. In fact, he intended to make "plastic mixtures" that could be molded and "heated under pressure" to yield stable *non*-plastics.[23] At the end of the lab entry for June 20 he listed more than twenty potential filling materials, ranging from cotton to bronze powder to slate dust, and dashed off a list of applications ranging from linings of chemical apparatus to insulating varnish for dynamos to floor tiles "that would keep warm in winter time."[24] Although months of refining the process remained, he had defined Bakelite enough to consider sending this unprecedented new material out of the laboratory into the world.

Before doing so, Baekeland reflected on his work. On July 11, after mailing his heat and pressure patent application, he confided to his diary, "Unless I am very much mistaken this invention will prove important in the future." With his family on vacation he enjoyed "the luxury of being allowed to stay home in shirt sleeves." While "bakalizing" such materials as meerschaum, cotton, lamp black, graphite, and glass wool, he contemplated the plight of "slave-millionaires in Wall Street who have to go to their money-making pursuit notwithstanding the sweltering heat." Still alone on July 23, the twenty-third anniversary of his doctorate, he experienced "great excitement in opening all my tubes with Bakalite compounds" and congratulated himself on his good fortune. "I am again a student," he declared, "and a student I will remain until death calls me again to rest." Actually his days as a chemist were numbered. He was about to become a businessman, an entrepreneur, an associate of the "slave-millionaires" he so blithely dismissed.[25]

ANNOUNCING BAKELITE

The marketplace began to define Bakelite before anyone, including its creator, knew what it was. Baekeland in 1907 had no more idea about its chemical composition than had Baeyer or Kleeberg. With a nod to the pure chemistry of his youth, he sometimes attempted theoretical descriptions, but on one such occasion he dismissed his own presentation as "a mere matter of conjecture" based on "insecure theoretical notions."[26] If later polymer chemists claimed to produce made-to-order designer materials for specific markets, that was not true of Baekeland. Many of the uses he envisioned in June 1907 did prove successful—a varnish for electrical coils, for example. Others, such as floor tiles, never quite worked. Defining Bakelite through its practical uses proved as uncertain as the search for a chemical definition. For every application Baekeland demonstrated for potential customers, another came to him from an outside enthusiast. This element of unpredictability eventually generated an aura of the miraculous for plastic, an aura that first emerged as Bakelite defined itself in the minds of manufacturers and consumers who encountered and transformed it.

As Baekeland realized some of the potential of his discovery, he began "thinking about the best method for developing this into a substantial business that would not involve me too much."[27] At first hoping to avoid raising capital and taking in partners, he planned to patent specific applications, grant licenses, collect royalties, and neither manufacture nor sell the material himself. Toward the end of 1907 he applied for patents describing refinements of the basic process and use of the material as a varnish, as a binder for abrasives, and as a coating for electrical coils.[28] A few days before the formal announcement of Bakelite in 1909, protecting his turf from spoilers, he submitted specifications for using the material in self-lubricating bearings, as a molding compound, and in gaskets of resin-impregnated fiber.[29]

Owing to the difficulty of the phenol formaldehyde reaction and the inexperience of manufacturers accustomed to working with other materials, his non-involvement plan immediately faltered. The commercially diffident chemist soon dismissed as "erroneous" his optimistic prediction "that almost anybody would be able to make the new material easily, himself."[30] Baekeland discovered that not only would he have to make and sell molding compounds and varnishes ready for use by others; he would also have to offer laboratory services and on-site technical assistance to avoid the bad reputation their mistakes otherwise would generate. Eventually the use of trained engineers and chemists as salesmen became standard throughout the plastic industry. In the autumn of 1907, however, as Baekeland set out discreetly to publicize his invention, such a system would have seemed monstrous to him.

While continuing to experiment with catalysts and fillers, Baekeland began telling a few chemists and potential customers about his discovery. In August 1907 he first demonstrated his work for an outsider, Charles F. Burgess, chairman of chemical engineering at the University of Wisconsin. As Burgess examined lengths of Bakelized wood and smooth hard rods of cast Bakelite, some a transparent amber, others a patchwork of lines and seams owing to their fillers, he probably considered the material's use in the dry cell batteries on which he was working.[31] Two months later, an annual meeting of the American Electrochemical Society at the Chemists' Club in New York gave Baekeland an opportunity to discuss his work informally and distribute samples to a few chemists and engineers. Among those responding enthusiastically were Ferdinand G. Wiechmann, a prominent sugar chemist who soon became involved in developing Bakelite, and an engineer with the Norton Company, a manufacturer of grinding wheels. Within a week the latter had written to say he was already discovering "many new applications for Bakelite."[32] Rumors spread quickly, and by the end of the year Baekeland was swamped with inquiries ranging from the esoteric (electrical anodes) to the predictable (billiard balls).[33]

Baekeland quickly realized it was not enough simply to offer Bakelite to the right people. Norton's engineers reported after many tests and much correspondence back and forth that they had "not succeeded in making [grinding] wheels hard enough" with Bakelite as a binder; in fact it was "not as hard as shellac." Convinced of Bakelite's superiority to shellac, Baekeland concluded it was "best to carry our experiments out in our [own] lab." His assistant Thurlow began traveling back and forth from Yonkers to Worcester, a shuttle that became more complicated as other companies found they could not adapt to Bakelite without help.[34]

A typical pattern of cooperation emerged in Baekeland's long relationship with the Boonton Rubber Company, the first of the "custom molders" in the plastics industry—a company that bought molding compound from a "material supplier" to make parts for a product's primary manufacturer.[35] Located in Boonton, New Jersey, the company was founded in 1891 to recycle rubber from bicycle tires, with some of the output vulcanized for molded electrical insulation. A new general manager, Richard W. Seabury, had heard of Bakelite from Baekeland's former employer Anthony, who happened to be an officer of the rubber company. Ambitious to transform his marginal business, Seabury visited Snug Rock in January 1908, practiced molding the material, and returned to Boonton with samples. Within a month Baekeland had offered Seabury and Anthony an option on an exclusive license for Bakelite for molded electrical insulation. As proposed by the optimistic inventor, Boonton would order thirty tons of Bakelite A in 1908 at $500 a ton or forfeit up to $6,000 in penalty; by 1911 they would purchase 150 tons or forfeit $30,000. But it was not that easy. Four days later Baekeland spent an unsuccessful day tinkering with the molding process at Boonton and arrived home around midnight "very tired and fagged out."[36] His assistant Thurlow traveled frequently to Boonton throughout the spring with all thought of options and exclusive licensing abandoned. On March 28 Baekeland jubilantly reported "beyond doubt" that "pulverized solid A properly mixed will allow molding in a few minutes, less in fact than shellac." By June 6, when Seabury turned up at Snug Rock "with a set of the nicest Bakalite insulators," the chemist's optimism seemed justified. Two months later, after receiving an order for three hundred pounds of Bakelite A (nowhere near the projected thirty tons), Baekeland concluded, "It looks as if Bakelite was not going to disappoint me." Success seemed assured at the end of the year when the New York Central Railroad ordered 100,000 third-rail insulators to replace ceramic on electrified tracks. A year later Boonton built a new plant exclusively to mold Bakelite; within four years only 20 percent of their business came from rubber.[37]

This development work marked Bakelite's emergence as a plastic as the word is now popularly understood. The Boonton Rubber Company molded solid one-piece objects in hydraulic presses from a mixture of powdered Bakelite A and chopped asbestos fibers or wood flour. With application of heat and pressure to a cast-iron (later steel) mold, the Bakelite binder liquified, impregnated and surrounded the asbestos filler, circulated through the mold's enclosed volume, and solidified as Bakelite C—with a film of pure resin curing last, against the mold's surface, to yield a smooth, uniform finish. The processing technology came not from celluloid but from hard rubber. Even without considering inherent qualities of the materials, Bakelite offered a distinct improvement over both. Finished celluloid had to be cut, formed, pressed, or machined, and then fabricated—all steps requiring skilled labor. And hard rubber emerged from the mold in a rough outline of its final form and had to be machined to specifications that remained approximate at best. With Bakelite anything seemed possible once the molding process was perfected. Boonton's very first commercial Bakelite moldings, carried out in 1908 for Weston Electric, were tiny bushings for a precision electrical instrument. Remembered in the folklore of the plastic industry as "the size of mustard seeds," they were molded—not machined—to a tolerance of plus or minus one thousandth of an inch.[38] Three years later, when Boonton published a brochure for electrical manufacturers, the company guaranteed all work "accurate to dimension, every piece alike, ready to be taken from the stock bin and used." As Seabury's brother-in-law and fellow custom molder George K. Scribner recalled, "millions of parts could be duplicated cheaply," a feat contributing to expansion of the automotive and electrical industries and representing for Scribner "as much of a contribution to the stream of civilization as anything else we've ever seen."[39]

Boonton's success brought other molders and manufacturers to an appreciation of Bakelite. Most required as much assistance as Seabury. Baekeland frequently complained that "people do not know the technique of Bakelite and I have to teach them."[40] One customer admitted later he had begun using the molding compound without understanding "just what the nature of Bakelite was or what it was composed of, or the process that was followed in making [it]."[41] Customers blamed problems on the material itself, not realizing that "proper grinding, good dry materials, good mixing and specially good pressure and good application of heat at the right time during the pressing" also affected results. Writing to a chemist whose company was having difficulties, Baekeland claimed he had samples of moldings from "two different manufacturers, using the same mixtures, [that] varied so much as to give totally different results."[42] Despite his experience with Boonton, he found that "people who were proficient in the manipulation of rubber, celluloid or other plastics were the least disposed to master the new methods." Only those "who were not engaged in plastic before" were flexible enough to deal with a new situation. Most customers failed at first, and demands for assistance forced Baekeland to work on technical aspects of processing the material. He hoped to perfect those processes "beyond the stage where chemical knowledge or too much experience is required," thus minimizing damage done by hasty workers concerned only with final products.[43]

Early in 1908 Baekeland converted Snug Rock's barn into an expanded laboratory with office and library in the loft. He installed a hundred-gallon iron autoclave for "cooking" Bakelite A—looking like an elongated diving bell about five feet from top to bottom, equipped with a porthole and an agitator driven by a steam engine taken from an old automobile. A secondhand hydraulic press and Westinghouse air compressor enabled Thurlow to perfect molding techniques on site, and the hiring of chemist August H. Gotthelf freed Baekeland to handle business demands. Without realizing it he was constructing a pilot plant.[44] By this time, on the eve of the formal announcement of Bakelite, its creator clearly understood its innovative essence. Writing to a cutlery manufacturer, Baekeland introduced himself as "discoverer and inventor of a new material called Bakelite which is neither a compound nor a mixture, but a new synthetic chemical individual, an improvement on nature's synthetic methods for the building up of resinous substances." The material's "field of useful applications in the most varied branches of industry" was so "unusually large" that he was proceeding carefully "from fear of being swamped with helter-skelter, hasty trials which might work to a disadvantage in keeping up the [product's] good reputation." By manufacturing the raw material himself and licensing others to use it under his patents, he hoped to avoid the sort of "ponderous company organization" that would "paralyze my initiative or my freedom of action." His "research laboratory" was already "rigged up to deliver as much as one ton at a time," and he would soon be able to "produce Bakelite in any quantities."[45]

Baekeland first publicly revealed his phenol formaldehyde investigation and discussed its commercial potential at a meeting of the New York Section of the American Chemical Society on February 5, 1909.[46] The night before, as he sat in his study towering over Snug Rock, polishing his formal paper—"The Synthesis, Constitution, and Uses of Bakelite"—he felt anxious at "thus sending my work boldly into publicity." Arriving early at the Chemists' Club, he unpacked his samples and admired "a splendid framed exhibit" of Boonton moldings Seabury had brought—valve parts, bobbin ends, small casings, a spool for an electric coil, and miscellaneous industrial parts (Figure 2–3). The audience was larger than usual. The opportunity to confirm rumors had drawn businessmen as well as chemists. Old friends Anthony and Sperry were joined by such new associates as Joseph Steinberger of the General Insulate Company. The chemical elite had turned out in force, including Charles Baskerville, a leader of the American Association for the Advancement of Science, and Wilder D. Bancroft, president of the American Electrochemical Society, who had traveled from Cornell University for the occasion.[47]

Baekeland spoke informally, illustrating points with several chemical demonstrations. The talk divided naturally into four parts, a review of the work of Kleeberg and others, a presentation of his own investigation and process, a discussion of commercial applications of Bakelite, and, as a nod to the purists, a tentative chemical analysis based on Gotthelf's work.[48] Summarizing the material's qualities, Baekeland described it as "perfectly insoluble, infusible, and unaffected by almost all

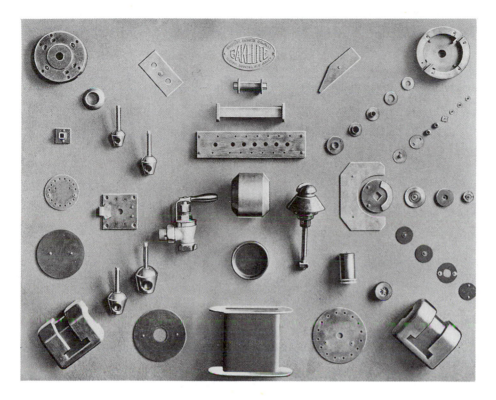

2-3. Bakelite Moldings by the Boonton Rubber Manufacturing Company, 1908–1909, displayed at Baekeland's public announcement of his invention

chemicals, an excellent insulator for heat and electricity." It was also so hard it
could not be scratched by a fingernail, a characteristic making it "far superior to
shellac and even to hard rubber." Its most serious fault was "lack of flexibility"; all
the same, Baekeland concluded, "as an insulator, and for any purposes where it has
to resist heat, friction, dampness, steam or chemicals it is far superior to hard rubber, casein, celluloid, shellac and in fact all plastics." Bakelite was also (stretching
the truth) "splendidly" competitive in price. He next discussed the material's major
applications—as a molding compound, as a protective coating, as an adhesive or
binder, and as a varnish for impregnating soft wood or cheap cardboard (yielding a
"hard resisting polished material"). Although it made admirable pipe stems, billiard
balls, knobs, buttons, and knife handles, he preferred developing the "so many more
important applications for engineering purposes." After closing by acknowledging
the help of Thurlow and Gotthelf, Baekeland received an enthusiastic ovation, flattering comments from Baskerville and Sperry, and a general round of informal
congratulations.

News of the announcement spread quickly by word of mouth and through the
New York papers.[49] Requests for information "came pouring in," and the chemist
often spent whole days "answering another batch of inquiries about Bakelite."[50] At
least one job-seeker asked to "join [the] Bakelite enterprise." Despite Baekeland's

response that he "did not intend to form a company," Hylton Swan did not give up. A nephew of Edison's lighting rival Sir Joseph Swan, he soon earned Baekeland's gratitude by introducing him to John Wesley Hyatt and eventually won a position.[51] Responses to the publicity perhaps led Baekeland to suspect his life was about to change. At this point he understood his material's potential and was in control of its development. But he failed to understand that his announcement, by expanding Bakelite's "multiplicity of applications," would trigger what he most feared—a "squandering of efforts in too many directions."[52] More to the point, he would lose control of the definition of Bakelite to outsiders, to custom molders and electrical manufacturers intent on their own ends, and to infringers of his patents whose so-called Condensite and Redmanol by their very names diluted the significance of Bakelite. Over the next fifteen years he struggled to maintain control of his invention despite centrifugal forces of business expansion, in-house technical improvement and diversification, patent infringers, larger technological innovations, and the public's growing awareness of the material. This struggle slowly generated a cultural product. By 1924 the meaning of Bakelite had come to a focus and was merging with a general definition of plastic as an embodiment of twentieth-century modernity.

GOING INTO BUSINESS

After the formal announcement Baekeland struggled to decide how best to exploit his patents, of which several, including the basic heat and pressure patent, were granted on December 7, 1909. A stream of information seekers encompassed an expanding range of industries. The S. S. White Company hoped to substitute molded Bakelite for hard rubber in denture plates, while the International Steam Pump Company wanted to line mining pumps with Bakelite varnish. The National Conduit Company discussed making pipes from impregnated paper tubes. Westinghouse Electric, soon to become his best customer, sent an engineer to learn "how to use Bakelite for impregnation of coils," and within a week General Electric sent someone on the same errand. Even manufacturers of "fancy goods," for which Baekeland expressed disdain, flooded the chemist with inquiries. Leopold Demuth, for example, who manufactured smokers' goods, approached him about using cast Bakelite to replace natural amber in pipe stems. By the end of 1909 hundred-pound orders from Boonton and Westinghouse were not uncommon, but at thirty-one cents a pound Baekeland was not going to make a fortune at Snug Rock.[53]

Clearly it was time to expand if only to ensure a comfortable living. Although Baekeland had continued to consult at Niagara Falls while investigating the phenol formaldehyde reaction, he was living on capital, often selling stocks and bonds for ready cash. He could not satisfy demand for Bakelite by making it on his estate, nor could he personally finance a manufacturing plant. The daily tedium of promoting and demonstrating various applications was wearing him down. "Sometimes I wish there was no Bakelite," he complained in a typical diary entry; "sometimes I wish the whole Bakelite matter was out of my hands." It was a measure of his own unclear definition of Bakelite that he spent much of 1909 pursuing the Protal Company as the major vehicle of his material's promotion and his own deliverance. Protal was the brainchild of sugar chemist Wiechmann, whose "greatest drawback," Baekeland later said, was "childish ignorance of the chemistry and technology of plastics, coupled with a misdirected imagination and inability to face facts as they are." Baekeland's own childish ignorance reflected a willingness to accept the claims of anyone who promised to take Bakelite off his hands.[54]

As an expression of turn-of-the-century enthusiasm for improving on natural materials, Protal was an unlikely substance. Conceived by Wiechmann even before he learned of Bakelite in 1907, Protal was intended as a substitute for hard rubber. This new molding compound included a binder of ivory dust, sulphur, and casein (a natural plastic derived from milk), and a filler of ground vegetable ivory. The latter came from the corusco nuts of the tagua palm, harvested in Ecuador and brought into the United States at about three cents a pound. Whatever the value of vegetable ivory as a filler, Wiechmann's binder had proven intractable, so he tested Bakelite after its 1909 announcement and eagerly ordered a hundred pounds of Solid A. The Protal Company quickly became so serious a prospect that Baekeland granted Wiechmann exclusive use of Bakelite with vegetable ivory. But even the optimistic Baekeland reacted with shock—"I guess this is some bluff or too much enthusiasm!"—when he learned that Protal planned to order five to ten tons of Bakelite every week. All the same, he loaned Thurlow to Wiechmann for days on end and referred promising job applicants to Protal—including two who later came back to work for him, Hylton Swan and engineer Herbert S. May.[55]

Despite lingering technical problems with Protal-Bakelite, Wiechmann publicly announced it in January 1910 as a rubber substitute compatible with all processing machines. He sent samples to eighteen electrical and communications companies, many of which had already dealt with Baekeland or with Boonton Rubber Company. In April Baekeland confided to Wiechmann his intention of forming a "General Bakelite company" based around Protal with its existing factory, laboratory, chemists, engineers, and workers.[56] That plan began to disintegrate over the summer, however, as Thurlow and Swan reported continuing difficulties with Protal-Bakelite moldings. The personal friendship of Baekeland and Wiechmann dissolved in disputes and legal threats. While Wiechmann claimed that variations in different batches of Bakelite made it impossible to produce moldings of consistent quality, Baekeland pointed to such factors as the expense of grinding corusco nuts, the inferiority of vegetable ivory to common wood flour as a filler, the incompetence of Protal's plant manager, and financial mismanagement. As Swan recalled, Protal engineers reverted in desperation to wood flour and asbestos as fillers—with "a 'pinch' of Protal so as to save the name." Baekeland soon regretted this flirtation and even feared that the Protal connection might "knock to pieces my whole prospect."[57]

47

If the Protal connection seemed insensitive in joining the fortunes of a new synthetic material to something as limiting as corusco nuts, then Baekeland recovered his senses when he formed the General Bakelite Company in September 1910. The main investor was Jacob Hasslacher, a partner in the Roessler & Hasslacher Chemical Company, which produced the formaldehyde used at Snug Rock and also imported phenol from Germany. Bakelite was thus clearly identified with its chemical sources and as a synthetic substance gained a measure of independence from the natural materials for which it substituted. Basing exploitation of Bakelite within the chemical industry also initiated a trend in the emerging plastic industry of clear separation of material suppliers from fabricators of finished products. Although Baekeland sometimes toyed with setting up his own molding operations, and a few suppliers of plastic materials also engaged in molding or fabricating, General Bakelite set the dominant pattern in 1910.[58]

Although General Bakelite remained linked to Roessler & Hasslacher, sharing a business office in Manhattan for six years, Baekeland did not escape the administrative routine he dreaded. He objected to being made president but finally acquiesced—so long as he was "not bothered with . . . office routine"—because Hasslacher assured him his presence would give the company "more prestige."[59] After all, Bakelite's inventor could best proceed with its commercial development. If such arguments paled, his own financial interest spurred him to accept the burden. Although Hasslacher was the major shareholder, subscribing $230,000 out of an estimated $330,000 raised, Baekeland received shares valued at $200,000 in consideration of his patents.[60] The day when members of the company paid in their subscriptions and the chemist assigned his patents seemed the "happy termination" of a long, painful process; within a few weeks, however, he was in place on the treadmill, complaining of the task of editing a promotional brochure that left him so dizzy he "had to give it up." The General Bakelite Company was in business.[61]

The thirty-four-page text of *Bakelite: Information Number One* introduced the company's products and their uses in a straightforward manner, even including plans for constructing a Bakelizer. It opened on a grand note by proclaiming just "What Bakelite Is." Using familiar phrases, Baekeland described it as a "new substance" offering "the advantages of hard rubber, Japanese lacquer, celluloid," and surpassing their properties. Not only did Bakelite improve on natural products, however; it also improved on nature's very own processes. "In the Synthesis of BAKELITE," he wrote, one finds "a laboratory reaction, which seems to run along lines parallel to those of the delicate physiological changes which occur in plant life." But the Bakelite reaction was "quicker and less expensive" than the painfully slow growth by which nature prepared her materials. By referring to the chemical process as an improvement on nature's, he suggested that the chemist approached the miraculous in his work and hinted at millennial prospects for the material realm. And by revealing a multitude of applications, he discarded imitation of any particular natural substance as a goal and invited people to make of Bakelite whatever they wanted.[62]

Although Baekeland oversaw promotion during the early years, his associates took over much of the business of the General Bakelite Company. Thurlow continued as technical troubleshooter for customers, while Gotthelf remained as chief chemist at Snug Rock for more than ten years—the inventor himself only rarely donning a lab coat. May served as business manager, Swan as sales manager, and Louis M. Rossi came from Roessler & Hasslacher to run the plant. The plant was installed first in a building leased from R & H's Perth Amboy Chemical Works and in 1913 moved to a new three-story building on adjacent land purchased from R & H.[63] Baekeland supervised the first batch of material at Perth Amboy on February 11, 1911, with results that hardly boded well. Intending to make two barrels (150 liters) of varnish, he found everything at the plant "wrong and unfinished." Workmen were inexperienced, electric lights went out, the steam engine was not running properly, chains kept jumping from gear wheels, and faulty agitation made the reaction potentially explosive. After losing his temper several times, he "finally managed to pull through without a complete break down" (whether of himself or the process was unclear), and they drew off the two barrels. As a finale, a workman peering from a platform into the open still with a kerosene lantern succeeded in igniting the fumes and burning off all his hair.[64] This incident reflected start-up chaos rather than standard operating procedure. Production rose sharply from about 100,000 pounds in 1911 to more than two million pounds in 1916. The General Bakelite Company prospered.[65]

Even so, a dispute with the Boonton Rubber Company in 1912 suggested Baekeland was losing control of his invention as a larger plastic industry emerged. Until then Boonton had functioned as a semi-official development arm of General Bakelite, a place for testing new mixtures and processes. For years Richard Seabury filled his lab notebook with experiments involving Bakelite compositions, fillers, additives, and dyes.[66] Given this relationship, Seabury was angry when some of his molding customers installed their own presses and bought material directly from General Bakelite. The issue came to a head after Baekeland returned from visiting the Remy Electric Company of Anderson, Indiana, makers of automotive electrical parts. There he had toured a "well equipped plant of considerable size" and admired its "nice equipment for molding." Seabury showed up at New York a few days later complaining that Remy had canceled all its molding orders with Boonton.[67] Caught in a "rather delicate position," Baekeland lamented Remy's failure to honor a supposed promise that it would only gradually establish its own molding department. In addition he warned Remy, with a carbon copy routed to Seabury, that he had seen "evidence of porosity and insufficient flowing" in some of their moldings and reminded them that Boonton was still "the pioneer in the art of Bakelite molding."[68] Seabury continued to grumble about industrial espionage and complained that his reward for doing Bakelite's "missionary work" was a loss of business "as fast as we get [it] established on a profitable basis."[69] Able to sell resin one way or the other, Baekeland advised Seabury to rest content in the industry's general expansion. No matter how many orders Boonton lost, there would always be new customers to keep Boonton's proposed new plant operating to capacity.[70]

This episode marked the emergence of a characteristic tension in the plastic industry. Custom molding companies, which remained small businesses with small profit margins, were often run by skilled craftsmen. They resented such material suppliers as Bakelite, which belonged to an increasingly monolithic chemical industry. Large suppliers often relied on custom molders for technical innovation but gave price breaks to in-house molding departments of large corporations that generated high-volume orders. As a postscript to the Boonton-Remy dispute, Westinghouse Electric announced only a few months later that it, too, formerly a Boonton molding customer, was "equipping a plant for making molded goods."[71] At that point, however, General Bakelite hardly figured as a chemical giant. Baekeland had already learned that he too was not immune from competition. Despite the patent wall he thought he had erected, other chemists had recognized Bakelite's potential and sought alternate processes for achieving the same end. By 1912 they were threatening Baekeland's enterprise. His major reason for turning his back on Seabury "in direct violation of the spirit of [their] agreement," as the latter complained, was the success of Condensite, a so-called "infringing material" whose availability freed Remy and Westinghouse from dependence on Boonton even if General Bakelite refused to sell directly to them.[72]

FIGHTING THE PIRATES

The Condensite Company of America was incorporated in 1910 to commercialize phenolic materials and processes of Jonas W. Aylsworth, chief chemist of Edison's laboratory at West Orange, New Jersey.[73] Unlike Baekeland, who considered Bakelite a material of diverse uses, Aylsworth sought in the phenol formaldehyde reaction a solution to a specific problem: a hard material for phonograph records. Learning of Baekeland's announcement in 1909, Aylsworth rushed to develop somewhat different processes for reaching the same end. His first patent applications, filed less than three months after Baekeland's presentation at the Chemists' Club, finessed the heat and pressure method in a number of ways. Two applications specified rotating molds to generate centrifugal force that would prevent the foaming that disfigured so many phenolic products.[74] Others covered a more fundamental chemical process, the so-called "two-step wet process," which Aylsworth conceived of as distinct from Baekeland's method.[75] In the first step Aylsworth reacted solutions of phenol and formaldehyde to form a permanently fusible, soluble product, similar to Baekeland's B but capable of being heated to drive off water at a temperature so high it would have turned B into a porous, worthless C. In the second step Aylsworth combined his fusible product with a wet solution of hexa in a reaction that absorbed no water and gave off no gas. Although the final product was a hard, infusible substance chemically similar to Bakelite, Aylsworth emphasized that his process required no countervailing pressure and therefore differed fundamentally from Baekeland's heat and pressure process. Not only that; the resulting product contained no water, not even the traces found in Bakelite, and promised to provide better electrical insulation.

When it became obvious Edison was not interested in exploiting phenolic plastic beyond phonograph records, a group of investors joined to commercialize Aylsworth's work. Among them were businessman Kirk Brown and Frank L. Dyer, president of the Edison Company and a patent attorney who vouched for Aylsworth's originality. With Brown as president, the Condensite Company came into existence in September 1910, six days before the dilatory General Bakelite Company. It quickly installed a plant in an Edison building at Glen Ridge, New Jersey (moved to Bloomfield in 1914). Two months later Dyer paid a visit to Baekeland, who already knew what was afoot. It was small comfort to learn Condensite was "not yet doing any business, only experimental." For a while Baekeland heard little more, but he chafed at having to teach customers to use Bakelite. "Afterwards," he lamented, "some pirates will have the benefit of my educational work!" During 1911 and 1912 evidence of Condensite's threat trickled in. Seabury reported being offered Condensite molding compound at a competitive price. General Electric's plant at Pittsfield, Massachusetts, began purchasing Condensite at a price Baekeland could not match. Even worse, Condensite was proving technically superior to Bakelite for some uses.[76] In a series of skirmishes with Aylsworth in *The Journal of Industrial and Engineering Chemistry*, Baekeland proved himself an aggressive combatant. He argued that Aylsworth's intermediate product was a common novolak, that his own specifications covered hexa as an intermediary formed by formaldehyde reacting with the catalyst ammonia, that Aylsworth's final product could not be cured without pressure, and that Condensite was not as hard or infusible as Bakelite. Having nothing to lose, Aylsworth rested his case with a clever insult, noting that his non-pressure method required no "vulcanizer or equivalent apparatus" as in "the rubber industry."[77]

Negotiations to reach an accommodation with Condensite began in November 1911 but languished until June 1912, when a patent infringement suit was decided in Baekeland's favor. Four months later Condensite agreed to a licensing arrangement and finally signed in May 1913. But Baekeland had won a questionable victory. Although he received royalties for every pound of Condensite, its very existence diluted his own material's value and questioned his claim to a fundamental innovation. He continued to lose sales to Condensite when customers found it technically superior or less costly. Even more threatening was the specter of other "infringing pirates," a fear that materialized in October 1912, even before the Condensite compromise, when he learned of another phenolic substance that eventually came to market as Redmanol.[78]

The third major phenolic contender derived like Condensite from research into a specific application. Adolph Karpen, president of S. Karpen & Brothers, a furniture company in Chicago, sought a synthetic varnish more durable and less expensive than those derived from natural products. Using the program of industrial fellow-ships at the University of Kansas, Karpen in 1910 funded research by a young Ca-nadian, Lawrence V. Redman, assisted by Archie J. Weith and Frank P. Brock. Like Aylsworth, Redman eagerly read reports of Baekeland's work as a solution to his own problem. But Redman found Bakelite varnish unacceptable because its water content raised the grain of wooden furniture. A professor at Kansas who had inves-tigated the embalming of corpses suggested Redman abandon aqueous solutions of formaldehyde and experiment instead with hexa. Success would mean an alternate process not protected by Baekeland's patents and a similar but superior product.

Redman and his associates spent four years perfecting a phenolic varnish based on a so-called "one-step dry process" in which phenol containing no water was heated with an expensive dry form of hexa. The reaction gave off ammonia gas and left a porous intermediate product that could be ground up and then dissolved in alcohol as a varnish, or combined with more hexa to produce a syrup for casting or a compound for hydraulic molding. Both products cured without pressure, if some-what slowly, and Redman's process avoided Baekeland's problem of separating out large quantities of water. Careful to distinguish his process, Redman claimed it avoided "serious difficulties" of "the old wet formaldehyde process."[79] Early in the development process, in June 1910, Redman filed an initial patent application that was unsuccessful; a subsequent application in February 1911 yielded the first of many patents on August 18, 1914.[80] By then Redman and his associates had left Kansas for Chicago to work for Karpen's newly organized Amberoid Chemical Products Company. A trademark conflict resulted in a new name—the Redmanol Chemical Products Company.

Baekeland confronted his latest competitors and presumed infringers in Septem-ber 1913 at an American Chemical Society meeting at Rochester where Redman, Weith, and Brock all presented papers on what he dismissed as "a shabby impudent copy" of his own work. Although he enjoyed describing Weith and Brock as "two clumsy hayseeds" in his diary,[81] he had to face Redmanol's competition in 1914 and soon became involved, as with Condensite, in a grinding battle conducted both in *The Journal of Industrial and Engineering Chemistry* and in court. Baekeland's opening salvo hinted at the rage provoked by a patent system he had trusted to protect him. Replying to a paper published after the ACS meeting, he began not by addressing his competitors' scientific claims but by ridiculing their academic pretensions. Al-though they signed their paper from the University of Kansas, their work resulted from "no more nor less than one of the so-called industrial research fellowships . . . paid for in the interest of purely commercial enterprises." To a disingenuous Baeke-land that fact "offset somewhat any prestige the mention . . . of the University of Kansas may bring about." Moving on, he accused them of disguising the old familiar novolak with the absurd term "phenyl-endika-saligeno-saligenin." Despite all the fuss about "wet" and "dry" processes, he concluded, tests revealed the "absolute identity" of their products. In other words, Redman was making Bakelite.[82] Subsequent exchanges became increasingly bitter, with Redman's group referring to a benighted Baekeland as "the Doctor" and he in turn applauding their work "as an act of loyalty to the dry State of Kansas."[83]

Debate centered around Baekeland's specious claim that their process was derivative. He maintained that in his process ammonia (an alkaline catalyst) reacted with formaldehyde to form hexa, which then reacted with phenol. As "any chemist knows," he lectured, ammonia and formaldehyde on the one hand, and hexa on the other, were as alike as "tweedledum and tweedledee." Baekeland had thus used hexa from the beginning, even before Aylsworth arrived on the scene, and Redman's process was nothing new. In fact the Redmanol process *was* innovative, and Redman demolished Baekeland's argument by printing in parallel columns quotations from the latter's writings showing definite changes in his understanding of the role of hexa since 1909, when he had actually warned that hexa's presence must be limited.[84] Baekeland had the last word in 1921, however, when U.S. District Court Judge Thomas I. Chatfield ignored the sparring over process and accepted Baekeland's insistence on the "absolute identity" of the final products. In the meantime, however, he continued to worry about losing control to the "pirates" of Condensite and Redmanol.

Baekeland especially feared losing an important new market for phenolic varnish—as an impregnator and binder of laminated sheet materials—that was growing faster than consumption of molding compound.[85] The chemist had experimented early on with "polished Bakalite fiberboard," "Bakalized pulpboard," and "composite cardboard"; by 1910 Westinghouse Electric was using Bakelite varnish as a substitute for shellac in its Micarta insulation board.[86] In a process devised by Westinghouse engineer Daniel J. O'Conor, a continuous ribbon of kraft paper passed through a bath of Liquid A and then through a drying chamber before being cut into resin-impregnated sheets. Stacked in layers and compressed between the heated plates of a hydraulic press, where final curing occurred, they emerged as a single rigid laminated sheet with the chemical and electrical properties of Bakelite, ready to be cut into shapes for insulating panels and supports. This Bakelite-Micarta, which was a considerable improvement over shellac-impregnated Micarta, was soon joined by Bakelite-Micarta-D made from layers of cotton duck, used for blanks for cutting noiseless gears for automotive timers and machine tools.[87] Recognizing the potential of his process, O'Conor left Westinghouse and in 1913 began the Formica Insulation Company at Cincinnati. After a couple of lean years O'Conor and his partner Herbert A. Faber, a former Westinghouse salesman, established Formica as a viable competitor of Micarta for commutator rings, winding tubes, and mounting panels for radio sets—all at Baekeland's expense. Outraged by O'Conor and Faber's defection, Westinghouse threatened Baekeland with loss of its considerable business if he sold resin to the renegades. So he denied them access to Bakelite and forced them to use Redmanol varnish instead. As a result his diary was sprinkled with mutterings about "imitation processes . . . Redman Formica etc. etc."[88]

2–4. Autographic Kodak with end panels of "pressed Bakelized paper" (laminate), 1915

On a more positive note, evidence of success kept pouring in. Although Baeke-land's basic patents were surrounded by a network of interconnected applications patents, he could never have predicted how people would use Bakelite—and Condensite and Redmanol. Late in 1913 he proudly observed he was writing with his first Bakelite pen, a Parker model. Two years later George Eastman sent him the Model 1-A Autographic Kodak Special, whose bellows extended from a case with two side panels molded from Bakelite-impregnated paper (Figure 2–4). Like anyone encountering Bakelite at the annual trade shows, Baekeland noticed magnetos at the motor boat show, inspected self-starters, gearshift knobs, radiator and gas caps, steering wheels, and door handles at the automobile show, and bought several Bakelite-lacquered brass beds at the furniture show.[89] As a member of the Naval Consulting Board during the First World War, he noted "with pleasure" that Liberty airplane engines had "two well molded Bakelite caps covering the Bakelite insulated spark distributors." During the war phenolic plastic was used for uniform buttons, battery cases, airplane propellers of laminated wood, and as replacements for tin caps of shaving cream tubes.[90] Earlier, in 1915, touring the Delco plant at Dayton with Charles F. Kettering, inventor of the automotive self-starter and later General Motors director of research, Baekeland proudly observed "several of our steel barrels of Bakelite being rolled in." Astonished at the array of molding presses, he declared it "truly a wonder" that in so short a time "the automobile industry of the United States should have developed to the point that one single accessory . . . should necessitate several plants like this (now 1,500 employees)." Bakelite, he implied, had contributed to that expansion.[91]

Despite the inroads of Condensite, Redmanol, and Formica, despite the increasing importance of Westinghouse, General Electric, Eastman Kodak, Delco-Remy, and other large independent manufacturers, not to mention custom molders such as Boonton, Kurz-Kasch of Dayton, and Northern Industrial Chemical of Boston, all defining and redefining Bakelite's uses and meanings, Baekeland was acknowledged as the industry's central figure. When he received the Willard Gibbs Medal at a meeting of the American Chemical Society in May 1913, the presenter observed that "the chemical inventor . . . fills a new want as often as he does an old one . . . or he may even create a want where there was none before." That recognition meant much, coming from colleagues and associates who appreciated his work even when concealed within power stations or motor cars. Public recognition came more slowly, and he derived as much pleasure from a brief private encounter as from any professional ceremony. While boating on Lake George in the summer of 1918, he stopped at a country store. As he exchanged names with the garrulous proprietor, Baekeland remarked that his own would prove hard to remember. "Oh no," said the storekeeper. "I have a way of remembering names . . . for instance when I think of your name I shall think of Bakelite." Not wanting to spoil the moment, the chemist did not reveal himself but noticed the man was "smoking a pipe with a Bakelite bit."[92]

If Bakelite, a prototype for plastic in general, did not yet radiate an aura of good or ill, at least it was filtering into public consciousness. But to become a household word it had to emerge from industrial and electrical uses into the realm of "fancy goods" for ordinary consumers. Bakelite also had to assert its identity against the centrifugal forces of Condensite and Redmanol in a struggle that ended with con-solidation in 1922. Endowed with the strength of synergy, the new company then set out to present a positive public image for phenolic materials, which Baekeland himself described as "peculiar for the reason that their main qualities are so-called negative qualities, their inertness."[93]

THE BAKELITE CORPORATION

Consolidation emerged from a patent infringement suit filed in 1917 by General Bakelite against General Insulate of Brooklyn, a molding company that had aban-doned Bakelite for Redmanol. Baekeland and his attorney Townsend were suing General Insulate as a means of getting at Redmanol, whose Illinois location made legal action difficult. If successful in winning a decision for priority of their basic patents over Redman's, they intended next to sue Formica for infringing Baekeland's "composite cardboard" patent by using Redmanol. The trial ran for a month in the spring of 1919 at the U.S. District Court in Brooklyn. Baekeland enjoyed the pro-ceeding because it forced him to return to the laboratory to prepare demonstrations. His "short review of the art of plastics" proved so compelling to Judge Chatfield that when the defense moved to strike it from the record, he countered that it ought to be "copyrighted and printed."[94] While Redman argued the distinctiveness of his hexa process, Baekeland argued the virtual identity of the final products Bakelite and Redmanol. Although a quick decision was promised, Chatfield delayed two years until August 1921. In the meantime Baekeland's courtroom euphoria vanished and he entered a "doldrums" so severe that he came to "hate the whole Bakelite enterprise" with its "infringers" and "pirates." When the judge finally pronounced in his favor, Baekeland personally wrote the press release announcing victory and carried it by hand to chemical industry editors.[95]

A number of factors disturbed the precarious balance of the phenolic industry at the time of Chatfield's decision. The economy had entered a recession so severe that General Bakelite's sales plummeted. In 1921 the firm experienced the first major decline of its existence. Actual poundage of material sold decreased by 57 percent from 1920; gross income declined even more, by 68 percent, owing to a drop in average price from sixty-five to forty-eight cents per pound.[96] The business slump added to Baekeland's depression over the delay in settling the patent situation. For several months before the decision he toyed with selling out to the Hercules Powder Company or to Union Carbide, or with joining Condensite under Kirk Brown's presidency. Immediately after Chatfield's decision, Baekeland and Brown began negotiating a merger.[97] They had the phenolic business legally tied up. No competitor could move without infringing the patents of one or the other, shared through cross-licensing. Baekeland failed to reckon with Adolph Karpen, however, whose investment in Redmanol was rendered worthless by Chatfield's decision. Knowing Baekeland would never license anything to Redmanol, Karpen secretly took an option on a controlling interest in Condensite and in November 1921 announced he had obtained access to all of Condensite's Bakelite licenses. Baekeland fumed that his "relations with Condensite Co under former control, and Condensite in the hands of [his] bitterest enemies were two different things," but he had no choice other than to reach an accord.[98] Unwilling to allow Karpen and Redman to continue sniping at him as licensees, he proposed consolidating the three companies.

During protracted negotiations Baekeland came to distrust his former ally Brown as "petty and selfish." Karpen, on the other hand, seemed "a dominant personality . . . desirable as a guide and a counselor in any business," and the former "pirate" Redman became a "real asset," someone with whom Baekeland could "get along splendidly" because their "tendencies and views" ran "very much along [the] same lines."[99] Despite continuing problems with Brown, who seemed to realize he had not as much to gain as the others, they devised a formula for determining the relative worth of each company, signed a secret agreement in December 1921, held the first board of directors meeting on May 8, 1922, and then publicly announced formation of the Bakelite Corporation as a holding company for its three operating branches. Once again unable to shake responsibility for his invention, Baekeland was elected president, with Karpen, Brown, and Philip Schleussner of R & H serving as vice presidents.[100]

A holding company did not guarantee true consolidation, however, nor even a single corporate identity. Each branch retained its own officers, plants, products, and trademarks, comprising a welter of different names for substances a federal judge had declared identical. Not only did Bakelite, Condensite, and Redmanol vie for attention. So did Westinghouse's Bakelite-Micarta, Continental Fibre's Bakelite-Dilecto, Diamond State Fibre's Condensite-Celeron, and Formica Insulation's Redmanol-Formica—all laminated sheet materials. At least one custom molder devised its own name and trademark, with Boonton offering a "Boonton Bakelite" brand distinguished by mirror-image B's back-to-back inside an oval. Baekeland hardly improved matters when he suggested calling the new corporation Synthoplastics or Neoplastics. His colleagues deferred to his inventive priority, however, and allowed Bakelite to stand. Shortly after incorporation Redman even suggested (in what Baekeland called a "very sensible letter") that they simplify their "complicated advertising by concentrating on [the] Bakelite trade mark and omit Condensite and Redmanol whenever desirable."[101] The three constituents formally dissolved themselves and made the Bakelite Corporation an operating company during the summer of 1923.

Full consolidation could not have come at a better time. By 1923 general prosperity had returned. The Bakelite Corporation was benefiting from an increase in automobile sales and a large phenol surplus left from the war.[102] Even more important was radio's popular emergence. Military demand for radio sets during the First World War had strained manufacturing capacity, but the real boom came between 1920 and 1924, when civilian enthusiasts began assembling sets at home and scores of companies supplied ready-made equipment. Most sets were assembled on laminated phenolic mounting panels or chassis that insulated coils, tuners, tubes, and other parts from each other. A typical radio set also possessed a front panel on which its molded Bakelite dials and knobs were mounted—an expanse of black Bakelite or Formica whose flawless gleaming surface derived from polished stainless steel plates used in the material's final compressing. The Bakelite Corporation's sales nearly doubled from 1922 to 1924, when the total topped seventeen million pounds; Formica's annual sales jumped from $400,000 in 1920 to $3 million in 1924. In April 1922 both Rossi at Perth Amboy and Redman at Chicago sought Baekeland's approval for new stills and grinding mills to meet the demands of the "Radiophone" industry. A laminate salesman later recalled the "great radio 'do-it-yourself' boom" by exclaiming, "What a difference, no peddling, just taking orders." Baekeland returned from a radio exposition in 1923 to report that Bakelite was "used almost everywhere, and prominently advertised in the printed matter of almost all its users." As a publicist wrote in 1935, Bakelite "became a household term" with the advent of radio.[103]

By coincidence the Bakelite Corporation was consolidating its products and creating a unified image just as radio was domesticating phenolic plastic, admitting it into the American parlor as a visual component of an innovative technology and form of entertainment. Bakelite thus shared in the aura of the new. It became easier to use Bakelite in other consumer goods that frankly announced their material independence, their rejection of the imitative tradition of plastics like celluloid. Bakelite's success in radio contributed to a public willingness to accept other uses unprecedented in visual appearance, tactile qualities, or function. That willingness had to filter into consciousness and be recognized by promoters of Bakelite, however, before it could be presented to manufacturers and consumers in an organized manner.

BAKELITE TILL KINGDOM COME

The new Bakelite Corporation began to emphasize advertising, public relations, and image building—in effect to define Bakelite's innovative distinctiveness. Baekeland never quite understood how or why that had to be done, but he recognized that General Bakelite had not effectively presented its products to customers. Back in 1921, while negotiating with Brown of Condensite, he confessed "that our present management is not good as far as the selling or introduction end is concerned." Sales manager May had disappointed him, and it was clear that neither his son George nor son-in-law George Roll, both employees of the company, saw things as he would have liked. Brown, on the other hand, had three sons, Sandford, Allan, and Gordon, all involved with Condensite. Baekeland told Brown he had "no objections if one of his sons, perhaps two," took positions in the new company.[104] As it turned out, all three became central figures even though their disruptive father was simultaneously forced out. Although Baekeland's "two Georges" received administrative positions, while Redman became vice president for research, Rossi vice president for manufacturing, and Swan sales manager, the Brown brothers transformed the way the company did business. Sandford managed a former Condensite subsidiary, Gordon worked with Swan in sales, and Allan, who had learned every aspect of the phenolic business from lab work to production to sales and accounting while still in high school, became director of advertising and public relations. As a Bakelite chemist recalled years later, Allan Brown did "a bang-up job of it"; another oldtimer, a trade editor, stated simply that "Allan Brown was the sparkplug at Bakelite."[105]

When Brown assumed responsibility for promotion, most people thought of Bakelite as an industrial material used for electrical insulation. Almost immediately he disengaged it from that limiting connection and created an open-ended identity. Within a year Bakelite had become "The Material of a Thousand Uses," as announced in a booklet illustrating pipe stems, cigarette holders, necklaces, earrings, a candlestick, a pen, a shaving brush, and other "art and jewelry applications" in rich colors of jet, amber, emerald, and carnelian. Only the back cover depicted a "strictly utilitarian use," a view of acid-resistant lab equipment whose elegant simplicity reminded readers of the material's synthetic origin without detracting from the unearthly beauty of the other objects. The slogan was not new, having occasionally cropped up over the years. Baekeland had once referred to "the thousand and one . . . articles" that might be made of Bakelite. More to the point, Du Pont had run an advertising campaign in several trade journals in 1920 referring to Pyralin celluloid as "The Material of a Thousand Uses." That campaign lasted no more than two years, and Condensite picked up the slogan in 1923.[106] From 1924 onward it appeared in every advertisement or publication of the Bakelite Corporation.

Along with the slogan appeared a distinctive trademark devised by Brown: the letter B above the mathematical symbol for infinity, symbolizing "the infinite number of present and future uses" of the company's products (Figure 2–1). At first a vertical oval surrounded these two graphic elements. Within a year the oval became an outline formed by three half-circles. To most people that "three-sided circle" seemed no more than a graphic convention; to insiders it symbolized a long struggle whose outcome united three companies and gave them a new identity. That image of unity, based on a concept of continuous innovation, was essential to the company as it defined itself in the face of competition that would threaten only a few years hence, in 1926, as Baekeland's basic patents expired.[107] Brown hoped the trademark would become a seal of approval used by manufacturers on consumer goods made of Bakelite. Satisfied consumers would then recognize, prefer, and demand Bakelite products—making it impossible for anonymous competitive phenolic plastics to gain a foothold. As a booklet aimed at the radio market explained in 1924, "Both manufacturer and user profit by the use of Bakelite—the only material which may bear this famous mark of excellence."[108]

Brown's sales effort addressed two groups whose members did not overlap with the company's prime customers. Most Bakelite was sold directly to six laminating companies and about thirty molding operations, some of them custom molding companies like Boonton and others captive departments of large corporations like Westinghouse or General Electric. Thanks to early efforts of Baekeland, Kirk Brown, and Redman, those companies knew and used Bakelite; but they had to be convinced to keep using it after patent expiration brought other material suppliers into the market. To exert pressure on molding companies by expanding demand for Bakelite, Allan Brown focused on two other groups. The first was composed of manufacturers who might be convinced to abandon rubber, wood, or brass in favor of Bakelite in their own products. The second encompassed the general public who formed the ultimate market for Bakelite products.

Brown often reached manufacturers through trade journals like *Electrical World*, which carried an annual advertising insert with full-page ads for Bakelite and for other companies using Bakelite in their products. When the first issue of *Plastics* appeared in 1925, Brown bought the front cover for an advertisement offering "the cooperation of our engineers and laboratories in adapting Bakelite to your needs."[109] After Brown's appeals stimulated inquiries, his brother Gordon and Hylton Swan supervised a team of engineer–sales representatives who not only explained Bakelite's technical advantages (dimensional accuracy and stability, durability, reduced labor costs for assembly and finishing) but also engineered products, designed molds, and produced custom mixtures for special jobs. As Allan Brown recalled, the task was not easy. Sales representatives spent days "overcoming the obstacles in adapting our materials to old methods and overcoming old prejudices." The company often broke into an industry by working with small upstart firms unburdened with heavy investment in obsolete tools and training; their successful competition later pressured larger companies to adopt Bakelite.[110] In his most extravagant gesture, foreshadowing the company's later linking of plastic and design, Brown organized a Bakelite Caravan of Ideas in 1927. Addressing the problem that Bakelite looked "like gunpowder" and had "scarcely any display possibilities," Brown filled two moving vans with two thousand Bakelite products and sent them on a busy itinerary with stops at more than two dozen industrial cities. Installed in hotel ballrooms and university auditoriums, the exhibition was open only to local executives and their engineers.[111]

Proselytizing among end-use manufacturers had served Baekeland well as early as 1907. But he never quite understood Brown's second strategy: direct appeal to the general public. In 1924 Brown organized a national campaign with advertisements in major newspapers and free displays for dealers of products containing Bakelite—radios, cars, appliances, costume jewelry, smokers' accessories. The general impression came through clearly in A Romance of Industry, a booklet small enough to fit into a woman's purse.[112] The cover, with a genie's lamp yielding a slender houri, feminized the creative power behind Bakelite. "What is this wonderful material," asked the text, and "what makes possible its wide variety of uses?" With a nod to "our modern magician—the chemist," it advised that by "reading this little booklet, you will find perhaps to your amazement, the number of places you are likely to come into contact with this unique chemical creation." After a progression of illustrations of domestic scenes and products, it admitted coyly that "in factories you will find Bakelite in still other forms." Mostly fluff with enough detail to whet the appetite, the booklet encompassed the poles of Brown's endeavor. Placed in the hands of a middle-class woman, the "great national purchasing agent,"[113] it was intended to promote Bakelite as a material to seek out when shopping. Taken home to a husband in manufacturing or retailing, it exerted the indirect pressure that all brand-name promotion of Bakelite had as its goal.

Above all, Brown strove to define Bakelite as more than a substitute for shellac and hard rubber, more than an obscure electrical component. One could hardly imagine a more celebratory effort than The Story of Bakelite, a little volume that created an almost millennial aura and set the tone for subsequent discussions of plastic. Sometime in 1923, as part of the new public relations campaign, Brown commissioned a writer named John Mumford to produce a book about Bakelite. When Mumford came to Snug Rock in July to gather information, Baekeland reported an "interesting talk about many subjects." As an experienced popularizer of industrial topics Mumford worked quickly. In September Baekeland was already correcting the manuscript, and later the two met to discuss the work.[114] Published in March 1924 without any hint it was a commissioned work, it was well received by Edwin E. Slosson of Science News Service, who told Baekeland it demonstrated "how chemistry can be enlivened by the introduction of historical, biographical and artistic elements." Slosson also reviewed it in his *Science News Letter*, recommending it to "teachers and librarians throughout the country" for "information they cannot get from the textbook."[115] Despite the praise, which helped distribute twenty-five thousand copies over the next three years, Baekeland seemed embarrassed by the book. When chemical consultant Arthur D. Little wrote him late in 1924 seeking material for a talk on industrial research, Baekeland sent him a copy of *The Story of Bakelite*. Distancing himself from it, he explained that "while I was in Africa, Jack Mumford wrote the story the way he saw it. . . . The book was finished when I came back here." Although Mumford had "undoubtedly . . . taken great pains to collect scattered information" from factories and customers, there was, in Baekeland's opinion, something not quite right about it.[116]

Mumford opened the story with the creation of life, when nature first began storing up "waste heaps" of dead organic matter from which research chemists were later to derive "colossal assets." Bakelite appeared as "a wonder-stuff, the elements of which were prepared in the morning of the world, then laid away till civilization wanted it badly enough to hunt out its parts, find a way to put them together and set them to work." Mumford marveled at Bakelite's "Protean adaptability to many things," at the ease with which it could be molded. He found even more miraculous the way it froze "in the very heat which gave it birth" to form "a solidity that mocks at the disintegrating forces of heat and cold, time and tide, acid and solvent." It would "continue to be 'Bakelite' till kingdom come." Out of the very stuff of death, Mumford suggested, Leo Baekeland had created an indestructible material possessed of immortality. And the chemists who followed him would indeed "make a new world" out of "marvels which earth had never known." They would "create new substances, and create them out of anything." Although Bakelite was already found "wherever people are living, in the Twentieth Century sense of the word," it was impossible to predict "all that it will or will not do." Bakelite remained "at once a contradiction, a mystery, a tireless factotum."[117]

In among the breathless rhetorical phrases, Mumford offered an accurate history of Bakelite—early research into the phenol formaldehyde reaction, Baekeland's discovery, subsequent commercialization. Why then did the chemist object to Mumford's book to the point of apologizing for it? The answer may lie in the writer's comparison of Baekeland to Columbus. The navigator had sought a passage to the Indies, the chemist a substitute for shellac. As it turned out, "each discovered a continent"—far more than expected. No other synthetic material could compare "in the universality of its uses, present and potential, with the thing that Baekeland created in his little Yonkers workshop, after the best minds in Europe had given up."[118] But Baekeland remained no more than a Columbus of synthetic material realms. Capable of crossing interpretive borders, of abandoning one frame for another, he never quite succeeded in visualizing his discovery's ramifications, nor in giving articulate expression to its developing paradigm. Nor did he understand the metaphorical flights of a John Mumford or an Allan Brown, which rendered him uncomfortable even as they defined his work's cultural meaning. Over and against the traditional techniques and imitative strategies of celluloid, Bakelite's emergence marked a triumph of the frankly artificial celebrated for its own sake, a perfect machine-age material evocative of precision, rational control, unearthly beauty. It seemed true, as another promotional booklet intoned, that "there is no real substitute for Bakelite."[119]

VISION AND REALITY IN

THE PLASTIC AGE

*P*lastic became a household material during the years between the world wars, a period that divides too easily into decades of boom and bust, of hedonistic excess followed by social reconstruction. A different perspective reveals a single era when varied trends converged to form a consumer culture. Sophisticated advertising reached a public whose experience of the automobile, radio, and movies disposed it to a universal identity based on mass-produced goods rather than on regional tradition. Electrification triggered demand for domestic appliances on easy credit terms. Technical improvements merged with annual styling changes to create a sense of unending material progress—slowed but not halted by economic depression. Roosevelt's New Deal aspired to guarantee these advances to all citizens by means of a managed economy. Modernity's rise seemed inexorable, and "machine-age" became an adjective referring to just about anything. Plastic, its forms and colors proliferating as the consumer culture emerged into consciousness, embodied that culture and visually distinguished it as unique. In the words of a designer who shaped products and environments, "new materials" spoke "in the vernacular of the twentieth century."[1] That vernacular changed as the century progressed, but plastic remained an emblem of modernity, a focus of praise and blame, an artificial substance whose inexpensive raw materials and processes enabled continuous expansion of consumer goods and defined their limits.

At the beginning of the interwar era, plastic's potential appeared uncertain. American shoppers recognized celluloid as a material that had long imitated ivory and tortoiseshell. Hardly a modern material, celluloid seemed old-fashioned after 1924 when bobbed hair allowed women to abandon ornate combs and hairpins;

annual production of the material fell by more than a third. Its use in movie film brought it a certain glamour—rendered ambiguous, however, by hints of superficiality as "celluloid" became a metaphoric adjective.[2] Bakelite seemed no more promising. Before the rise of radio and Allan Brown's publicity campaign, it remained an industrial material, usually black, often resembling hard rubber, which itself had been around for over seventy years. Apart from electrical outlet plates, gearshift knobs, and other mundane objects, Bakelite's most visible consumer applications were pipe stems—where it mimicked amber.

By 1940 the situation had changed entirely. Celluloid and Bakelite, so different as not to seem related, were joined by many new substances, each with unique qualities, creating a varied plastic spectrum. Colorful cast phenolic resin provided an alternative to Bakelite's industrial black or brown. Cellulose acetate, a nonflammable celluloid cognate, and urea formaldehyde, a molding compound similar to Bakelite, offered bright colors and pastels for consumer goods. Vinyl polymers appeared in experimental uses that revealed their warmth and yielding resilience. The lightweight crystalline transparency of acrylic promised an unbreakable glass substitute. Polystyrene offered similar glasslike transparency through an injection molding process quicker and less expensive than compression molding. Also appropriate for cellulose acetate, injection molding liberated plastic from Bakelite's industrial phase and made it the material of choice for inexpensive toys and gadgets. Between 1921 and 1937 annual production of coal-tar resins rose from 1.5 million to 141 million pounds. Urea formaldehyde entered government statistics in 1932 when two million pounds were produced; by 1937 production reached twenty-one million. Cellulose acetate, not monitored in 1929, climbed from three million pounds in 1933 to nineteen million in 1937. By the end of the era these materials appeared so prominent that some journalists abandoned "the Machine Age" as a slogan and proclaimed "the Plastic Age."[3]

Something of the confusion about plastic surfaced in *Fortune*'s review of the industry in 1940. The editors seemed uncertain how to present these new materials, whether to portray plastic as an extension of natural materials or as an intoxicating disruption of the natural order. These contrary interpretations emerged not in the article's text, which offered clear explanations of processes and applications, but in two illustrations, each a two-page full-color spread, each so bizarre in its own way, so rationally unwarranted, as to suggest an intrusion into consciousness from a site of unresolved psychological conflict.[4] The more easily grasped of these illustrations was a photograph of a relief map of an imaginary continent called Synthetica (Figure 3–1). Descending from an isthmus in the northwest, it resembled South America; a western bulge gave it the general outline of Africa. Rising from a blue ocean rippled with waves, the continent was divided into several countries, each cast from a differently colored plastic—in shades of black, brown, purple, green, lavender, orange, yellow—by Ortho Plastic Novelties, Inc.[5] Rivers (the Great Acetylene), mountain ranges (the Crystal Hills), and lakes (Acetic Acid Lake) contributed

3–1. "Synthetica: A New Continent of Plastics," *Fortune*, 1940

an illusion of geographic verisimilitude. Names of countries reflected chemically distinct types of plastic (Coumarone, Casein); those of capitals and major towns indicated different brands (Plaskon, Paraplex). Off the coast lay the island of Nylon. To the northwest ran the isthmus of Natural Resins; to the west at the edge of the map rose a land mass of indeterminate size, Inorganica, with its provinces of Steel, Plaster, and Putty barely visible.

Cleverly conceived, fabricated at some expense, and splendidly reproduced, the map of Synthetica served no obvious purpose other than to suggest by spatial arrangement the chemical groupings of various plastics. As if aware of Synthetica's physical inertness, *Fortune*'s editors provided a long evocative caption that amplified the exercise without explaining it. Synthetica, according to this caption, extended "right out of the natural world—that wild area of firs and rubber plantations, upper left—into the illimitable world of the molecule." Although it floated on the "Sea of Glass, one of the oldest plastics known," the continent was only recently discovered. "New countries, like Melamine, constantly bulge[d] from its coastline," and its boundaries were "as unstable as the map of Europe." Already possessed of its own Ruhr district, known as Phenolic, "a heavy industrial region of coal-tar chemicals fed by Formaldehyde River," Synthetica also boasted the "more frivolous and color-loving state" of Urea and the "glittering night life" of the resorts of Rayon Island, as well as "the Alkyd country, a great swamp of bright, impervious plastic paints, varnishes, and lacquers" creeping "like an implacable sargasso" into the sea to the south. And so on.

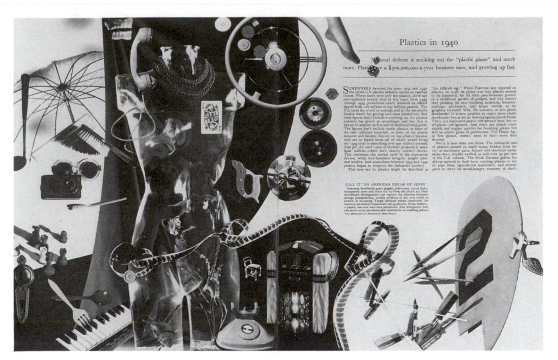

3–2. "AN AMERICAN DREAM OF VENUS," *FORTUNE,* 1940

The map of Synthetica firmly rooted plastic in the extractive materials culture of the past. The conceit of a map implied an odd denial of organic chemistry's promise to free the human race from geographic accidents of scarcity and supply. The very outline of the continent suggested the tropics, civilization's colonial preserves exploited for ivory, rubber, gutta percha, copal, and other natural resins and substances. Despite the map's bright spots, areas of dark viscous flow predominated, evoking images of coal tar, asphalt, petroleum, all products of the planet's own dawn-of-time organic waste. Or, shifting perspective a bit, the map resembled a medical professor's anatomical model, a graphic representation of earth's body opened up to reveal viscera of unnerving hues. *Fortune*'s cartography offered plastic as the product of a heavy industrial process dependent on earthbound materials, a process whose development followed that of the founding and growth of civilizations, a process that lent itself, finally, to human control.

By contrast, the article's other full-color two-page illustration created a bizarre sense of airy transcendence, of otherworldly escape from the heavy realm of natural materials. A breathtaking photomontage portrayed what *Fortune* described as "An American Dream of Venus" (Figure 3–2). Against a deep bluish-purple backdrop of vinyl sheeting appeared the pellucid outlines of a stylized armless female torso rendered in acrylic. Within its reflective hollows, visible through thick, light-concentrating surfaces, exotic goldfish swam and marine plants floated. Outside this otherwise undefined goddess, suspended about her in an immaterial medium, appeared the diffuse plastic accoutrements of her artificially colorful machine-age existence. Three strands of deep red vinyl wire woven into chains encircled her neck; the Queen of Diamonds from a pack of cellulose acetate cards covered her heart. A white telephone handpiece rested on her collarbone, its cord snaking down to a gleaming set at the photo's lower edge. Around her appeared a score of other objects rendered in rich hues with no regard for relative size or perspective: dentures, buttons, dice, sunglasses, steering wheel, transparent umbrella, amber hairbrush, red and white chessmen, black camera, Hammond organ keyboard, yellow and green picnic flatware, golf club, luminous jukebox, chair of crystal acrylic, red saw handle, nylon-stockinged foot, pink toothbrush, and a twisting loop of cinema film enclosing the lower field of her realm. Outside lay scattered signs of her male suitors: pens, mechanical pencils, pocket knives, screwdrivers, an airplane's rudder. With the exception of her own fishy depths, nothing suggested connection to living nature. It all implied radical divorce from the limits of a drab pre-plastic existence.

More evocative than Synthetica's ponderous geography and possibly more representative of emerging reality, the plastic Venus and her realm required less explanation. After listing a few of the artifacts "ris[ing] up from the plastic sea," a brief caption admitted in provocative language that "only surrealism's derangements can capture the limitless horizons, strange juxtapositions, endless products of this new world in process of becoming."[6] With the exception of a few visual associations, the photomontage offered a perception of total disjunction, both of present from past and of all present phenomena from each other. Objects of plastic simply appeared from nowhere with an explosive insistence that defied reason. Their significance derived not from any functions they might fulfill but from their own artificial, near alien agenda. Witty, playful, catering to the lighter, more transient side of contemporary life, plastic had evaded human planning and transcended control. Unlike the map of Synthetica, which postulated a land mass ripe for systematic exploitation, the plastic Venus promised an American dream of shifting shapes, an irrational phantasmagoria of ungrounded images, all in brilliant synthetic colors, a carnival of material desire.

Coming at the end of the era, neither of these visions quite reflected popular understanding of plastic during the 1920s and 1930s. But they did encompass two extremes of actual response to plastic. Each reaction made much of plastic's utopian potential as a substance capable of transforming the material conditions that had always limited human life. A conservative vision that remained dominant into the 1940s considered plastic as a vehicle of controlled social stability. Inexpensive plastics provided by organic chemists would foster true democratization of society by ending the strife generated by scarcity and replacing it with universal material abundance. This dream of equilibrium attained visual expression in the streamlined forms of consumer products molded during the 1930s, their rounded aerodynamic lines suggesting frictionless motion through undisturbed currents of time. At the other pole, a more radical vision considered plastic as a vehicle of proliferating transformations, of continuous transcendence. More congruent with a culture of consumption, of waste and inflation, this vision of endless disruption did not become dominant until the 1950s and 1960s.

Oddly enough, and perhaps only by coincidence, this polarization of visions extended even to the particular plastics typical of each era. Bakelite and other phenolic and urea resins, by far the most prominent of the interwar period, were *thermosetting* plastics; that is, their formation under heat and pressure was an irreversible process. They became stable, inert, nearly "immortal" as John Mumford had implied, during the chemical reaction that formed them. The significant postwar plastics, on the other hand, were such *thermoplastics* as polyethylene and polystyrene. Unstable, capable of being melted and reshaped, they reflected an expansive culture of impermanence similar to that projected by *Fortune* in its "American Dream of Venus." Suggestively prophetic as a surrealist vision in 1940, the magazine's photomontage hardly reflected the plastic utopianism of the 1920s and 1930s, whose publicists and designers favored a culture of stability.

MODERN MIRACLES

The apotheosis of conservative plastic utopianism appeared in 1941 in *Science Digest*, a magazine devoted to popularizing science and technology. V. E. Yarsley and E. G. Couzens, two British applied chemists, contributed an impressionistic survey of the life of Plastic Man, "a dweller in the 'Plastic Age' that is already upon us." He would be born "into a world of color and bright shining surfaces, where childish hands find nothing to break, no sharp edges or corners to cut or graze, no crevices to harbor dirt or germs." He would experience crib, teething ring, bottles, spoons, and toys of plastic and would be "surrounded on every side by this tough, safe, clean material which human thought has created." Later the young Plastic Man would be educated in "a new kind of schoolroom with shining unscuffable walls," sitting at "a molded desk, warm and smooth and clean to the touch, unsplinterable, without angles or projections." At home, "the universal plastic environment," he would live within walls covered by synthetic veneers, with molded furniture, plastic bathroom fixtures, and "bouquets of flowers miraculously preserved in geometrical masses of brilliant, clear, transparent plastic." The two chemists described Plastic Man's personal airplane as "the motor-car of the future . . . mass-produced from reinforced plastic." Unfortunately plastic could not prevent "the inevitable end." But old age would be cheered by plastic teeth, eyeglasses, playing cards, and chess pieces before Plastic Man's inevitable descent into the grave "hygienically enclosed in a plastic coffin."

This prophecy seemed a caricature of the world-of-tomorrow magazine piece, but the two chemists took their vision seriously. The emerging Plastic Age offered humanity an opportunity to gain freedom from chance and from nature's imperfections. The world of Plastic Man would be "brighter and cleaner" than any previously known, "a world free from moth and rust and full of color." Reversing the main economic fact of all history, it would be a place of abundance rather than scarcity— "a world in which man, like a magician, makes what he wants for almost every need, out of what is beneath him and around him, coal, water, and air." The new civilization—"built to order" by industrial chemists and technocrats—would embody "the perfect expression of the spirit of scientific control." For Yarsley and Couzens plastics were miracle materials from which the human race would mold the precise contours of a desired—and stable—future.[7]

A utopian aura had long gathered about the chemical industry. As early as 1907 a journalist informed readers of *Everybody's Magazine* about the "miracle-workers" of synthetic chemistry. Scientists would never be satisfied until they could fabricate "a loaf of bread or . . . a beefsteak" from "a lump of coal, a glass of water, and a whiff of atmosphere." Other articles in the popular press echoed this modern-miracles refrain, especially after the First World War. Stung by criticism of lack of preparedness in the face of Germany's chemical power and by accusations of war profiteering, industrial chemists and their employers launched a "chemical crusade" to promote recognition of their contributions to contemporary life. Major participants included the American Chemical Society, Du Pont, and the Chemical Foundation, a semi-public organization that funded its promotional activities by selling licenses for confiscated German patents. If the war had focused public attention on "the power and importance of chemistry" in such destructive areas as explosives and poison gas, chemists now had to embrace a "more congenial task of contributing to national health and prosperity."[8]

No one expended more energy promoting industrial chemistry as an engine of utopia than Edwin Slosson. After serving thirteen years as a one-man chemistry department at the University of Wyoming and seventeen years on the staff of the progressive journal *The Independent*, Slosson assumed the directorship of the Science News Service in 1921. As he wrote to Baekeland, he believed that "silent revolutions of science are often more important than the noisy revolutions of politics." But convincing people was hard. They were attracted by "freaks of nature" and ignored its "ordinary processes." To reach them he admitted casting science into a "sensational form," an approach he used in *Creative Chemistry*, a collection of his articles with an introduction by Julius Stieglitz, past president of the American Chemical Society. Stieglitz wasted no time getting to the point. The "wizardry" of chemistry "permeate[d] the whole life of the nation as a vitalizing, protective and constructive agent." Slosson himself also went straight to the point. The human race possessed "the power of reducing a substance to its constituent atoms and from them producing substances entirely new." Medieval alchemists, though "foolish and pretentious," had "aspired at least to the formation of something new." Fulfilling their ancient dream of transformation, "Man the Artifex" would "gradually . . . substitute for the natural world an artificial world, molded nearer to his heart's desire."[9]

Despite this preference for the artificial, Slosson waffled over celluloid's imitative use. While praising its exact reproduction of "the clear luster of amber, the dead black of ebony, the cloudiness of onyx," he lamented that it "had to resort to *camouflage*" to gain acceptance. Chemists had wasted tremendous effort discovering how "to imitate the imperfections of nature." Eventually, however, people would recognize plastic's superiority and allow this "chameleon material" to assume "forms and textures and tints that were never known before." The issue of imitation was resolved in plastic's utopian role as the material basis of democracy. Slosson described plastic as "a mechanical multiplier" whose potential for inexpensive mass production would eliminate the material scarcity that had fostered inequality and war throughout history. Imitation took on emblematic significance because inexpensive reproductions of rare materials that were formerly "confined to the selfish enjoyment of the rich" were now "within the reach of every one." A universal "state of democratic luxury" based on synthetic chemistry was at hand.[10]

Bakelite most fully reflected Slosson's utopian vision. The human race stood at the beginning of unprecedented material abundance made possible by coal—dirty, black coal—one of the world's most aesthetically unappealing substances. This limitless "scrap heap of the vegetable kingdom" provided chemists with "all sorts of useful materials" because coal contained in condensed form "the quintessence of the forests of untold millenniums." From the "black mass" of sticky, evil-smelling coal tar, formerly discarded during the coking process, chemists synthesized not only dyes in "all the colors of Joseph's coat" but also the phenol that went into Bakelite. From out of waste products came the materials of abundance.[11] If Slosson, like the alchemists, seemed to be suggesting getting something for nothing, he was also arguing for efficient conservation and transformation of waste. This synthesis of order from chaos yielded a metaphor that exemplified the social meaning of his chemical utopianism. Slosson compared chemical "polymerization" to the process of a financial "merger." The resin formed by condensation of phenol and formaldehyde was a "molecular trust" that had remained "uncontrollable and contaminating [to] everything it touched" until Baekeland imposed order on it. Just as governments had learned to put trusts and mergers to good use by controlling and channeling their energy, so too had chemists "learned wisdom in recent years."[12] It was all very well to praise the chemist as an "agent of applied democracy," but Slosson's metaphor suggested that a well-ordered society would have the molecular inertness of Bakelite.[13]

Other popular journalists frequently echoed Slosson's major themes—the magic of chemistry, the creation of something from nothing, the democratization of everyday life, the triumph of the artificial over the natural, and the engineering of social stability. When Baekeland received the Perkin Medal, for example, a profile in *The World's Work* celebrated his achievement as "rivalling the legendary magic of the medieval alchemists." For *Barron's* in 1927 the figure of the chemist was a "modern magician"; even celluloid was "a sort of Aladdin's Lamp product." For *The Saint Nicholas Magazine* in 1928 the chemist was a "creator" at whose "magic touch" the very "structure of molecules becomes plastic." Four years later the *New York Times* de-

scribed new plastics as "conjured out of laboratory test-tubes." Other articles referred to an "era of make-believe" in which "the magic of modern chemistry" seemed
"much like Arabian Nights tales" or "stranger than a fairy tale." A skeptic writing for
Business Week attributed all the hoopla to public ignorance, a result of "the difficulty
of translating into common terms the mysterious ways in which chemical processes
move, their wonders to perform."[14]

Most excursions into plastic utopianism quickly passed beyond alchemy and
magic to consider social benefits of new materials. Some observers discussed plastic
in the light of diminishing natural resources; others more positively by celebrating
its potential for democratizing abundance and wealth. *Barron's* suggested in 1927
that Du Pont's Fabrikoid "leather substitute" had proven indispensable because "all
the cows on earth could not supply the demand for leather" for upholstering furniture and automobile seats. According to *The Saint Nicholas Magazine*, the chemist's
"most useful work" lay "in the realm of substitutes": "as Nature's storehouses of
supply become exhausted or inadequate for world needs, synthetic substitutes must
replace them." Many of those substitutes in turn came from "by-products" that otherwise would have gone to waste, as *The Index* observed in 1936. Chemists had
succeeded "in transforming almost useless materials into materials of eminently
practical value"; the plastic industry thereby became "quite literally a creative contribution to the resources of civilization."[15]

These issues received perceptive treatment in 1930 in a book by Pauline G. Beery
with the fascinating title of *Stuff*. Her opening sentence asserted that life revolved
around "Strife for stuff!" Trade resulted from "a desire to obtain new kinds or greater
quantities of stuff from other lands." Wars were fought "to gain possession of regions
where certain important stuffs can be secured." When it came right down to it, "the
chief difference between the hungry and the well-fed, the homeless and the well-
housed, the poor and the rich—is simply stuff." Beery devoted most of her attention
to natural or traditional materials, but plastic and other synthetics received their
due. In a chapter on rayon, for example, she argued that "by placing artificial products on the market at very low prices," the chemist had "already done more than
any other single agency toward making a democracy of all the peoples of the world."
Pessimists, she observed in a final chapter on "Stuff of the Future," might worry
about using up natural resources, but optimists believed that "no matter what happens, the chemist can conjure new types of stuff from the air, or the ocean, or the
moon, like a magician taking rabbits out of a borrowed hat."[16]

C. C. Furnas, a professor of chemical engineering at Yale, talked about "Outdoing
Nature" in a volume of scientific prophecy in 1936. "Mother Nature did rather
poorly in the matter of plastics," he declared. *Fortune* concluded more radically that
same year that "only man can make a plastic." Hyatt had broken "the divine monopoly on primary substances" when he "brought forth the first plastic . . . a new
substance under the sun, a material that was not to be found in nature and that
could not be converted back again into the substances out of which it was made."

This desire to transcend nature's limits became for Lewis Mumford the motivating force of all technological change and thus of all history. His influential book on *Technics and Civilization* discussed synthetic materials along with lightweight alloys and clean hydroelectric power as crucial elements of a humane neotechnic phase that civilization was on the verge of entering. In Mumford's opinion technological development was marked from the beginning by an "unwillingness to accept the natural environment as a fixed and final condition of man's existence." From the seventeenth century onward "the machine" had functioned as "a counterfeit of nature, nature analyzed, regulated, narrowed, controlled by the mind of man." Its development had triggered "displacement of the living and the organic" in many areas of human life. Mumford hoped the neotechnic phase would restore equilibrium to natural and social environments ravaged by the industrial revolution but placed his faith in more—not less—control over nature.[17]

Those who felt ambivalent about replacing natural materials with synthetic plastic often had trouble articulating their concern. Furnas, for example, maintained that plastics of high quality and low price would take society "one step closer to a completely synthetic environment from the underwear next the skin to the roof covering on the houses." Such a development would "add variety," he confessed, but for unspecified reasons "might not be entirely desirable." Beery ended her chapter on "synthetic stuff" by describing some Bakelite furniture and concluding that "our descendants . . . may never see any wooden tables or chairs except in museums." Unlike Furnas she implied no criticism; yet her comment seemed a curious note on which to end. It may have seemed that plastic, despite its basis in organic chemistry, was too inorganic, that it encroached too much on life. Certainly one recoils even half a century later from the enthusiasm of Pierre S. du Pont, former president of the chemical company, celebrating his Lucite acrylic dentures in a letter to a friend who was "still plodding along on nature-made teeth" even though chemistry offered "a great improvement on nature's manufacture." Du Pont confessed he could not see "pictures of grinning girls with a full set of teeth" without imagining they owed their "perfection" to "teeth by Du Pont." Even perfect teeth implied victory over nature's imperfections, over time's vicissitudes, over death itself—the final insult of the organic. A degree of control so complete that it aspired to the rigidity of death while apparently defying it achieved expression in *Fortune*'s enigmatic observation that "a plastic is anything *but* plastic, once it has hardened and taken shape." That statement encapsulated the "thermoset" phase of plastic utopianism.[18]

But how seriously did the public respond to these promotional writings as they encountered the actual materials? In 1936 *Fortune* declared, "The layman has been taught to believe that an age of plastics is at hand." Some leaders of the industry blamed journalists for spreading too many wild tales. The editor of *Plastics & Molded Products* complained in 1930 about mass-circulation articles putting "fantastic notions in the minds of the public." Worried about disillusioned consumers, he blamed "over-harassed chemists" who told reporters what they wanted to hear just to get rid of them. Six years later, the editor of *Modern Plastics* complained that "nothing is more confusing to a manufacturer, an engineer or designer who is considering the use of plastics . . . than the half-truths and essentially misleading tidbits which he picks up in the course of his everyday reading." As late as 1939, when synthetic materials were more common, a contributor to *Dun's Review* thought it necessary to announce that "no miracles are performed with plastics." Not so concerned about the impact of exaggerated claims on consumers, he feared that businessmen, "befuddled into expecting and demanding all sorts of impossible uses," would be angered to learn many of their demands were "not within the realm of possibility."[19]

All the same, plastic industry pioneers could not fairly blame unrealistic expectations on harassed chemists or sensation-seeking journalists because they had also fostered a utopian mood. Two years into its existence as the industry's trade journal, in 1927, *Plastics* posed the rhetorical question, "Is it a Plastic Age?" Of course it was, came the rhetorical answer. In fact it was the editor's "mission" to "tell the world it is."[20] If Baekeland had discovered the new continent of Synthetica, then the editor played the role of a land-company agent recruiting immigrants from the old country's realms of rubber, wood, and metals. From his perspective, the industry looked to horizons as inviting as those facing the explorers who opened up the American continent. The realm of plastics was still a "wild and wooly West" of "wide open spaces" waiting to become "bee-hives of industrial endeavor." As this landscape became more artificial, it would fill with furniture of resin-impregnated wood scraps, extruded window frames, walls and flooring of synthetic materials, unbreakable dishes, molded cars and airplanes. "Truly," he concluded, intoxicated by his own vision, "if man at any time in his career has had the actual shaping of his future in his own hands, now is that time."[21]

There was something refreshing about this naive optimism of the 1920s and 1930s. People like Slosson and the editor of *Plastics* sincerely believed new synthetic materials would transform the world from a crude, uncertain place into a stable environment of material abundance and startling artificial beauty. If this transformation had not yet occurred, its promise seemed all the more pure, not yet sullied by doubts about waste, toxicity, ecological destruction, or the ease with which plastic might be molded into tasteless gadgets. Plastic's contributions to radio and the automobile already suggested it would become a universal requirement of contemporary existence—with its qualities of malleability and artificiality mirroring the essence of that existence. A genuine excitement motivated those who entered the field as chemists, investors, executives, molders, designers, and promoters. When a young chemist named Gordon M. Kline began tracking synthetic resins and plastics for the National Bureau of Standards in 1930, he had only to read three books and three journals (two of them foreign) to catch up with the state of the art. "It was a good place to start," he recalled, "because it was right at the beginning."[22]

It is impossible to state with certainty that such an expectant mood had an impact

on the general public, on ordinary people not caught up in the plastic industry for
their livelihoods. More important was their direct, perhaps often subconscious, experiencing of real plastics. Next to this tactile or physical exposure, journalistic accounts seem like epiphenomena. But we cannot measure these personal reactions or
even verify their existence. Marshall McLuhan once addressed a similar problem
when he suggested that any new communications medium profoundly disturbs a
culture by elevating one sense over the others and destroying a balanced sensory
ratio.[23] A new material might likewise disturb a culture by changing the ratio of
types of materials in the everyday environment. With the ratio of artificial to natural
materials rising constantly throughout the twentieth century, American consciousness might have experienced a subtle but continuous disturbance. Unable to verify
this theory, a historian of plastic can only extrapolate connections among three quite
distinct areas: the culture's major themes, the statements promoting new materials,
and the proliferating array of new materials with varied colors, textures, and functions. Let us turn now from the intellectual framework that accompanied plastic's
introduction to consider the stuff itself—what people did with it, and how, and why.

ALL THE COLORS OF THE RAINBOW

Plastic did not emerge from a dark industrial underworld into the more colorful
realm of consumer goods until the end of the 1920s. In theory, phenol formaldehyde could assume any color, but only black or dark brown pigments concealed the
fillers that gave Bakelite its strength as a molding compound. Even cast phenolic
proved notoriously unstable in color, as Baekeland learned through years of trying
to commercialize it. In 1928 a manufacturer complained that his blue pens "spon-
taneously changed to green, and not a nice green either."[24] For a while a wide range
of colors did not seem so important. Radio not only boosted consumption of phenol
formaldehyde but also seemed likely to domesticate black Bakelite as a material for
consumer products. A gleaming front panel of black phenolic laminate was a source
of pride for amateur radio technophiles. As radio ownership became more common,
however, conservative listeners often chose a dark brown panel or a swirl of brown
and black so as to suggest wood. In 1927 manufacturers began enclosing radio sets
in wooden cabinets harmonizing with traditional furniture. The move threatened to
force Bakelite back to its prior status as a hidden industrial material at a time when
bright color was becoming central to selling consumer goods. "The Anglo-Saxon is
released from chromatic inhibitions," enthused *Fortune* as Americans consumed yellow and blue gasolines and installed bathroom fixtures of lavender and cobalt blue
during the final years of prosperity before the crash of 1929. Unable to take advantage of the trend, the Bakelite Corporation suffered even more from competition of
new varieties of plastic formulated to respond to the demand for color.[25]

The rush to color occurred as Bakelite faced increasing competition owing to loss
of patent protection. Several of Baekeland's patents, including the heat and pressure

process, expired in December 1926; others followed in 1927. Within two years Bakelite was forced to share its market with a host of new phenolic materials graced with such imaginative names as Duranoid, Durez, Kellite, Lacanite, Lennite, Makalot, Neolith, and Textolite. Not all the hopefuls made it, but the survivors were soon joined by Albertol, Carboloid, Crystillin, Durium, Joanite, Marblette, Moldarta, and Resinox. Some were marketed by major electrical or chemical corporations, others became significant in their own right, and the rest soon disappeared. Although Bakelite remained the largest producer of phenolic resins and opened a $3 million plant at Bound Brook, New Jersey, in 1931, competition meant the end of artificially controlled prices. Lower prices enabled end-use manufacturers to use phenolic resins for less costly applications, and it became feasible to put chemists to work on new types that could be independently patented.[26]

The prospect encouraged newcomer Harry M. Dent to brave legal action several years before Baekeland's patents expired. Dismissed from Du Pont during the recession of 1921, Dent founded the Durez Company at North Tonawanda, New York, and began selling phenolic resin on such a small scale that he personally delivered sample barrels in his car. The venture succeeded. By 1924 Baekeland was complaining that a Michigan molding company had threatened to "go over to Durez infringing material" unless he cut the price of Bakelite. In 1926 Dent expanded his operations by equipping a grandly renamed General Plastics Company with resin kettles large enough to hold a Model T Ford. Durez proved so superior to Bakelite for some uses that someone at Bakelite hired Pinkerton detectives to steal samples from the General Plastics plant—a scheme that backfired and forced Baekeland to drop an infringement suit to avoid publicity. General Plastics was solidly entrenched long before a chemical trade show of 1931 where the company exhibited 750 objects molded by some two dozen custom molders from Durez, "the new-age material."[27]

Like Bakelite, however, Durez remained an industrial plastic. Dent was no more successful than Baekeland in devising a bright, colorful phenolic plastic that would not discolor, fade, or reveal its filler. A partial solution came in Germany, as chemists at Herold AG devised a phenolic resin impervious to ultraviolet light. A wide range of coal-tar dyes could be dissolved in the resin to color it throughout. Varying the proportion of water yielded opacity, translucency, or transparency as desired. Unfortunately, however, the resin could not be used as a molding compound because Herold's chemists were unable to incorporate fillers without destroying the material's clarity. Instead it was used to cast various shapes—tubes, rods, slabs, sheets— from which final products could be cut, turned, and fabricated. The Herold process for making color-fast cast phenolic came to the United States in 1928 when the American Catalin Corporation announced "a new plastic material" with "a depth of color" that was "almost unbelievable." The substance had "to be seen to be appreciated." It was a "gem of modern industry" combining "the sparkle and lustre of a precious stone with the toughness and strength of metal."[28] After a period of legal wrangling American Catalin licensed its imported process to such firms as Marblette, Joanite, Fiberloid, Du Pont, and even Bakelite, which in 1933 claimed its own cast resins as a new development. Bakelite emphasized the material's democratizing potential. Capable of "simulat[ing] nature's gems at a price within reach of the average woman," cast phenolic could "satisfy woman's craving for fine things" whether she lived in "the simple cottage of Main Street or the pretentious mansion of Fifth Avenue." Colorful cast resins offered a means of overcoming the "curse of sameness" that too often afflicted mass-produced machine-age products.[29]

Despite all the talk of gems, Catalin and its competitors proved most useful for making durable novelties with small power tools in low-volume fabricating shops. Denser, tougher, and more stable than celluloid, cast phenolic widened the range of inexpensive applications for plastic. Material suppliers stocked rods, tubes, and blocks in standard sizes and colors (Plate 4). Given the cheapness of the lead molds used for casting, they also had no trouble supplying custom shapes on short notice. Fabricators sliced off poker chips or checkers from solid rods, milled them as required, and then tumble-polished them. Specially-shaped tubes yielded belt buckles for dresses or napkin rings in the outline of stylized rabbits and birds. A rod of narrow width yielded blanks for chess pieces whose carving by machine gave them a modernistic appearance. The bracelets that attracted collectors of so-called Bakelite jewelry fifty years later were sliced off from cast tubes, machined for ornament, and then polished. Catalin's color range and ease of fabrication lent itself to rings, buttons, earrings, and brooches. In 1934 more than half the costume jewelry sold in New York City was made of bright plastic; at some shops the proportion reached 70 percent. Cast phenolic was so successful that *Modern Plastics* feared popular jewelry—and the material from which it was made—was being "ford-ized." Although even lower-income women could now enjoy "multiple sets of costume decorations," the journal warned that a "cheap and shortlived fad" would reflect poorly on all consumer uses of plastic.[30]

Catalin and other cast phenolics arrived in time not only to take advantage of the rage for color but also to stimulate it. The uncompromising artificiality of Catalin's synthetic colors—bright, clear, uniform, reflecting a depth beyond that of any painted surface—contributed to an emerging commercial aesthetic of modernity. But cast phenolic did not solve the problem of combining bright colors and strong molding compounds—an essential objective if plastic was to expand more fully into consumer goods. The eventual solution was announced in 1921 by an Austrian chemist, Fritz Pollak, whose urea formaldehyde condensation process was analogous to Baekeland's phenol formaldehyde process. Heating urea and formaldehyde in the presence of a base, Pollak obtained a colorless liquid that became rubbery with further heating and finally, under heat and pressure, became "a hard, infusible transparent mass, insoluble in acids and alkalies." His urea formaldehyde resin worked with such fillers as wood flour and could be used to impregnate wood, paper, or fabric. Most important, a "water-white" transparency enabled it to assume any desired color from light pastels to rich primary hues. While Pollak perfected his methods, chemists at the British Cyanides Company devised a parallel process whose products first reached Anglo-American consumers as molded tableware.[31]

Variously marketed as Bandalasta, Beatl, and Beetleware, the new plastic dishes were introduced to the British market in 1925 but did not catch on until Harrods mounted a major display late in 1926. Christian A. Kurz, Jr., an American custom molder traveling in Great Britain, was entranced by the novel effects possible with the new material. He began importing it for use by the Kurz-Kasch Company at Dayton and eventually convinced the American Cyanamid Company to take out a license for its manufacture. One of the first American descriptions of Beetle urea formaldehyde molding compounds appeared in *Plastics & Molded Products* in 1927. Noting that Beetle contained no coal-tar chemicals, the article described it as color-less in its pure state, light-fast, nonflammable, and capable of being boiled and im-mersed in cold water without crazing or cracking—perfect, in other words, for cups, saucers, plates, and flatware handles. Above all, its "many beautiful translucent shades" offered "the possibility of reproducing artistic shapes in delicate shades and colors." Especially in its incarnation as Bandalasta, urea formaldehyde plastic pro-jected an aura of unearthly delicacy. Unlike melamine dishes of the 1950s and 1960s that often resembled heavy stoneware, Bandalasta was molded with extremely thin walls whose translucence enhanced fantastic swirling patterns of creams and pinks, light blues and greens. Less expensive Beetleware exhibited in solid pastels the same graceful lines and simple forms, the same tendency to translucence. According to *Business Week* in 1930, it "looks and feels like fine English porcelain but, unlike it, is practically unbreakable."[32]

Early applications of urea formaldehyde molding compounds in the U.S. encom-passed more modest things (Plate 5). In 1928, for example, Kurz-Kasch molded drinking cups for the Marshall Field department store in Chicago and offered a pro-prietary line of solid-colored Beatl Duroware. The Northern Industrial Chemical Company of Boston began molding translucent lamp shades and electric switch plates for use in residential interiors. In 1933 American Cyanamid exhorted manu-facturers to use Beetle for cosmetic jars. Packaging a facial creme in Beetle instead of glass offered an inexpensive means of putting "new life in old lines with gay colors that beckon to buyers."[33] More important to American Cyanamid was the market for premiums, which the company hoped would familiarize the public with Beetle and speed its acceptance. As the company predicted in 1933, *"every fourth home"* would "soon have a Beetleware tumbler, bowl or other useful household item." In that year more than sixteen million "Beetleware bowls, mugs, shakers, tumblers, spoons, measuring cups, [and] scoops" were given away with brand-name foods, mouthwashes, cleansers, and so on—mostly from commissions solicited by Ameri-can Cyanamid and farmed out to custom molders whose marks appeared as free advertising on each piece.[34] Most prized by American children were some 1.5 mil-lion Ovaltine shakers—each a large cream-colored cup with a solid red or blue lid and a full-color decal of Little Orphan Annie. Beetle had abandoned the aristocratic pretensions of Britain's Bandalasta to embrace a more popular spirit. But it had not abandoned quality. Bright-colored, durable, rust-free, hygienic-looking, and easy to clean, Beetleware premiums and cosmetic jars not only promoted plastic as a material "in all the colors of the rainbow" but also reinforced Bakelite's prior claim that plastic was a substance of permanence even when used for something as ephemeral as packaging.[35]

BEYOND CELLULOID

Not all new plastics came from phenol formaldehyde or cognate processes. The chemistry of cellulose yielded a new variety, cellulose acetate, even as celluloid was beginning to recede into the past. Meanwhile the new field of polymer chemistry became more theoretical in approach. Trial-and-error entrepreneurs like Baekeland yielded to younger chemists employed by large chemical companies—Union Carbide, Rohm and Haas, Du Pont, IG Farben in Germany, and Imperial Chemical Industries in Great Britain. Their somewhat more informed discoveries led to acrylic plastics, to an array of vinyl compounds, and to polystyrene—all known experimentally before the Second World War but not fully exploited commercially. Journalistic accounts, world's fair exhibits, and occasional trial products made the public peripherally aware of them and left an impression of an expanding array of new plastic materials.

Cellulose acetate resulted from a long search for a nonflammable celluloid substitute that would be safer to manufacture, store, and fabricate. Although cellulose acetate was intended as a direct analogue of celluloid, both in chemical composition and in processing, it revolutionized the molding industry. Its use initiated a shift of attention from slow, heavy, compression-molded thermosets to quick, light, injection-molded thermoplastics. Little more than a technical concern through the 1930s, this transformation eventually affected plastic's cultural meaning. As the material itself became cheaper and less durable, and as objects of plastic became lighter and more plentiful, the common perception of plastic moved beyond the permanence of phenolic to a sometimes disruptive sense of impermanence and evanescence (Plate 6).

Chemists had long known that cellulose could be combined with acetic acid (the main component of vinegar) as readily as with nitric acid. Early in the twentieth century the resulting compound became a prime candidate in the search for a "nonflammable celluloid." It was an elusive goal because no plasticizer equal to the camphor of celluloid could be found. Chemists in several countries announced one promising but flawed solution after another. Among the first to succeed were Swiss brothers Henri and Camille Dreyfus, who sought not a plastic but an artificial fiber. Their invention came just in time to provide a cellulose acetate "dope" for coating and strengthening the fabric wings of British and American airplanes during the First World War. Later they returned to their original goal and perfected a cellulose acetate yarn introduced in the United States in 1925 by the Celanese Corporation. Expanding into plastics, Celanese in 1927 purchased a controlling interest in the Celluloid Company and began marketing sheets, rods, and tubes of Lumarith cellulose acetate, followed two years later by molding powders. In the meantime Eastman Kodak perfected a "safety film" of cellulose acetate on which it had been working since 1906. About 1929 its subsidiary Tennessee Eastman began selling sheets, rods, tubes, and molding powders of Tenite cellulose acetate. Other celluloid producers followed suit, with Du Pont Viscoloid marketing Plastacele, and Fiberloid offering Fibestos and Torteloid, the latter a mottled sheet product substituting for celluloid in imitating tortoiseshell.[36]

Like its older cousin, cellulose acetate was tough, flexible in thin sections, capable of assuming a wide range of colors, easily machined when cold, and easily manipulated by hand when softened by heat. At first it seemed little more than a nonflammable celluloid. That virtue was considerable. One veteran of celluloid manufacture recalled the safety measures of a small Leominster shop where he worked sawing teeth in comb blanks while "stripped to the waist under streams of water." Fabricating techniques for cellulose acetate remained as laborious as those of celluloid—but without the continuous shower. As with celluloid, making a comb from cellulose acetate sheet stock required "up to twenty-five mostly hand procedures" for "cutting and shaping the blanks, sawing the teeth, rounding edges and corners, polishing and finishing."[37] Even innovative uses of cellulose acetate at first merely replaced the older material, as in the inner layer of safety glass, where it was used—both for auto windows and those of the Empire State Building—because it did not yellow as quickly as celluloid.[38]

But cellulose acetate, a direct descendant of horn and tortoiseshell, soon took on a radical identity. Leapfrogging phenolics and ureas, it became the most advanced mass-production plastic of the 1930s. Cellulose acetate owed success to its nonflammability. Because it could be melted without danger of fire, it was a prime candidate for molding. Because it was a thermoplastic, melting and rehardening without undergoing a complex chemical reaction, it was easier to work with than phenol or urea formaldehyde. With no tradition of cellulosic molding to rely on, the earliest molders used compression techniques learned by working with phenolics. Within a short time, however, most adopted a new technique, injection molding, that took full advantage of cellulose acetate's thermoplasticity. Compression molding entailed applying intense heat and pressure to a mold until the material within hardened to its final chemical state, and then cooling the mold until it could be opened and the finished product removed. Injection molding, by contrast, began with cellulose acetate already in its final chemical state. Chopped up and dumped into a hopper, molding powder flowed in measured amounts into the chamber of an injection molding machine. In a single continuous motion a plunger thrust this charge through a heated cylinder, rendering it molten, and then injected it into a cooled mold where it instantly solidified. Unlike compression molding, which wasted time and energy in heating and cooling the mold during each cycle, the injection process maintained a hot cylinder and a cold mold. Injection molding of thermoplastics offered the additional bonus that scrap and wastage could be ground up, mixed with virgin powder, and dumped back into the hopper. Years later the significance of injection molding became ironically clear when anyone who had anything to do with it laid claim to having been first. As one molder recalled, however, most of them were so used to compression molding that they missed out at the start, but "after we did get religion, then we had a field day."[39]

As with so many other developments between the wars, Americans imported the technique of injection molding from Germany. Hyatt had received a patent in 1872 for "a vertical extruder or stuffer" for celluloid, but its use never extended beyond a few machines in his own plant owing to danger of fire. Die-casting of metals required a parallel technology, but it did not spread to plastic until about 1919 when A. Eichengrün introduced a cellulose acetate molding powder and Hermann Buchholz invented a hand-operated injection press to mold small pieces with about a teaspoon of material per charge. Attracted by rumors of these German developments, a number of American companies began to experiment with injection molding. Probably the first to become involved, William D. Grote in 1922 imported several Buchholz machines (and powder to go with them) for the Grotelite Company of Bellevue, Kentucky. The following year Buchholz applied for a U.S. patent for his injection process (granted eight years later), and Grote obtained an exclusive American license. Four years later, in 1927, the Celluloid Corporation obtained two Buchholz machines made by Eckert & Ziegler of Germany and began experimenting with their own new molding powder Lumarith. The primitive machines were operated with a wheel that resembled a railway brake and heated by a gas flame that often went out. Within a few years Celluloid had relegated them to a storage basement as a failure and was promoting compression molding of cellulose acetate.

Unaware of these experiments, in 1931 Samuel Foster, Jr., of Foster Grant, a small celluloid fabricating company in Leominster, obeyed a partner's demand that he go to New York to examine some moldings imported from Germany—combs, brushes, cutlery handles, toys. Impressed by the obvious improvement over celluloid, he ordered ten Eckert & Ziegler injection molding machines. Sometime before the first of them arrived, Foster heard about the failure at the Celluloid Corporation. It must have been with great anxiety that he unpacked the first of three machines to arrive before Hitler's rise to power led him to cancel the remainder of the order. Foster found, like Celluloid's engineers, that the Eckert & Ziegler machines, which were intended for use with a German polystyrene costing nearly two dollars a pound, did not work well with cellulose acetate. With too much at stake for his small company to write off the investment, Foster and several mechanics spent two years modifying the machines and began commercial injection molding of cellulose acetate in 1934.

While Foster worked in secrecy, others learned of German developments and sought to exploit or improve them. Grote, who was more an equipment builder than a molder, had already in 1929 perfected a partially automatic machine that used compressed air and in 1933 introduced a hydraulic injection press. That same year Du Pont Viscoloid purchased a Buchholz license from Grotelite and soon bought at least a dozen Grotelite machines at about $3,000 apiece. By then several other equipment manufacturers and molding companies were violating the Buchholz patent by experimenting with homemade presses of pirated design—the primary goal being to increase the size of the charge, which went from half an ounce at the beginning to sixteen ounces by the end of the decade. Hoping to increase demand for its Tenite molding powder, Tennessee Eastman subsidized this rapid development by promising to defend anyone sued by Grotelite for patent infringement. Injection molding reached its prewar climax in 1935 when Joseph F. Geers of the Index Machinery Corporation of Cincinnati imported the first fully automatic press, an Isoma manufactured by Franz Braun AG. Geers had witnessed an Isoma in action at the Leipzig Fair of 1934 and later recalled his astonishment at seeing how it "automatically molded and ejected a finished aspirin box at every stroke of the machine."[40] American molders quickly recognized its superiority, not only over earlier injection presses but over compression molding as well. It was hard to miss the advantage of being able to mold in fifteen seconds something that would have taken fifteen minutes if made from compression-molded phenol or urea formaldehyde. Geers distributed more than fifty Isoma presses before the war began. By then the American press manufacturers Reed-Prentice and Watson-Stillman had designed automatic machines and were moving injection molding beyond the experimental stage.

The desirable qualities of the material and the method appeared in a pamphlet called *Molding with Lumarith* published in 1936 by the Celluloid Corporation. Not only was cellulose acetate visually attractive and available in more colors than any other plastic. It also had "touch appeal." It did not feel "stony" or "cold" but "rather warm and pleasing to the touch." To sum it up, cellulose acetate offered "life, tone, vividness," all made inexpensively possible by means of injection molding.[41] Already it had appeared in adding machine keys, knife handles, bobbins and spools, bushings and radio dials, eyeglass frames, automobile door handles and interior trim, steering wheels and horn buttons, combs and buttons, telephone desk sets, and, perhaps most typical of all, in the fluted handle of the common screwdriver—whose metal shaft was magically restrained without fastener or seam in a hard liquid golden depth of wondrous clear transparency, a tool that fit naturally, smoothly, warmly in the hand and left grizzled carpenter and young boy alike pondering, "Now how did they do that?"

That inseparable mingling of surface and depth, that near identity of appearance and function—the sign of an apparently unquestionable rightness or fitness— seemed to express the meaning of cellulose acetate and of plastic in general during the 1930s. As Tennessee Eastman summed it up in a booklet touting Tenite, cellu-

lose acetate was "a plastic of superior strength and beauty" with "a hornlike toughness, uniform texture, high luster, and an unlimited range of transparent, variegated, and opaque colors." And of crucial importance to a manufacturer seeking to
contain costs, it could "be molded into finished products at the fastest speed ever
attained with plastics"—"molded complete in only a few seconds, eliminating hand-
carving, machining, and polishing operations." A plastics executive whose career
spanned much of the century later concluded that injection molding had triggered
"explosive growth" by "taking plastics out of the gadget and novelty class and
launching plastics, not as substitutes, but as basic materials in their own right." In 1939,
however, he might have analyzed the scene with more pessimism. American molders owned only one injection machine for every four compression machines, and
the sheer poundage of phenol and urea plastics produced that year outweighed cellulose acetate by three to one. Even so, while phenol formaldehyde remained the
industrial workhorse, production of cellulose acetate rose from 2.8 million pounds
in 1933 to 20.3 million, and its price fell from about $1.75 a pound in 1923, when
statistics were first reported, to about thirty-five cents. If one considered other new
thermoplastics—the vinyls, polystyrene, and the acrylics—not to mention those
rumored over the horizon, such as polyethylene, then "explosive" seemed an apt
description.[42]

NEW THERMOPLASTICS

The newer plastics appearing just prior to the Second World War marked a fundamental shift in the way new materials came into existence. Such figures as Hyatt,
Baekeland, and the Dreyfus brothers set about in practical ways seeking substitutes
for specific materials. Small entrepreneurial companies focused on supplying a limited range of chemical materials ready for molding or fabricating by others. The new
generation of chemists, on the other hand, worked outward from chemical discoveries to the marketplace. When they found something of interest, they looked for
ways to commercialize it. They were driven not so much by market demand as by
the pressure of supply, an overabundance of chemical raw materials waiting to be
exploited. This shift derived in part from the greater sophistication of chemists, who
were beginning to devise general theories governing polymerization of long-chained
molecules, but it owed as much to the rise of large vertically integrated chemical
companies whose influence spanned many industries. In 1907 Baekeland had
thought nothing of starting out as an individual to develop a specific plastic based
on a specific chemical process. By the late 1930s the push to exploit basic chemical
raw materials that derived from coal, natural gas, or petroleum had submerged such
individualist efforts. Development of new plastics occurred within a complex world
of international corporate rivalries, parallel processes, cross-licensing (sometimes
among three or four firms), informal exchanges of information among the industrial
chemical fraternity, and agreements that traded rights to one plastic for those to

another—as if Du Pont, Union Carbide, Rohm and Haas, Imperial Chemical Industries, and IG Farben were carving the map of Synthetica into colonial spheres of influence.

One of those new territories was Vinyl, which *Fortune* in 1940 located along the Amazon-like Great Acetylene River, running down from the territory of Chloroprene (or Synthetic Rubber) just to the south of Petrolia.[43] Ethylene, found in coal gas and natural gas, was the basic raw material for vinyl chemistry, which yielded several compounds of potential significance to the plastic industry. One of them, vinyl chloride gas, was first reported in 1835. In 1872 a German chemist first made its solid polymer, polyvinyl chloride (later referred to as PVC, or by the general public as "vinyl"). During the 1920s chemical companies focused on the fact that vinyl compounds, unlike celluloid, could assume a range of textures and densities, from soft and flexible to hard and resilient, depending on the plasticizers used with the polymer. The trick was to find a plasticizer that would not evaporate, leaving the material weakened, cracked, or shrunken. Researchers also had to find stabilizers that would prevent vinyl compounds from releasing roiling clouds of hydrochloric acid when heated for processing—a tendency that sometimes exploded injection machines and forced workers to wear gas masks.[44]

By the end of the 1920s B. F. Goodrich, Union Carbide, and Du Pont had all expressed interest, though the latter gave it up as a hopeless cause. Waldo L. Semon of Goodrich began working with polyvinyl chloride in 1926. His discovery that the compound became a rubbery solid after being dissolved in a solvent led to the rubber substitute Koroseal in 1931. Used initially for seals for shock absorbers, Koroseal soon appeared as a coating for fabric raincoats and shower curtains, as a flexible tubing for bottling plants, in suspenders and watch straps, and as a coating for General Electric's Flamenol insulated wire. Koroseal became a visible consumer material at the end of the 1930s with raincoats and shower curtains cut from an extruded PVC film. It was unlike any other waterproof film or coating, a far cry from earlier oilcloths, rubberized fabrics, and artificial leathers. Warm, smooth but yielding, like a baby's skin to the touch, the film projected an otherworldly aura with its diffused gray translucence like a dense fog yielding to soft transparency when something directly touched it from the other side. A publicist described the effect as "like billowy curtains of mist." Unfortunately PVC sheeting did not always live up to its appearance. Poorly compounded film became tacky or brittle and exuded a chemical odor as plasticizers volatilized into the air—among the first indicators that so-called miracle materials might not always be so.[45]

Union Carbide's interest in vinyl compounds represented a convergence of its overall strategy as a large chemical company with a search for a particular market substitute. Organized in 1917, the Union Carbide and Carbon Corporation brought together three smaller companies moving toward exploitation of petroleum and natural gas—an agenda that involved seeking uses for acetylene and ethylene. As a small part of that strategy, Joseph G. Davidson began in 1926 to explore use of

3–3. Artifacts molded of polyvinyl chloride, early 1940s

polyvinyl chloride as a substitute for tung oil. Pressed from seeds harvested in China, the oil was subject to irregular supply and price fluctuations; it also caused a skin disease among those who handled it. Davidson soon found that polyvinyl chloride turned brittle and black when exposed to sunlight. Rather than seeking plasticizers and stabilizers, he turned to other vinyl compounds and eventually settled on a copolymer of vinyl chloride and vinyl acetate. Announced in 1930, Union Carbide's Vinylite followed an uncertain path to viable applications. No substitute for celluloid, Bakelite, or cellulose acetate, Vinylite turned into a gummy mess when boiled. When tried out for toothbrush handles, the material gummed up the bits that were used to drill holes for bristles. Belts and suspenders sold well until unscrupulous fabricators cut the material with other oils and gave the product a bad reputation. Success came only toward the end of the decade with raincoats, insulation, toys (such as soft dolls' heads), insulation for wiring, long-playing phonograph records, and an inner lining for beer cans (Figure 3–3).[46]

Union Carbide's slow progress did not keep the company from introducing Vinylite to the public at Chicago's Century of Progress Exposition in 1933. On the first floor of the constructivist Hall of Science, the company's Age of Plastics exhibit joined those of other industrial companies in revealing their contributions to material progress. Described for the public as odorless, tasteless, colorless, inert, and nonflammable, Vinylite could assume any color from opaque to transparent (the latter an exaggeration). On display was an array of experimental products, many of questionable practicality. Brightly colored tumblers and clock cases that were rigidly molded with few plasticizers and many fillers marked no improvement over urea formaldehyde. Such artifacts paled next to the exhibit's central attraction, a three-room Vinylite House that offered fairgoers a temple of plastic utopianism.

Co-sponsored by the John B. Pierce Foundation, a serious promoter of prefabricated housing, it was a simple one-story structure. Living room, kitchen, and bath were enclosed within modular walls made of two layers of aluminum coated with Vinylite lacquer surrounding an inner insulating layer of Vinylite. Interior doors were of Vinylite panels on pressed board. Vinylite floor tiles (actually 75 percent slate powder embedded in vinyl resin and plasticizers) proved so durable that they were taken up when the exposition closed and installed at Union Carbide headquarters, where they remained more than twenty-five years. Dishes, towel racks, switch plates, door knobs, and toilet seats were of molded Vinylite; baseboards and sills of extruded Vinylite. Quarter-inch translucent Vinylite fixtures suffused soft light over it all. Despite its vibrant colors—lime-green living room walls with dark-green-and-yellow checkerboard floor, pink bathroom, orange doors—the Vinylite House seemed spartan, severely modernistic, too austere for average Depression-era Americans but appropriate for future visions nourished by Buck Rogers in the Sunday comics. A correspondent for *Scientific American* described it well as the "house that chemistry built . . . truly a dwelling that came out of a test tube."[47]

Uncompromising in revealing plastic's unprecedented difference in texture and color from traditional materials, Union Carbide's display exposed a dilemma that makers and fabricators of plastic eventually had to confront. When plastic, used frankly and directly, became the overwhelming material in a given environment, the result often seemed cold or inhuman. But modifying plastic to conform to traditional expectations of the material world coerced acceptance of environments that subverted not only the natural or traditional but also the synthetic or artificial. Such thoughts followed almost too predictably from a consideration of vinyl, with its smooth flexible surfaces simulating so closely yet incompletely those of organic life. The acrylics, on the other hand, possessed a transparency so brilliant that they seemed a direct replacement for glass—certainly the result of chemical magic but hardly capable of molding whole environments. Acrylic plastic was not seriously employed until the Second World War provided a need for strong, lightweight, optically perfect thermoplastic sheets that could be formed into aerodynamic shapes for airplane cockpit covers and gunners' enclosures, but it first appeared on the market during the 1930s. More than other new materials it depended on the complex web of international competition and cooperation that was beginning to dominate both the chemical industry and the increasingly captive plastic materials industry. The acrylics also provided another example of the fundamental shift in raw materials from cellulose and coal tar to petroleum and natural gas.

Although chemist Otto Röhm investigated polymerization of acrylic acid for his doctorate in 1901, he abandoned the subject to focus on the leather tanning chemicals that enabled him to start the German firm of Röhm and Haas in 1907. His partner Otto Haas came to the United States two years later to open an American branch in Philadelphia. The two branches became separate companies after the First World War, each headed by one of the original partners. Seeking to expand the

85

business, Röhm returned to acrylic research during the 1920s in time to get a head start on parallel work at Imperial Chemical Industries (ICI) in Britain and at Du Pont. In 1928 Röhm and his associate Walter Bauer developed a polymethyl acrylate interlayer for safety glass that was marketed by the American company as Plexigum in 1931. Better than celluloid, which yellowed, or cellulose acetate, which became brittle at low temperatures, Plexigum could not compete with polyvinyl butyral, introduced in 1936. In the meantime, however, Bauer and Röhm were experimenting with a related polymer, polymethyl methacrylate, that was transparent, glasslike, and capable of being cast, sawed, and machined. A happy accident in 1932 provided them with Plexiglas, an optically perfect transparent material cast in sheets. Seeking to improve on Plexigum, which had to be "buttered" on two sheets of glass that were then pressed together to form safety glass, they poured methyl methacrylate monomer directly between two sheets of glass. Polymerization occurred through exposure to light. Instead of cementing the two panes together as intended, the polymer separated cleanly from the glass in a strong solid sheet. For the next three years both companies worked to improve the material and to find applications for it. Plexiglas was offered for sale in both Germany and the United States in 1936.[48]

Although Du Pont and ICI had signed an agreement in 1929 to divide world markets and share information and patents, each company entered acrylic research unknown to the other. Du Pont's interest stemmed from a need to find an outlet for surplus isobutanol. Setting to work in 1931, D. J. Loder devised a means of obtaining methacrylic acid and thence "polymerized methyl methacrylate," which he intended to develop as a synthetic resin. In the meantime the company learned by routine report that ICI was applying for a patent on polymethyl methacrylate research. For the next two years both companies reported regularly on their progress. Despite the attractions of a strong lightweight substitute for sheet glass, Du Pont focused on acrylic molding powders and cast turnery shapes (rods, tubes, and blocks). By the end of 1934 a small "semi-works" was making a weekly delivery of 350 tapered hexagonal handles of transparent cast acrylic for use with Pyralin celluloid hand mirrors. Cast resin seemed perfect as well for automobile tail lights, and experiments continued for using polymethyl methacrylate for dentures, phonograph records, and safety glass. The company projected annual sales of a million pounds a year, three quarters of it going for cast turnery shapes; ironically only fifty thousand pounds seemed likely for such "miscellaneous small uses" as airplane windshields.[49]

As research converged at Röhm and Haas, ICI, and Du Pont, and commercial introduction of acrylics became imminent, the companies in 1936 concluded an intricate set of cross-licensing agreements to forestall expensive litigation. The German firm of Röhm and Haas traded licenses with ICI; so did the American firm with Du Pont.[50] Toward the end of the year, after Röhm and Haas's announcement of Plexiglas, ICI followed suit in Great Britain with Perspex and Du Pont began promoting Pontalite in October as "a new, water-clear plastic, strong as glass, flexible and non-shattering."[51] Lacking full knowledge of Röhm and Haas's method of casting large acrylic sheets until 1939, when a new license granted Du Pont half the annual cast-sheet capacity of the American firm Rohm and Haas, the larger company focused on novelty items. Even the choice of names revealed a conceptual distinction—Plexiglas suggesting the material as a flexible or shatterproof improvement over glass, Pontalite clumsily indicating crystalline optical properties that made it a unique material for toiletware, jewelry boxes, and other fashion items. Soon after the material's introduction Du Pont abandoned the name Pontalite to avoid the possibility that a trade name incorporating part of the company's own name might someday become a generic term. The new name, Lucite, was taken from the celluloid toiletware, translucent and solid-colored, that the company had begun marketing nine years before.[52] More than anything else, Lucite reflected a stylish frivolity as it became available to manufacturers who made jewelry from it or embedded flowers and seahorses in it to create desktop novelties.

Dave Swedlow, a west coast designer and fabricator, recognized acrylic's true novelty, that it could be softened with heat and twisted into compound curves. Liberated by isolation in Los Angeles from preconceptions of celluloid or Bakelite fabricators, Swedlow created bowls and trays by folding and crumpling sheets of acrylic into shapes varying from the representational to the vaguely biomorphic to the abstract. Loop-the-loop candle holders shared space in his catalogue with stylized barley-twist table lamps, all transparent and marked by the sharp brilliance of their transmitted light (Figure 3–4). An item of high fashion at the end of the decade,

3–4. DECORATIVE OBJECTS FABRICATED FROM TRANSPARENT POLYMETHYL METHACRYLATE, 1940

Swedlow's acrylic furniture appeared in a *New Yorker* advertisement that characterized Lucite as emerging "out of a test tube" to become "the most fantastic, fabulous
stuff since Cellophane." Provoking the sensory disjunction of surrealism, Lucite
looked "like limpid columns of water . . . light as blown glass . . . sturdy as wood."[53]
The material reached an apotheosis of frivolity with release of the film *Eternally Yours*
in 1939. Swedlow and eight workers created a Lucite chemical retort five feet
high from which Loretta Young emerged as a gorgeous being created or synthesized
by master magician David Niven. This scene, foreshadowing *Fortune*'s "American
Dream of Venus," relied on Swedlow's skill in heating and shaping relatively small
Lucite sheets over a form and then cementing the pieces together. In a short time,
as war approached the United States, he was applying his techniques to larger one-
piece airplane enclosures. Swedlow joked later that "Loretta Young deserves at least
some of the modeling credit for the B-17 nose bubble." Initially, however, only
Rohm and Haas possessed the technical capability for such functional applications.[54]

   The American firm of Rohm and Haas obtained a license for casting acrylic sheet
from the German firm late in 1935.[55] In January 1936 Otto Haas sent Donald S.
Frederick to Darmstadt for two months to familiarize himself with acrylic sheet
manufacture and fabrication. The chemist had joined Rohm and Haas in 1934 after
earning a doctorate at the University of Illinois under the supervision of Carl S.
Marvel, one of the two or three leading polymer chemists in the United States. Frederick gained firsthand experience of the "flexible gasket" technique for casting large
sheets. To overcome the effects of polymethyl methacrylate's tendency to shrink and
crack as it hardened, the two pieces of plate glass between which the resin was
poured were fastened at the edges with a flexible gasket that allowed the sheets of
glass to move closer together as the resin contracted between them. Frederick returned home unaware that the Germans had not revealed their blow molding technique for forming large bubble enclosures by softening a sheet of Plexiglas and then
distending it with forced air. As a result, Americans made do for years with hand
shaping of softened sheets over curved wooden forms until wartime demand forced
independent development of blow molding. From the beginning, however, Rohm
and Haas's prime market was aviation. An aggressive sales manager, Frederick demonstrated samples of Plexiglas on the west coast in September 1936 and proved the
material's superiority to celluloid and cellulose acetate, which were only available in
small twenty-by-fifty-inch panes, and to glass itself, which had to be used in flat
sections. That same year he won a decree from the Army Air Corps that polymethyl
methacrylate was the only plastic sheet material approved for use in military planes.
In 1937 Frederick sold $135,000 worth of Plexiglas; the following year sales rose to
$432,000 with more than 90 percent of the material going for aviation. Both Rohm
and Haas and Du Pont experimented with other functional uses for acrylic plastic,
including back-lit signs, retail displays, highway reflectors, clear instrument panels,
surgical instruments that piped cold light, and dentures.[56] With the exception of
dentures, none of them extended beyond the experimental stage before the war.[57]

Consumers came across acrylic plastic in precious few applications, and most of those were novelties, though Swedlow's designs for housewares hinted at plastic's potential for rendering the material world visibly insubstantial.

Another new thermoplastic polymer, polystyrene, completed the roster of plastics commercially available by the end of the 1930s.[58] It too was a product of the emerging chemistry of petroleum and natural gas. Ethylene and benzene combined to form ethyl benzene, which then yielded styrene. Heating the monomer under pressure produced polystyrene, a relatively hard, brittle material that could be ground into a molding powder. It possessed the transparency and light-conducting quality of acrylic, the injection-molding potential of cellulose acetate, and its own unique ability to reproduce the most finely detailed molds with precise accuracy. Although polystyrene was polymerized in the late 1920s by Herman F. Mark of IG Farben and by Iwan Ostromislensky of the Naugatuck Chemical Division of U.S. Rubber, both of whom were investigating synthetic rubber, the plastic was commercialized more quickly in Germany. The first American polystyrene was Naugatuck's Victron, introduced in 1933 at a prohibitive price of more than $1.50 a pound; the material also suffered because impurities prevented the company from providing a "water-white" or fully colorless variety. That problem was solved by the Dow Chemical Company, which introduced Styron in 1937 at sixty-eight cents a pound.[59] A year later both Bakelite and Monsanto offered versions of polystyrene. An advertisement for Monsanto's Lustron in 1941 suggested why polystyrene soon became a major plastic. Illustrating a set of "bright-hued" picnic forks and spoons of the sort that seemed disposable a decade later, the ad explained that similar sets "formed from sheet plastics" had been "best sellers in exclusive shops for two years" at a premium price of $2.50. Now that they were cheaply made by "one shot" injection molding of polystyrene, they were "selling like wildfire *in chain stores at popular prices!*"[60] To younger members of the plastic industry the future seemed certain. Even so, they could not have guessed that postwar thermoplastics would so transform the scene that even polystyrene, then a marginal experiment, would seem as common as Bakelite.

Celebration of an approaching "plastic age" reflected the euphoria of the industry's executives in the 1920s and 1930s. In a few short years they had witnessed their field's expansion beyond the limited materials and methods of celluloid. If phenolic plastic seemed full of promise at its announcement in 1909, it now yielded its aura, though not its economic position, to urea formaldehyde, to cellulose acetate, to the vinyls and acrylics, to polystyrene. Each issue of *Plastics* and its successors revealed new materials, new processes, new applications, or at least someone's hopeful projection of innovation. A *Plastics Directory* published in 1929 listed eighty-four trade names for American plastics, some of them basic materials from major companies, others infringing materials from shoestring operations, still others in fact applications like film or laminated sheets. Four years later a similar list contained twice as many names. A severely edited list from 1939, which eliminated obsolete and marginal names, included nearly 250 different plastic trade names.[61] Each supplier hoped to make the general public aware of its own individual brand, but few succeeded beyond Celluloid, Bakelite, Formica, and Plexiglas. To expect people to distinguish Lumarith from Lustron was asking too much. But they could not help noticing plastic's expansion into everything from jewelry to auto parts, from toys and novelties to radios and telephones. The proliferation of names and types, of colors and textures, of new materials and novel applications, lent an air of reality to the visions of plastic utopians. Most people did not spend their days marveling over modern miracles. But they would have skimmed right over a journalist's reference to an approaching "plastic age" without questioning it. The many shapes of things to come were already manifest; they could be seen, touched, desired, possessed. If only to control this unprecedented expansion, to order the chaos of new materials, people ignored trade names and learned to refer to them all as plastic. The industry itself, as an economic endeavor, proved to be equally chaotic and uncertain of definition. Only gradually did a sense of common purpose emerge within the industry, a self-consciousness appropriate to the vision that chemical promoters had already created.

4

AN INDUSTRY TAKES SHAPE

*D*espite a welter of formulas, brand names, and specialty resins, a few giants of the emerging chemical industry dominated production of plastic resins. By the end of the 1930s Du Pont was making celluloid, cellulose acetate, and polymethyl methacrylate, and was developing nylon fibers and molding powders. Monsanto had acquired Fiberloid (the third largest producer of celluloid and cellulose acetate) and Resinox (the third largest producer of phenolics). Celanese soon absorbed the industry's founding company by extending its control over Celluloid, and Tennessee Eastman and Dow continued their expansion into plastics. American Cyanamid, which had introduced Beetle urea formaldehyde to the United States, eventually consolidated its interest by buying Plaskon, another producer of urea formaldehyde. And in 1939 Baekeland relinquished personal control of the Bakelite Corporation by selling out to Union Carbide to avoid having to enter the chemical business to produce his own raw materials.[1] Even so, plastic was not yet a major industry in the general scheme of things. Although annual production increased tenfold between 1921 and 1939, reaching 255 million pounds, with a sales value of $78 million, it totaled less than 1 percent of the annual poundage of steel.[2] And the molding and fabricating of final parts and products remained in the hands of several hundred plants and shops of varying size.[3] The increasing gap between material suppliers and processors created a tension that both enlivened the industry and made it difficult to control. Beyond perfecting their materials, providing technical assistance, and developing prototype applications, material suppliers relied on an army of restive entrepreneurs, engineers, machinists, designers, and visionaries to realize the potential of the plastic age. This motley crew, its personnel constantly shifting, encompassed an array of backgrounds and motives. At times offering brilliant in-novation or solid development, they just as often revealed ignorance or irrespon-sibility whose consequences reflected on the entire industry. Baekeland's hard-won insight that he could not simply release his wonder-stuff to a waiting world proved prophetic; nor, however, could he do it all himself.

A RISKY BUSINESS

The custom molder or fabricator between the wars was an independent business-man, often of working-class origin, who had started out in rubber processing, in celluloid fabrication, or in fields such as jewelry, comb, or toy manufacturing that later adopted plastic materials. Compression molding of phenolics attracted former molders of rubber and shellac, while celluloid fabricators tended to expand into cast phenolic or, during the 1930s, to shift into injection molding of cellulose acetate. Dependent on chemical companies for raw materials, the custom molder also feared consolidation of the molding process itself. Large corporations using large quantities of plastic parts often established "captive" in-house molding departments that com-manded discounts from material suppliers. General Electric's molding division ab-sorbed 15 percent of America's phenolic resin in 1930, not only producing plugs, switch plates, insulated panels, and half the nation's radio-tube bases but also engag-ing in custom molding for other firms. The company's million-dollar plant at Pitts-field, Massachusetts, had 324 presses, 900 workers, and an inventory of a thousand molds when it opened in 1938.[4] Large electrical and automotive companies oper-ated at such volume that they could afford captive plastic departments. That was not often the case with companies making umbrellas, alarm clocks, pencil sharpeners, drafting tools, fishing reels, office machines, cooking utensils, cocktail shakers, and dozens of other everyday items. Instead they turned to custom molders for plastic parts that would save them money by reducing assembly costs or eliminating metal parts. Such artifacts, extending beyond major industries and cutting across all areas of everyday life, contributed substantially to making the public familiar with plastic.

The custom molder was a jobber, dependent on others' successes for the repeat orders by which he made a profit. He was usually a craftsman who had mastered a difficult process by attaining an intuitive sense based on trial-and-error experience. He knew how to design a mold so material would flow smoothly to all sections and cure evenly to yield a uniform piece. He knew when to incorporate hidden ribbing to strengthen a molding, how to use decorative bosses and fluting for the same purpose, and how to disguise unevenly colored material by sandblasting a mold to create a textured surface. He knew how to insert flanges and other metal parts before molding so a piece would fit easily into its next larger assembly. He was a master of adaptation, modifying old equipment for new materials or processes and keeping up with such rapid innovation that one old hand joked of "a new kink every day." The presses he used were not precision instruments. More evocative of the age of

4-1. Compression molding press and operator, 1930

iron and steam than of anything that followed, compression presses were heavy, bulky, and oily, in marked contrast to the pristine artificial objects that emerged from them (Figure 4–1). Hydraulic systems, "tableting" or "pre-forming" of precisely measured resin charges, and use of mechanized pins to knock finished pieces from the mold before they cooled only began to automate a process that formerly required the custom molder to measure out molding powder by hand for each press cycle, to load the mold into the press, to remove the mold when it had cooled, and then to remove the piece. Curing times remained a matter of experience if not guess-work. Early injection presses required months of tinkering by machinists and engineers just to get them to work, never mind efficiently. The image of calm precision that recurred when journalists celebrated plastics occasionally approximated the reality of the chemical laboratory, but it never came close to describing the world of the custom molder.[5]

The custom molder's situation as an independent businessman emerged in a novel of corporate takeover published in 1955. The author of *Cash McCall* might have chosen any industry for his tale of a self-made entrepreneur who built a business and survived the Depression only to lose when the rules changed after the war. That Cameron Hawley focused on a custom molder suggested the marginality of the plastic industry between the wars, its sense of opportunity for individuals who defined it as they wished without having to mesh in some larger system. At a crucial point in the novel the founder of the Suffolk Moulding Company, Grant Austen, looks from the terrace of his house outside Philadelphia to his factory in the distance and recalls its scenes to memory as he ponders the question of selling out. While evoking the sensibility of a custom molder, this romanticized passage also suggests the brute physical history of those artifacts whose materialization seemed so mysterious to promoters of modern miracles. Looking down at an "old mill building" that once housed his first presses but now serves only for storage, Austen imagines himself walking through the tool shop, "bright with sunshine through the sawtooth roof, pungent with the musk of hot oil, alive with the nerve-tingling sound of steel cutting steel." On he goes "into the tool storage shed, seeing the molds stored behind the wire-fenced walls, the whole history of the business written in those chalked numbers on the oak-plank shelves, dates that went as far back as 1923, the male and female steel that had given birth to radio tube bases and bottle caps and switch assemblies and drill housings and rouge boxes and tool handles and watch cases and a thousand other things molded out of plastics, some remembered, more forgotten." Finally Austen reaches the pressroom, "brown-dusted and carbolic-odored," and imagines himself "hearing the steam-hissing suck of the opening presses, feeling the jarring thump of the hydraulic rams, following the conveyor line through the firewall door into the sharply contrasting quiet and cleanliness of the warehouse and the shipping room, out into the yard, past the high-stacked power-house, up the back steps into the Administration Building."

While Austen contemplates selling his company, a potential purchaser obtains an objective account of its history from one of Austen's former associates. Suffolk, he explains, is "one of the oldest molding plants in the country," engaged in "what we call *custom* molding—making molded plastic parts that are used by manufacturers of finished products." One of the best in the industry, "in the old days Suffolk got most of its business on difficult jobs that required a lot of development work—the kind of molding problems that no one else would tackle." Even the "big electric companies" with their own molding departments would give their business to Suffolk "if they'd run into something that was off the beaten track." Often that meant "working out some entirely new production technique or inventing a new kind of a mold" or even "building a special press." Although the special presses were actually constructed by "Hartzell-Bauer" (a reference to Watson-Stillman or Reed-Prentice), Suffolk "did all the engineering." Austen himself, "a good engineer and a clever development man," was "the sparkplug" behind it all. One of the major chemical companies had used his plant "almost as if it were their own laboratory—trying new materials, working out the processing bugs, setting production standards, that sort of thing." Hawley's fictional description was so attentive to realistic detail that any of half a dozen molders would have recognized his own company. Even its phrases echoed those used by molders, as when George Scribner invited customers to "use your molder as a consulting engineer," or Elmer E. Mills referred to his shop as "the production laboratory for many materials suppliers."[6]

The number of molders and fabricators grew steadily between the wars. In 1929 there were about sixty custom molders, virtually all engaged in compression molding of Bakelite or other phenolics. Ten years later, in 1939, there were about 170 custom molders. Half limited their work to compression molding, about sixty had branched out into injection molding but continued to operate compression presses, and another thirty engaged only in injection molding. Some of these companies were also active in fabricating products from cast or laminated sheets, rods, and tubes; another sixty-five firms engaged only in fabricating. Two years later, before wartime expansion, American molders (captive as well as custom) were operating about nine thousand presses, some of them twenty years old and fitted with makeshift improvements, others equipped from the factory with increased capacity and advanced control devices. By 1946 there were some 370 molding and fabricating companies. Even then, with 40 percent of the industry's capital postdating Pearl Harbor, only fifty companies were worth more than a million dollars. Two-thirds of the companies were worth less than $125,000; nearly half were so small they escaped notice of financial rating services. Throughout the prosperity of the 1920s and the economic collapse of the 1930s, plastic molding remained an endeavor open to small businessmen and skilled machinists rising from the ranks of labor (Figure 4–2).[7]

Despite success, custom molders chafed under what they perceived as unfair

4–2. PLASTIC MOLDERS' MARKINGS OF THE 1930S, REVEALING THE VARIETY OF CUSTOM MOLDERS

exploitation by material suppliers. Most bitter was Seabury of the Boonton Rubber Manufacturing Company, who had worked closely with Baekeland early on. Seabury complained that his company was "responsible in no small degree for establishing and maintaining a reputation for Bakelite." In return he had watched as Baekeland allowed Delco, Atwater-Kent, Westinghouse, and other companies to establish captive molding departments at an annual cost to Boonton of $230,000 in lost business. In 1926 Seabury tried to enlist several other custom molders in a declaration of war. "For the past seventeen years," he argued, none of them had "made any money, due to the rotten policy of the Bakelite Corporation," which had "made millions of dollars by exploiting the trade molders." Noting that the oppressor's patents were about to expire, Seabury proposed that six leading custom molders combine in a General Molding Company to force price reductions from Bakelite and gain the advantages of rationalized administration, production, and distribution. Nothing came of the idea, which would have united Seabury's company with such competitors as Kurz-Kasch and Scribner's spin-off Boonton Molding Company. Custom molding remained such a risky business that an officer of Kurz-Kasch told a group of foreign visitors forty-five years later that his plant was "antiquated" because "intense competition" never permitted American molders "sufficiently high profit on sales dollars or on investment to keep our physical plants updated." Seabury and Scribner would have appreciated his joking that all new jobs came from "mistakes on the estimates." Only those molders able to squeeze profits from an impossible job could hope to survive.[8]

Companies that survived in this cutthroat environment usually benefited from shop experience running back to the early days of the industry and a solid knowledge of chemical engineering. A good example was Kurz-Kasch, which began molding Bakelite in 1916 as the Dayton Insulating Die Company. Its founder, Christian A. Kurz, Jr., was a chemical engineer in his early thirties who had earned a degree from Purdue University before setting up a consulting laboratory at Dayton. Attracted to plastic molding by the possibility of military contracts, Kurz began making buttons for army uniforms on ten small gas-heated, hand-operated presses. Success enabled him to move to a larger building in 1919, but he also sought the assistance of Henry J. Kasch, a toolmaker whose technical experience Kurz lacked. Born in 1885 at Newark, Kasch attended a vocational school and learned his trade in the machine shop at Thomas Edison's laboratory before going to work as a toolmaker for the Celluloid Corporation. After four years at Celluloid he joined the Charles Burroughs Company, then the leading manufacturer of molding presses. During twelve years at Burroughs he cooperated frequently with Bakelite on phenolic molding and advanced to superintendent of the plant and then head of sales. Kasch's expertise so impressed Kurz that he convinced the toolmaker to become his partner in a reorganized Kurz-Kasch Company in 1921. Five years later an observer described them as "doing good business at low prices and sailing pretty close to the wind." Starting with fifty employees, the company grew to about four hundred by

1930. While Kurz kept up with chemical developments (such as introduction of Beetle urea formaldehyde in England), Kasch perfected technical innovations (such as preforming of mold charges and use of semi-automatic ejection pins). As a company executive later recalled, they succeeded by taking on every "'impossible' job that presented a challenge."[9]

Kurz-Kasch was typical of regional molding companies whose business came mostly from local manufacturers. Others included the Reynolds Spring Company of Jackson, Michigan, the Northern Industrial Chemical Company of Boston, and the Chicago Molded Products Corporation. At first glance the Boonton Molding Company seemed similar, but it transcended regionalism because it owed its existence to Seabury's early work with Bakelite. In 1908 Seabury had become general manager of the Boonton Rubber Manufacturing Company, whose founder, his father-in-law, Edwin A. Scribner, had died in 1898; Seabury gradually acquired control from an English investor who had bought out the Scribner family's holdings. Scribner had left a six-year-old son named George who later worked for his brother-in-law Seabury at Boonton during the summers while earning an A.B. from Princeton and completing two years of graduate work in electrical engineering at Columbia. When George Scribner went to work full-time for Boonton in 1915, he possessed both practical experience and formal training, as well as a liberal arts education that enabled him eventually to articulate the molding fraternity's goals and methods. Despite these advantages, Scribner lost his job with Seabury in 1920 when he returned from a six-week sales trip to the western United States without a single order. Out of "sheer desperation" at the prospect of bagging groceries at the local A & P, he started the Boonton Molding Company by installing a single press in a corner of a factory down the street from Seabury's plant. The new company grew so quickly that in 1926 Scribner's former boss Seabury included it in his proposal for a General Molding Company. The presence of two Boonton companies confused things until 1933, when the original changed its name to Tech-Art Plastics and moved to Long Island City. By then Scribner had expanded from phenolic to urea, he had "dabbled" with compression-molding of cellulose acetate, and he had begun to investigate injection molding. He had also become a leader whose experience, he liked to say, dated "back to the dark ages of molding."[10]

That comment came in 1933, when the Boonton Molding Company was barely a dozen years old. For the president of a young company, himself just turned forty, to speak of the "dark ages of molding" indicated an industry experiencing rapid transformations rather than a long, slow evolution. Coming to grips with increasing mold and press sizes, with a plethora of new materials requiring modifications in machines and routines, and with markets that shifted with each new chemical or technical innovation, Scribner realized the necessity of communicating clearly with customers who knew next to nothing about plastic. In 1933 he wrote a twenty-six-page handbook, *A Brief Description of the Commonly Used Plastics Compiled for the Guidance of Engineers and Buyers*, which doubled in size at its third revision in 1940 and went through six more revisions by 1950. The first edition offered "specific mechanical and electrical properties of the nine most common types of plastics with which the buyer or engineer is faced when the question of choosing a new material arises." Hidden in that measured sentence were the ramifications of the concept "new material," whose neutral tone masked the dismay and exhilaration of a manufacturer or engineer forced to do old things in entirely new ways or, more radically, to do entirely new things. After discussing various plastics in a carefully factual way, Scribner asserted that Boonton's staff, "born and brought up in the industry," had "overcome more 'things that can't be done' than the average group." Unlike unimaginative molders who relied on "routine quoting," Scribner looked to "development and application of new ideas" to create sales. He sounded as if he were promoting a pyramid scheme when he suggested that manufacturers who switched to plastic would gain an edge on their competitors. But the plastic pyramid never stopped expanding with new properties, new processing techniques, new applications. By 1940 the custom molder appeared as the manufacturer's "guide"—not through the uncharted wilderness of Synthetica that had confronted pioneers like Baekeland but through a more urban "merry throng of rapidly increasing plastics." Scribner was so confident of the brave new plastic world, it sounded hollow when he warned that no molder could take the "curious powder known as synthetic plastic molding compound" and "transform it into anything under the sun." They—and he—were too busy proving him wrong.[11]

ASSERTING CONTROL

At first glance "the plastic industry" of the interwar years hardly merits such a unifying phrase. The materials were too distinct to lump them together. Hard, dark, infusible phenolic bore little resemblance to bright, colorful, thermoplastic cellulose acetate, or to soft, sometimes tacky vinyl sheeting, or to hard, glasslike, formable acrylic. Different chemistries—those of cellulose, coal tar, and petroleum—divided the world of plastic. Different processing techniques—casting, laminating, compression molding, injection molding, extrusion of thermoplastics like vinyl, machining of cast phenolics and acrylics—came from several industrial traditions. Different applications appealed to a range of end-use industries—from radios and automobiles to toys, toiletware, furniture, and housewares—each with its own needs, preferences, and ways of doing business. A doll's head of blow-molded celluloid had nothing in common with a distributor cap of compression-molded phenolic; nor a cast acrylic airplane windshield with a thimble of injected-molded cellulose acetate. Material makers who were determined to distinguish Durez from Bakelite or Lumarith from Tenite only added to the confusion. So did custom molders who looked to the advantage of their own companies and regarded suspiciously any move to consolidate or standardize their activities. Even so, several institutions began to promote a sense of common purpose among those who made or used plastic. Uniting them was a recognition that they were engaged in transforming the world with radically malleable chemical materials unlike anything known to nature or tradition.

Nothing so clearly announced the new field of "plastics" as the first issue of the trade journal of that name in October 1925. Until then chemists and engineers relied for random coverage on journals broadly devoted to industrial chemistry and electrical engineering or narrowly focused on automobiles and aviation. Launched under the editorship of Carl Marx, a chemical patent attorney, *Plastics* served as a vehicle for material suppliers who bankrolled it with advertising revenue and plied it with articles touting their own products. Smaller advertisements promoted suppliers of chemical solvents and plasticizers, press manufacturers, mold makers, paper companies, and purchasers of plastic scrap. In line with Marx's opening promise "to help every fabricator and user of composition products," *Plastics* published technical articles on new presses, new molding techniques, properties of new resins, significant patents granted, and abstracts of relevant scientific articles.[12] At the outset, then, the journal put shop talk into print for people directly engaged in molding and fabricating operations.

Within a few years, however, the magazine was shifting focus from those within the industry to those outside, to potential users of plastic. The first sign came in May 1927 when the name changed to *Plastics & Molded Products* and the editor began covering new product applications as well as materials and processing techniques. Soon after the stock market crash the magazine was rescued from bankruptcy by Williams Haynes, a prominent chemical publisher whose involvement suggested recognition of plastic's rising importance and continued emphasis on technical matters. In January 1933 he announced a fundamental shift in editorial policy intended to stimulate the industry's economic renewal. Hoping for maximum impact among readers and casual browsers, Haynes placed his short manifesto on the front cover. For seven years, he told readers, the journal had "served the makers and fabricators of plastic materials" during a period of "rapid internal development." But the industry had matured. It had reached a technical plateau and now required "wider markets" to realize its potential. Because "thousands of potential users" had to be "educated" and "transformed into actual buyers," he had decided that "our next issue will become *Plastic Products*—a business magazine which will style and service all plastic materials not only for the Industry itself but for all Industrial Consumers."[13]

This shift in the journal reached a logical extreme in 1934 after it changed hands again. The journal became so closely identified with convincing manufacturers to use plastic in the design of consumer products that it served for several years as the nation's most important journal of industrial design. The new owner, Charles A. Breskin, already published *Modern Packaging*, a visually sophisticated magazine organized around the idea that new packaging techniques and materials could sway consumer choices in a depressed marketplace. A companion journal devoted to plastic made sense. Plastics had already appeared in packaging as closures and reusable gift boxes. More important, it was easy to think of molded plastic housings of electric shavers, clocks, and radios as packages because they all had to attract consumers at the point of purchase. Before taking over the magazine, which suffered from a no-nonsense layout and small grainy photographs, Breskin hired designer Joseph Sinel to create a slick mock-up inspired by *Fortune*. In the first restyled issue of the new *Modern Plastics*, for September 1934, Breskin and his editor heralded it as a "new type of industrial magazine" because it addressed an outside audience. Its new purpose was "to interpret to industry at large the growth of the plastics industry." Promising "to increase the use and consumption" of plastic, Breskin in effect took away the industry's only means of internal communication and transformed it into a medium of advertising aimed at convincing other manufacturers they could best attract consumers to their products by making them with plastic.[14]

Implementing the new policy fell to Earl F. Lougee, installed as editor shortly after the takeover. Until he retired in 1940, Lougee vigorously championed the cause of plastic with manufacturers and the general public. Although he claimed "a happily balanced diet of technical and non-technical editorial material," the journal in fact embarked on a promotional crusade. Lougee believed "the real job ahead of the industry" consisted of "education and publicity." Manufacturers had to be shown "how and where" to use plastic. Retailers had to be "made conscious of the profits" to be gained from plastic. And the public, at the end of the equation (or the beginning if one focused on desire and demand), had to be "apprised of these modern materials which are theirs for the asking."[15]

To reach the first two groups, Lougee ran a series of articles about new products that had succeeded commercially, in each case providing information about design considerations, choice of materials, processing techniques, and sales figures.[16] Other articles focused on specific plastics, explaining chemistry, physical properties, and processing techniques in simple terms meant for newcomers—hardly required reading for an experienced molder or fabricator. The most important series of articles, many written by Lougee himself, profiled designers who had championed the cause of modern design for mass production in plastic at a profit to their employers. Profiles of Raymond Loewy, Lurelle Guild, Ely Jacques Kahn, Jay Ackerman, Gilbert Rohde, and Belle Kogan, all published in 1935, promoted the status of industrial design as much as that of the plastic industry.[17]

To reach the general public as well as manufacturers who had not yet gotten the word, the magazine organized two "annual plastics exhibits," the first in December 1934, shortly after Breskin's takeover, the second in May 1935. Thousands of plastic products were displayed at the publisher's offices in cramped, somewhat dowdy surroundings.[18] In the following year Lougee initiated an Annual Modern Plastics Competition, lasting from 1936 through 1941. The expansion of categories in which products could be entered, from three in 1936 to nineteen in 1941, mirrored the industry's own expansion. Judges over the years included designers Loewy and Walter Dorwin Teague, architect Harvey Wiley Corbett, editors from such fashionable magazines as *Harper's Bazaar* and *Arts & Decoration*, and several product engineers. Lougee and Breskin invited entries from material suppliers, manufacturers, molders and fabricators, toolmakers, and consultant designers or engineers—anyone, in other words, who had anything to do with a product. Each year all entries went on display. The winners received full coverage in the magazine, complete with glossy photographs, and at least in 1938 toured department stores, public libraries, and engineering schools across the country.[19] A manufacturer impressed by what plastic had done for the Sears Silvertone radio or the Burton X-Ray Projector, two winners for 1936, could turn to a three-hundred-page "Catalog and Directory" issue, later known as the annual *Modern Plastics Encyclopedia*, for details about materials, techniques, suppliers, and processors. Although an advertising salesman for *Modern Plastics* recalled the competition as "a real 'scam job' whose prizes went to the biggest potential advertisers," even he admitted "the resultant publicity was good for the industry."[20]

More than a trade journal and less than a trade association, *Modern Plastics* functioned as both through much of the 1930s. It offered a unified identity glowing with modernity with which to face the outside world. Breskin and Lougee's awareness of design's potential for stimulating and satisfying material desires enabled them to establish plastic as a primary medium of the symbiotic relationship between manufacturer and consumer—a newly conscious element of twentieth-century culture. *Modern Plastics* seemed a creature of interests larger than the molders and fabricators whose expertise made the industry possible. By turning away from technical matters to address consumers of plastic, Breskin denied processors access to their only unifying forum. Almost as if aware of the loss, they acquiesced in the formation of a trade association, the Society of the Plastics Industry. Even then, their action represented a further surrender of the rough-and-tumble individualism that had formerly marked their endeavors. Another major element in the rationalization of a maturing industry, SPI owed the impetus for its formation to material suppliers—and in particular to Gordon Brown, sales manager of the Bakelite Corporation.[21]

Earlier attempts at uniting for efficiency's sake had come to nothing. In 1923, for example, Hylton Swan and Sandford Brown met with a dozen custom molders to form an association for "molded manufacturers" under the auspices of the American Society of Mechanical Engineers. Their plea to "let's all of us get back of this and *PUSH*" fell on deaf ears and the group soon dissolved.[22] Another quickly defunct group was a chapter of the National Electrical Manufacturers Association, for those who specialized in molding phenolic insulation. Custom molders were suspicious of the motives of material suppliers like Bakelite in pushing for organization. Their doubts were confirmed early in the Depression when suppliers treated the NRA's legislatively mandated Plastic Material Manufacturers Association as an invitation to legalized price fixing. This inauspicious record was on Gordon Brown's mind in 1935 as he embarked on yet another attempt to bring order to the industry. As sales manager at Bakelite, he had been collecting and collating sales figures from eleven phenolic molding companies since 1923 to achieve an overview of trends.[23] Even so, most custom molders continued to distrust Bakelite and other material companies. Aware of that fact, and inspired by the example of a Plastics Golfing Society in Great Britain, Brown arranged two days of golf and informal socializing at a resort in the Delaware Water Gap in July 1935. Although he phrased the invitation to "leave the impression that it was in answer to popular demand in the industry," the few molders he discussed it with in advance were skeptical, willing to attend only if he "would do all the work."[24] By inviting Bakelite's competitors, by acting as a private individual rather than a Bakelite representative, and by avoiding any hint of impending formal organization, Brown managed to entertain seventy-five molders, fabricators, and material salesmen so successfully that they clamored for another gathering. Almost before the rowdy, well-irrigated group knew what was happening, their golfing retreats yielded a trade association.

Several issues emerged immediately. Apparent dominance of material suppliers over molders and fabricators at first threatened to swamp the venture. When the Society of the Plastics Industry was formally organized in May 1937, forty-six individuals representing twenty-eight molding firms held a clear majority over twenty-six representatives of eleven material suppliers. At each of the next two meetings, however, in September 1937 and May 1938, the materials representatives gained a slight majority, probably because rival companies sent their own salesmen to ensure Bakelite did not enjoy an inside track to potential customers.[25] Following the recommendation of a program committee headed by Herbert S. Spencer of General Plastics, SPI solved the problem in October 1939 by giving each corporate member a single vote and granting five-dollar "social memberships" to individuals employed by member corporations. The committee also successfully maintained that because "the industry [was] growing rapidly and no one [could] predict its course," annual sales statistics should be collected and published for the guidance of members.[26] Other endeavors included attempts to standardize wage rates, to push for equitable freight rates, and to discourage unfair competition through price cutting. Referring to the latter as an issue "full of dynamite," Spencer observed that there seemed "no bottom to the price situation" with "some of the most outstanding [molding] organizations in the industry . . . quoting the lowest prices, claiming . . . they are making a profit on all work."[27] Less controversy surrounded a decision to lobby Congress on labor regulation. As SPI defined a serious agenda, its meetings took on a more serious tone. Despite some skepticism, most members stayed off the golf links long enough in May 1938 to attend a lecture on "super-heated hot water" in molding; two years later a slate of technical papers entirely replaced golf as an official activity.[28] By 1940 a paid staff of two was installed in an office on Madison Avenue. The Society of the Plastics Industry was definitely under way.

Almost as if intended to fill the vacuum left by *Modern Plastics*'s shift toward outside promotion, SPI ministered directly to needs of companies within the industry. But the trade journal and the trade association maintained close ties from the beginning. As a corporate member of SPI, *Modern Plastics* participated in its deliberations, served unofficially as a neutral arbiter of disputes, and reported on its activities. The two institutions even shared a key figure, William T. Cruse, in the early days. A graduate of the Wharton School, Cruse became sales director for cellulose acetate at the Celluloid Corporation in 1936—a position requiring him to study organic chemistry at New York University and mechanical engineering at Columbia in his spare time so he could "deal with problems in the field" without sounding "like a fool." When Lougee resigned the editorship of *Modern Plastics* in 1940, Cruse left Celluloid to take the position. His work at the magazine and as a "courtesy" member of the SPI board was so impressive that when SPI's executive secretary became ill in October 1941, the board asked Breskin to release Cruse from *Modern Plastics* so he could fill the SPI vacancy. For decades, until retirement in 1967, he served effectively behind the scenes and on ceremonial occasions.[29]

An agency of the federal government completed the trio of institutions that did most to define and unify the plastic industry. Unlike Breskin's magazine and SPI, both of which owed their existence to the industry's success, the National Bureau of Standards was independent of the shifting needs and profit motive of business. Established in 1901 under the Department of Commerce and Labor, the Bureau devised standards for industrial materials, tested products under consideration for federal purchase, evaluated products submitted to it by industry, and by 1920 had extended its mission to applied research in chemistry and physics on problems submitted by military agencies, engineering societies, and even trade associations. Much like everyone else who became involved with the evolving plastic industry, the National Bureau of Standards did so gradually. Around 1929 its Organic and Fibrous Materials Division set up a one-man "synthetic resins and plastics desk" to monitor the situation. The second occupant, appointed in 1930, was Gordon M. Kline, a chemist in his late twenties who had joined the bureau while working part-time on a doctorate at the University of Maryland. His career lasted as long, and was nearly as influential, as that of SPI's Cruse.[30]

Soon after Kline's appointment as chief of an expanded Organic Plastics Section in 1936, Breskin invited him to become technical editor of *Modern Plastics*. The bureau offered no objection because no one could accuse him of playing favorites by working for a magazine with no competitors. As Kline recalled, the position "established a direct contact with the plastics industry" that would not have existed had he remained "merely a chemist" at the bureau. He became acquainted with executives, research directors, and senior chemists under auspices that mitigated their distrust of sharing information and thus avoided "duplication and triplication of a great deal of research."[31] Even Kline's appearance inspired confidence. Always referred as "Dr. Kline," he was a thin, bespectacled, abstracted man, a scholarly expert rather than a worldly businessman. Owing to his connection with *Modern Plastics*, the bureau more directly addressed the industry's needs. At the same time, Kline provided an editorial counterbalance at *Modern Plastics* to the image-oriented

promotional slant Lougee had given the magazine. Kline served the journal by pry-
ing loose submissions from corporate chemists describing new materials or processes. Equally important, he served as the industry's bibliographer by publishing
long annual reviews of new developments and by editing the annual *Modern Plastics
Encyclopedia*. Owing to his visibility in both government and industry, Kline was a
founding member of Committee D-20 on Plastics of the American Society for Testing Materials, which began in 1937 to devise standard test methods.

Relations among the three institutional supports of the industry were so close that
when Cruse resigned as editor of *Modern Plastics* in October 1941 to become executive secretary of SPI, Breskin offered the job to his technical editor Kline at the National Bureau of Standards. Although the editing job was attractive, the bureau's
director refused to release Kline owing to the impending national emergency. All the
same, the offer indicated a fluidity of boundaries between three agencies almost
incestuously united in personnel and in common purpose, despite the distinctiveness of their specific agendas. That they all came into existence within three years
(counting the renamed *Modern Plastics* as a new entity) indicated an industry whose
"pioneer" days had ended, whose self-conscious process of definition had created a
stable identity, and whose members could have anticipated a period of slow growth
if wartime demands had not stimulated a forced technical advance whose phenomenal pace continued unbroken afterward. Despite the chronic plight of small molding
and fabricating companies whose interests clashed with those of expanding chemical companies, they all shared a sense of participating in the creation of a plastic age,
an era of new material possibilities whose horizons dwarfed those of novelties, or
toys, or electrical insulation, or housewares, or any other single business. This attitude came to a head during the war, when it became common to assert, as *Life* did
in 1943, that "the nation's biggest and most clamorous maker of gimcracks" had
become "a semi-secret, high-priority war industry."[32] By then, however, the concept
of a plastic age was already a common perception grounded in realities of sight and
touch, not just in the rhetoric of promoters like Slosson and Mumford. Journalists
had voiced prophecies of plastic utopianism, but the industrial designers of the
1930s actually embodied them in the artificial outlines, colors, and textures of
molded consumer goods.

DESIGN IN THE BAKELITE STYLE

At the end of the twentieth century it is impossible to recapture the physical textures
of everyday life into which plastic's material presence emerged during the Depression years. Although plagued by economic collapse and human misery, that era
appears as the last capable of supporting an optimistic, even utopian vision, innocent of the horrors of nuclear war, clinically efficient genocide, technological self-
pollution. A postmodern or post-progressive failure of nerve has invested those
years with a nostalgia that attains expression in the popular medium of film. Images

of that time come filtered through the haze of *Bonnie and Clyde* or *Chinatown*. The troubling future visions of *Blade Runner* or *Brazil* extrapolate from gleaming World's Fair pylons and bleak Hoovervilles. The ultimate urban hell is *Batman's* Gotham City, where darkly exaggerated Art Deco canyons reminiscent of Hugh Ferriss enclose stylized interiors whose cold pastels and unnatural highlights suggest the touch of the colorizer. As the millennium wanes, the sense of a material past is confused with these pastiches from the shards of the century's popular culture. When style and design become markers of a flawed historical awareness, it is not surprising that plastic becomes a near fetish, its history increasingly hard to recover. After all, plastic emerged during the 1930s as an artificial substance exploited by designers self-consciously intent on defining a new style for a so-called machine age.[33]

The difficulty of isolating plastic in the cultural landscape of the 1930s becomes apparent in a passage from the novelist James M. Cain, known for his precise, hard-boiled prose. The first paragraphs of *Mildred Pierce*, set in 1931 and published ten years later, focus on a suburban "Spanish bungalow, with white walls and red-tile roof . . . like thousands of others." Cain's description of its living room is crafted with a reporter's sociological instinct. The furnishings comprise a "standard living room sent out by department stores as suitable for a Spanish bungalow," with "crimson velvet drapes, hung on iron spears" and "a long oak table holding a lamp with stained-glass shade." On display is a single book, the "Cyclopedia of Useful Knowledge, stamped in gilt and placed on an interesting diagonal." Among several Western paintings is one of "a butte at sunset, with cow skeletons in the foreground." Two details remain to be mentioned, no more significant in Cain's ordering than any others: "one table, in a corner, in the Grand Rapids style, and one radio, on this table, in the bakelite style." Cain, never an esoteric writer, assumed that his readers in 1941 knew not only the material, and even the word, but also a particular bakelite style. But what, precisely, *was* a radio "in the bakelite style?"[34]

To a collector in the 1990s, the answer to that question seemed obvious. A "bakelite radio" was a sleek, streamlined table model with case, grille, and knobs fabricated from pieces of Catalin or Marblette cast phenolic in bright contrasting colors, an Addison or a Fada (Plate 7), one of those sold at auction in 1983 for $500 whose value then soared beyond the thousand-dollar mark. It was an object of nostalgic veneration of which a design writer observed that "you almost expect to hear Ronald Reagan's voice . . . announcing those old Cubs games."[35] But maybe a "bakelite radio" was a huge Pilot or a Sears Silvertone of 1936 (Plate 8)—its smooth one-piece cabinet molded of Bakelite, black or dark brown, round-edged but hardly streamlined, too plain to be celebrated as "machine-age," too industrial to blend in with ordinary domestic furnishings unless by fading into the obscurity of bad lighting. Or maybe it was a radio whose maker had wanted to offer color while avoiding both the labor cost of assembling Catalin castings and the material cost of urea formaldehyde, and so had devised the expedient of spraying an ordinary phenolic molding with cream or pastel lacquer in violation of standard rhetoric about bright crystalline

4–3. BELMONT RADIO CABINET
MOLDED FROM DARK PHENOLIC RESIN
AND LACQUERED IN A LIGHT COLOR, 1939

color embedded throughout (Figure 4–3).[36] Or perhaps Cain intended his phrase to suggest one of the earliest plastic radios, neither streamlined nor brightly colored but instead molded from a dark mottled phenolic in imitation of the "gothic ca-thedralette" style of large wooden radio cabinets. Fifty years later it was hard to reconstruct the diversity of Bakelite styles during the 1930s and inevitable that any discussion of the interactions of plastic and design would tend to generalization.

As Breskin and Lougee's strategy for transforming *Modern Plastics* suggested, the plastic industry and the new profession of industrial design developed together during the Depression years. Designers like Teague, Loewy, Henry Dreyfuss, Norman Bel Geddes, and Harold Van Doren spent the decade redesigning consumer products to make them more attractive. Businessmen hoped industrial design would overcome the problem of underconsumption to which they attributed most economic woes. Products from toasters to refrigerators experienced face-lifting operations in attempts to stimulate economic recovery. If, as industrial designers liked to say, their profession was a "depression baby," then the plastic industry was, in *Fortune*'s words, a "child of the depression." There was some truth to an engineering journal's assertion, "A Plastic a Day Keeps Depression Away"—a thought cynically echoed by Lurelle Guild when he noted that "plastics . . . offer amazing possibilities to both designer and manufacturer in tricking up design." Another designer, Peter Muller-Munk, recalled that "plastics became almost the hallmark of 'modern design' . . . the mysterious and attractive solution for almost any application requiring 'eye appeal.'" Many design decisions resulted not from the question "What can I do to improve [a product's] operation, maintenance, manufacture and appearance?" but from a questionable concern for "What do I have to do to make it of plastics?" The close relationship between plastic and industrial design became an unquestioned axiom, as *Business Week* indicated by asserting that "modernistic trends have greatly boosted the use of plastics," which in turn "by their beauty have boosted modernism."[37]

Plastic owed success as a material of mass production to a convergence of three factors. Economic motives that spurred plastic's advance from celluloid combs through Bakelite gearshift knobs became even more compelling in an era of scarcity. Manufacturers valued plastic's certainty of supply, its dimensional precision and stability, its relative light weight and ease of processing, and above all the savings obtained by minimizing costs of hand assembly and finishing. Secondly, the material benefited from public faith in technological progress as a counterforce to economic and social stagnation. Fascination with advancing modernity, with radio and electricity, with automobiles and aviation, with skyscrapers and streamliners, merged with more practical economic concerns. And finally, plastic benefited from a twentieth-century culture of modernism whose artists, designers, and visionaries were articulating the meanings of technical and industrial modernity. They not only celebrated plastic among other technological wonders but also showed manufacturers how to use plastic in idealized machine-age design modes.

New consumer plastics such as cast phenolics, cellulose acetate, and urea formaldehyde appeared on the market just as American commercial design experienced a renaissance stimulated not only by the recession of 1927 but also by an exhilarating mix of European modernisms. The extravagant novelties of Art Deco revealed at the Paris exposition of 1925 merged with the austere idealism of Le Corbusier's *Towards a New Architecture*. Secure in the knowledge that European artists were celebrating American progress, young designers reacted with equal enthusiasm to the sculptural expressionist fantasies of Eric Mendelsohn, the social responsibility of the Bauhaus, the abstractions of de Stijl, and the perceptual machines of Constructivists. More often a matter of gears, motor cars, and "machines for living," modernist visions also encompassed new industrial materials—plastic, aluminum, stainless steel—as the essence of modernity. Visitors to an avant-garde Machine-Age Exposition in New York in 1927, on viewing Naum Gabo's Constructivist sculptures of intersecting planes of transparent celluloid, realized that plastic, no longer an imitation of ivory or tortoiseshell, lent itself well to machine-age expression.[38]

Among the most vocal boosters was Paul T. Frankl, a designer of custom furniture who hoped to see high-art advances extended to the masses. He made a start in that direction in 1928 by designing celluloid comb-and-brush sets promoted by his client Celluloid for their "so-called modernistic art." With other emigré designers arrived in America since the First World War, Frankl shared a direct knowledge of European modernist trends and an enthusiasm for his adopted country's brash modernity. Although he identified himself as a philosophical disciple of Le Corbusier, his own design work revealed an eclectic expressionism in tune with American commercialism. In a book on contemporary design, *Form and Re-form*, which appeared in 1930, he devoted a chapter to "Materia Nova" considered as "expressive of our own age." These included such new materials as Bakelite, Celanese, Vitrolite glass, Monel metal, aluminum, linoleum, and cork. Despite the range of this list, all of which he was using in deluxe furniture and interiors, Frankl placed greatest emphasis on Bakelite and other synthetics. Warning about conservative clients who sought refuge in imitating traditional materials, he declared that "imagination is essential to visualize and to realize the potentialities of new materials—to treat them on their own terms, to recognize, in a word, the autonomy of new media." He challenged designers "to create the grammar of these new materials" that already spoke for themselves "in the vernacular of the twentieth century . . . the language of invention, of synthesis." Sounding much like Slosson or Mumford, Frankl exclaimed that "industrial chemistry today rivals alchemy" in offering processes by which "base materials are transmuted into marvels of beauty." Two years later, in another tract on modern design, *Machine-Made Leisure*, he extended his celebration of plastic in a similar chapter whose title announced that "A New Language Emerges." Describing new materials as the foundation of a new style, he again praised the "new alchemy" for yielding the "minute accuracy, strength, permanence, and stability" of molded plastic. With special praise for William Perkin, the chemist whose creation in 1856 of mauve, the first synthetic coal-tar dye, had "led the world into fields of loveliness undreamed of," Frankl observed that the future "promises to be a true Plastic Age."[39]

The earliest significant application of plastic to furniture and interior design involved phenolic laminates as protective surfaces for tables, shelves, cabinets, counters, and walls. Laminate had remained an industrial material except for that brief period in the 1920s when it invaded American parlors as the front insulating panel of radio sets. Anticipating a new market, the Formica Insulation Company in 1924 hired chemist Clarence M. Hargrave to improve on the standard expedient of using swirled or mottled mixtures of black, brown, and dark red shades of resin to approximate the appearance of wood. Hargrave worked for three years developing a process for incorporating a wood-grain lithograph as the top layer of paper in a Formica sheet. By the time his relatively crude process was ready, however, radio manufacturers had decided to domesticate the receiver by enclosing it in an ornamental wooden cabinet.[40] Other laminate uses revealed practical considerations. In 1927, for example, the Manger Hotel of New York provided each room with a wooden chiffonier whose fold-down writing surface was inlaid with black Micarta laminate—necessary for protection from alcohol stains because Prohibition had driven guests out of the lounge and into their rooms to drink.[41]

American modernists became aware of phenolic laminate as an industrial material reflecting the machine spirit they wanted to convey in custom furniture for Manhattan sophisticates. Although each designer displayed minor personal touches, their work collectively suggested an impersonal precision. Frankl began by crafting unique skyscraper-shaped pieces from elegant stained woods, sometimes trimmed with black lacquer, but gradually shifted to Bakelite laminate, at first as a substitute for lacquer trim, eventually as a material for surfacing entire pieces. Gilbert Rohde, who started as a designer of custom furniture and ended with commissions for mass-produced goods, produced tables for custom sale with Bauhaus-inspired tubular chrome-steel frames and Bakelite tops. Most active in spreading the use of plastic laminate was Donald Deskey, whose spare geometric furniture relied for effect on

4–4. Furniture of brushed metal and phenolic laminate designed by Donald Deskey for Ypsilanti Reed
Furniture Company, 1928–1929

the contrast of brushed metal supports with horizontal expanses of polished black
Bakelite so sheer as to encompass the concept of two-dimensionality (Figure 4–4).
Gaining experience from a room with gray Bakelite walls designed in 1929 for displaying prints in Abby Aldrich Rockefeller's apartment, Deskey went on to outfit an
austere modernist coffee shop for a Brooklyn department store, and in 1932 to
specify various laminates on furniture throughout the public rooms of Radio City
Music Hall.[42]

The only problem with phenolic laminate was its color—or rather its lack of
color. Only surface layers of black, brown, or other dark colors survived the dark
industrial resin that comprised the impregnating matrix of every single sheet. Although black was popular with trendy Manhattanites, for ordinary consumers it was
passing into oblivion along with the Model T Ford. Even *Plastics & Molded Products*,
whose editorial policy of 1928 echoed Henry Ford's stubborn technical orientation,
turned its attention in that year to the fad for boosting sales with bright colors.[43] The
Formica Insulation Company kept chemist John D. Cochrane, Jr., busy into the late
1930s seeking solutions to the problem. Eventually he developed a decorative Formica whose top sheet of paper was impregnated with water-white urea formaldehyde before being laminated under pressure onto sublayers impregnated with the
usual phenolic resin. The result was a hard, durable, fire-resistant, easily cleaned
surfacing material available in dozens of colors and finishes, light or dark, glossy or
matte, pearlized or marbled, wood-grained or linen-textured. The high cost of processing urea formaldehyde Formica prevented the material's true democratization,
however. Final curing required such precise temperature control that thermo-
couples had to be embedded in each sheet. While continuing to search for a cheaper
resin, Cochrane perfected a cigarette-proof laminate with a layer of foil embedded

4–5. Soda fountain with counter and trim of Formica laminate, 1930s

under the top layer of paper to disperse heat and prevent charring. Finally, in 1938, he successfully replaced urea formaldehyde with a chemically similar resin, melamine, developed in Great Britain as a material for unbreakable dishes. Within two years Formica laminate was inexpensive enough for surfacing mass-produced dinette tables. In the meantime, however, decorative laminate had emerged as the material of choice for surfacing Woolworth counters, diners, cocktail lounges, Checker cabs, and railway coach interiors (Figure 4–5). Wherever one looked among public spaces of the late 1930s, one found the unearthly flawless surfaces of smooth Formica, deep mirrors of jet-black or jade-green or synthetic hues reflecting a futuristic "polished orderly essential" that remained a dream to the real world of the Depression decade.[44]

The achievements of designers like Frankl and Deskey, in using laminates in custom furniture and interiors, alerted executives like Allan Brown of Bakelite that design could be a factor in promoting plastics among manufacturers who were reluctant to abandon tradition. Sometime toward the end of 1932, as Americans anticipated the inauguration of a new president with a new deal, Brown hit on his own recovery plan. Not only did it boost Bakelite's reputation as a machine-age material; it also gave tremendous assistance to the fledgling profession of industrial design and awakened designers to plastic's potential as a material of manufacture. In January 1933 Brown convened the first of several seminars to acquaint designers with Bakelite's technical and aesthetic advantages. Meeting over luncheon in a private room at a Manhattan hotel, Brown and his staff made an attractive proposal to a dozen of the most prominent designers. Deskey was there. So was Geddes, whose vision of a teardrop-shaped utopia was gaining attention in corporate board rooms and Sunday supplements. Also attending were Dreyfuss, Guild, Sinel, Lucian Bernhard, Helen Dryden, Gustav Jensen, Joseph Sakier, George Switzer, John Vassos, and Simon de Vaulchier. Others, among them Loewy, Teague, and Alfonso Iannelli, soon became involved. The proposal was simple. If the designers would use Bakelite in actual products created for their corporate clients, then the company would feature the designers and their products in an advertising campaign intended to promote industrial design as a strategy for economic recovery and Bakelite as a mainstay of industrial design. As an added incentive, the company's engineers would advise designers about molding techniques and technical limitations of Bakelite to ensure that the resulting products proved functional. Brown later recalled an expansive goal of "starting the business cycle moving in the opposite direction" and concluded that the campaign "helped rejuvenate interest in new and better products in general." As carried out, the plan publicized industrial design at its formative moment and established Bakelite—and plastic in general—as a material that expressed contemporary values while reducing manufacturing costs.[45]

At least ten designers signed up to appear in advertisements that ran throughout 1933 and 1934. A distinctive format gave the series a strong identity. Each ad focused on a single product and gave its manufacturer free publicity. Each contained a small photograph and capsule biography touting the designer as celebrity, and each quoted the designer on the virtues of sleek modern design—the message being that a product which looks better, sells better. Many of the sketches mentioned other companies a designer had worked for, thereby serving as a source of references for an executive thinking of hiring a particular designer. Those featured ranged from the raffish Iannelli, a goateed artist in shirtsleeves with a cigarette dangling from his lips, to the staid, bankerlike Teague, hat firmly on head, his appearance calculated to reassure the most conservative executive worried about trusting his future to a bohemian artist. An extensive range of products suggested that designing with Bakelite would prove a smart move in any industry. Personal accessories like barometers and telephone indexes, appliances like irons and washing machines, business equipment like mimeographs and soda-fountain dispensers—all were transformed by new materials and new styles. Through it all ran the message, as phrased in one ad, that "distinction in design has a very potent sales appeal." Or, as an ad for a sculptural electric iron phrased it, the product "sheds its Cinderella garb . . . takes a beauty treatment . . . and steps out to win housewives and sales."[46]

Appearing in the trade journals *Sales Management* and *Plastic Products*, neither of which boasted a particularly artistic layout, the well-styled advertisements reinforced their own point by contrast. Taken as a whole, the series comprised a primer on product design with an emphasis on plastic. Oddly enough, the texts of the advertisements said little about the actual style that plastic encouraged. Only the photographs suggested how plastic was visually transforming the things of the twentieth century. An alert observer would have discovered a significant trend. Plastic invited the shaping of artifacts as self-contained, irreducible wholes rather than as assemblages of parts. Objects as varied as a telephone index and a meat grinder appeared smooth, rounded, almost sculptural, and made it easy to forget they enclosed mechanical parts. Only one advertisement addressed the issue—and only

briefly. It featured a mimeograph machine redesigned by Loewy and included a small picture of the original for comparison (Figure 4–6). As exaggerated in description, the old machine was "a mysterious appearing assembly of wheels, gears and cylinders"; the redesigned model, on the other hand, was "a compact machine with operating mechanism concealed within attractive housings."[47] Such a degree of enclosure, reinforced by plastic's real and visual seamlessness, marked a break with much of civilization's prior material and mechanical evolution. Until the advent of plastic, most objects, even complex ones, appeared as rationally intelligible assemblages of various parts and materials; they did not seem smoothly, inviolably whole. But plastic contributed to a syndrome of ignorance about technological processes by enclosing them in irreducible molded forms whose deceptive simplicity found its clearest expression in the streamlining of the 1930s. It became easy to look no further than a flawless surface, whether geometrically precise or sensuously flowing, and to assume that beneath it lay an ideal state of perfection. Ironically, artifacts shaped by plastic during a self-conscious machine age indicated a willingness to ignore mechanical complexities, to abdicate responsibility for understanding or directing them, to assume that beneath those pristine surfaces everything was well under control. Style, it turned out, could be quite substantial.

The campaign to link Bakelite's fortunes to industrial design succeeded. Free publicity and direct experience with new materials inclined designers to use plastic in

the future. Manufacturers who employed designers proved receptive to their suggestions that plastic be adopted because the advertisements had shown what it could
do. At times the strategy backfired, as when manufacturers of two of the campaign's
star products abandoned Bakelite and began molding the same products from Durez
phenolic. Unlike most American industries during the Depression, however, the
plastic industry accommodated competition by continuing to expand. Whether
used to reduce costs or to appeal to a public seeking intimations of future progress,
the combination of plastic and design proved irresistible. The Bakelite Corporation
continued to promote plastic as a consumer material by publishing a portfolio on
design and designers, possibly a compilation of the advertisements. And the *Bakelite
Review* began printing names and addresses of designers who had demonstrated
adequate knowledge of the material's technical and aesthetic potential; in 1939 the
company consolidated these names in a list distributed on request to manufacturers.
Annual sales of Bakelite molding compounds in the meantime rose from 8.3 million
pounds in 1931 to 30.4 million in 1937, and that division of the company emerged
from red to black in 1935. If the design campaign was not entirely responsible, it
certainly contributed.[48]

Other plastic companies followed Bakelite's lead in embracing modern styling.
Some hired industrial designers to develop new applications or to work with specific
clients. Van Doren assisted Toledo Synthetic Products in promoting urea formaldehyde, and Rohde consulted with Rohm and Haas about introduction of Plexiglas.
Even custom molders sometimes employed designers. Kurz-Kasch maintained a
long association with the design partnership of Carl Sundberg and Montgomery
Ferar, who provided prototypes for convincing reluctant customers to switch to
plastic. James Barnes and Jean Reinecke enjoyed a similar relationship with Chicago
Molded Products beginning in 1935. Reinecke later recalled that they knew a bit
more than their customers about what was still an esoteric subject. "The fact that we
could glibly say 'methyl methacrylate,' as well as 'Bakelite,' was very authoritative,"
he recalled, leading not just to success with Chicago Molded Products but also to
commissions for "various 'how to' articles" for trade journals that in turn generated
further design commissions in plastic.[49] Nothing better signified the plastic industry's shift from technical to aesthetic matters than the appearance late in 1934 of the
transformed *Modern Plastics*, providing designers as much space as plastics executives or engineers. By then the new editorial policy hardly seemed innovative; instead it reflected an attempt to catch up with Bakelite's lead and reap the benefits for
the entire industry.

At the end of the decade Allan Brown decided to stimulate demand among the
general public with a Bakelite Travelcade, a touring exhibition devoted to "Modern
Plastics for Modern Living." The Travelcade was installed at company expense at the
Franklin Institute in Philadelphia, the New York Museum of Science and Industry,
the Buhl Planetarium in Pittsburgh, and the Chicago Museum of Science and Industry, in addition to serving as the company's exhibit at the New York World's Fair.

113

Visitors walked past a series of thematic panels framed by streamlined pylons, telling the Bakelite story from a diorama of the inventor's laboratory onward. If that seemed too static, they could gaze at a hydraulic molding press turning spoonfuls of powder into souvenir spoons at a steady clip. Equally dynamic was *The Fourth Kingdom*, a film about Bakelite's processing and applications. Narrated by famous journalist and broadcaster Lowell Thomas, it explored "how chemical research has taken the three kingdoms, Vegetable, Mineral and Animal, and created a fourth kingdom whose boundaries are unlimited."[50] Despite this geographic metaphor, soon to be echoed in *Fortune*'s map of Synthetica, the fourth kingdom marked a victory of synthesis over extraction, of the artificial over the natural. The Travelcade served as a vehicle of self-congratulation for the Bakelite Corporation and for manufacturers who had already proven Bakelite a material of more than a thousand uses. Visitors to the Travelcade needed no persuasion; they had already arrived at the fourth kingdom and had one foot in the gate. They were already acquiring plastic objects whose streamlined forms suggested the outlines of a future soon to be envisioned by a science fiction magazine as "a city of seamless, cast and rolled plastic materials of brilliant and beautiful colors . . . a city of curves and streamlines, of sweep and rounded beauty."[51]

Such visionary rhetoric obscured a tension between the natural and the artificial that remained central to the process of determining forms, textures, and colors of plastic objects in the interwar years. While a publicist might declare that a certain dresser set made from Amerith celluloid "makes one lose faith in the natural, so handsome is the synthetic plastic material," a photograph revealed a surface imitating the rich colorful patterns of onyx. As late as 1933 an advertisement in *Plastic Products* declared that "Discriminating Manufacturers Insist on Real Marblette," a cast phenolic illustrated by pictures of three apparently marble ash trays.[52] An executive of General Plastics began speaking out against such imitation midway through 1932, even before Bakelite launched its design campaign. Franklin E. Brill was more aggressive than anyone else in promoting plastic as plastic rather than as a simulation of something else. New materials and their uses, he argued, ought to be "frankly expressive of the machine-age which made them possible." Until recently, he and his colleagues had been "mis-using the gifts of our new synthetic age." Without considering plastic's unique possibilities, for example, makers of small electric clocks had released a "flood of imitation walnut Gothic Cathedralettes" so cheap in price and appearance that the market faltered. Seeing bargain counters "loaded with walnut and mahogany moldings" that could hardly be given away had convinced him of the "sheer folly" of "imitating older materials." In his opinion "the consumer subconsciously resent[ed] the manufacturer's low estimation of her intelligence" as revealed in imitations of other materials and simulations of hand craftsmanship. Designers of plastic moldings would succeed only by treating plastic as an artificial material, by "using simple machine-cut forms to get that verve and dash which is so expressive of contemporary life." Above all, they should abandon ornament—"time-worn motifs" and "fussy, odd shapes"—and instead develop plastic's

4–7. Radio cabinet molded of Plaskon urea formaldehyde, designed by Harold Van Doren and John Gordon Rideout for the Air-King Products Company, 1933

uniqueness by thinking sculpturally in terms of "contours." It was necessary "to give our materials an identity" to "make them desirable to the consumer and therefore to industry-at-large."[53]

Consumer demand was not the only force moving design in plastic beyond imitation to machine-age forms. The high cost of machining molds also put pressure on manufacturers and designers to adopt simpler designs. Complex "scrollwork" and "gothic" style yielded first to vertically oriented "skyscraper" compositions like Telechron's Electrolarm of 1930 and Van Doren's Air-King radio of 1933 (Figure 4–7). Nearly as complex visually as gothic moldings, though defiantly modern, even modernistic, they were cheaper to manufacture because zigzag setbacks "could be machined in the mold with the minimum of hand labor," as Van Doren pointed out. By the last half of the decade, economic pressure to further simplify mold-making coalesced with larger cultural concerns as manufacturers abandoned angular Art Deco for streamlining's sweeping curves (Plate 9). A journalist actually claimed in *The Atlantic Monthly* that streamlining as a design motif owed its existence to the example of plastic cases and housings. It seems more likely that plastics and streamlining came together by happy coincidence.[54]

"Oddly enough," Brill observed in 1935, "the streamline shape comes pretty close to the ideal contour which designers have always claimed that plastics should take." For the next ten years, nearly every authority on plastic product design praised streamlining's virtues (Figure 4–8). A rounded or streamlined mold could be cut and polished by machine, but the angles of a mold with sharp edges and corners

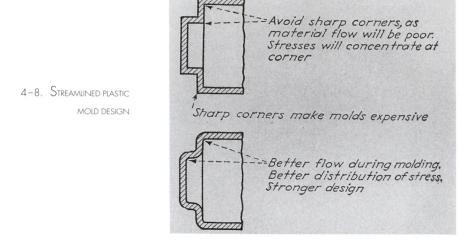

4–8. Streamlined plastic

mold design

Avoid sharp corners, as
material flow will be poor.
Stresses will concentrate at
corner

Sharp corners make molds expensive

Better flow during molding,
Better distribution of stress,
Stronger design

required expensive hand finishing. A rounded mold also permitted smooth, even flow of molten plastic over every surface and through every volume. Copying the wind tunnel tests of aerodynamic engineers, plastics engineers used special dyes to trace the flow of material through variously shaped molds. Such experiments led to a conclusion that "streamlined flow should be designed into both the inside and outside" of a part to avoid formation of gas pockets, uneven flow of material, and other flaws that could weaken a molding. Rounded corners used less molding compound than a right-angled part of similar dimensions, thus offering substantial savings over a long production run. And after a plastic housing for a radio or an electric shaver was out of the mold, assembled, shipped to stores, and in the hands of consumers, rounded edges and corners afforded greater protection from breakage than thinner, sharper angles and corners would have done. In the final analysis, rounded contours brought out the beauty of glossy plastic, which looked best when reflecting "at least one highlight from any angle." Since plastic lent itself to flowing shapes, everyone agreed it should be used sculpturally rather than in misguided imitation of other materials with wholly different characteristics. Streamlining as a design mode in plastic thus satisfied economic and technical constraints while taking full advantage of the material's unique qualities.[55]

Plastic's expansion into consumer goods occurred just as designers and architects were developing streamlining as a machine-age design mode expressive of Depression aspirations. As Muller-Munk argued, with the emergence of plastic's "large flowing curves," "the purely facade type of design" was "giving way to a really plastic conception of the machine as an object to be seen from all angles at last."[56] Flowing, sculptural, and evocative of speed, streamlining reflected desire for frictionless flight into a utopian future whose rounded vehicles, machines, and architecture would provide a visually simple, protective environment—closed off from the Depression's economic and social dislocations and marked by a static perfection. As a form of cultural expression generated by aerodynamic research and by the prominence of the automobile, streamlining did not depend on plastic molding techniques for its development. On the other hand, a molded plastic radio physically embodied a revolution in communication similar in scope to that effected by the automobile in transportation. While ordinary people might look longingly at Airflow Chryslers and Lincoln Zephyrs, they could actually afford a Sears Silvertone, a Pilot All-Wave, or a two-tone Fada. The plastic radio was the most democratic exemplar of stream-lining, a machine-age icon for the living room of everyman and everywoman. And molded phenolic, the decade's most typical plastic material, evoked utopian per-manence with its hard smoothness, its flawless surface, its quality of being "anything *but* plastic, once it has taken shape."[57] However accidental the convergence of a particular material with a larger cultural expression, the relationship became as sym-biotic as that of the plastic industry with industrial design.

Even so, to associate plastic with a particular design style possibly obscures the material's cultural meaning during the interwar years. James Cain and his readers in 1941 may indeed have understood "the bakelite style" as a unitary concept encom-passing Gothic Cathedralette as well as streamlined Silvertone or Fada. To group together a random collection of plastic moldings of all shapes, sizes, styles, and colors from those years is to see how much they all had in common and how much they differed from any previous artifacts (Plate 10). In terms of formal qualities, not much united a flat rectangular cigarette box with sharp-cut Mayan lines and a squat soap case with rounded edges flowing into a distended lid. Nor did a gothi-cized Lincoln electric clock with stylized vaulting have much in common with a low horizontal Telechron with digital counters and a seven-ribbed band sweeping up one side and down the other. Nor did a boxy Kodak Baby Brownie camera with molded-in speed lines share the aesthetic of a sculpturally contoured Bell Telephone hand-piece. But formal eclecticism did not disguise the fact that certain common characteristics marked them as distinct from other objects of everyday life, as new things made from a new material. The architectonic quality of some, the cigarette box or the gothic clock, referred visually, playfully, to earlier objects that had been constructed or assembled. However, each of these artifacts announced its integrity, physically and visually, as a single housing or shell, molded rather than put together, created instantaneously in a manner outside the experience of carpenter, welder, or machine-tool operator. From that fact came the appeal of alchemical imagery. Out of chemicals and powders, raw materials beyond the ken of ordinary folks, each of these objects had emerged with a single stroke of a hydraulic press. No clear series of human actions, indeed no human action at all, lay between conception and final form. Never mind that molds had to be designed, cut, and polished, or that most moldings had to be finished and assembled—flash ground off, surfaces buffed, works installed, and so on. Their appearance was of seamless unitary creation. If they aped earlier forms or modes of construction, they did so only to enable users to integrate them into the known world.

Much the same was true of colors and surfaces. Logically one expects a gulf be-tween objects whose materials ape the traditional and those that frankly express artificiality. No such visual gulf existed. Few imitations molded during the 1930s in fact resembled what they purported to copy. Some imitations of marble were credible—if viewed from a distance. But most imitative moldings were meant to simulate wood, and their solid browns and dark mottles and swirls proved wholly inadequate to the task. The smooth dark brown surface of a gothic clock functioned as a stylized reminder of what it once would have been. Stylization domesticated an object by enabling it visually to blend in with traditional surroundings but did not disguise its radically unnatural surfaces. Brown, black, swirled, mottled—it made little difference. All evoked the darkness of earth, of coal, of oil, of the substances from which they came. Too smooth and uniform to be products of natural processes, they suggested an unprecedented act of instantaneous transformation. All those dark brown and black phenolics had more in common with the unearthly bright yellows and reds of cast phenolic, or the rich synthetic cream of urea formaldehyde, or the occasional deep green or blue of molded phenolic, than with any natural substances. Viewed against an irregular surface of grained oak, worn and polished with age, a collection of typical moldings stood out as so starkly unnatural that they suggested a quantum shift in materials. Viewed on a surface of polished black Formica, by contrast, they seemed to emerge smoothly, without interruption, almost naturally, as artificial flora of the "fourth kingdom" of chemical synthesis. Designers who used plastic to create forms unknown in any other material not only gave physical embodiment to their era's utopian aspirations. They also gave plastic its identity. If there was in truth no uniform "bakelite style," the stuff certainly projected a distinctive aura.

PLASKON: A CASE STUDY

When Baekeland sold out to Union Carbide in 1939, not much remained of the scene that had confronted the struggling General Bakelite Company in 1918 as canceled military contracts seemed to threaten bankruptcy. Those two decades had brought continuous expansion. Demand for plastic increased with society's reliance on electricity, on the automobile, on radio, all of which exploited phenolic moldings for insulation. Consumer goods manufacturers turned to new plastics like urea formaldehyde and cellulose acetate for bright colors and ease of manufacture. The new technique of injection molding, still largely experimental, promised accelerated rates of production of ever cheaper goods. As all this happened, competitive material suppliers and their molding customers formed a new industry. The trade journal *Modern Plastics* and the Society of the Plastics Industry defined it and channeled its energy. Product designers chose plastic for an aesthetic expressing the faith in progress of a nation exhausted by economic chaos. By 1939, the outlines of the continent Baekeland discovered were largely filled in. Visitors to the New York World's Fair, dedicated to "building the world of tomorrow," found it was built of plastic—murals of cellulose acetate, statues and signs of Lucite or Plexiglas acrylic, washable Formica walls, displays, and dioramas. At least temporarily, a visionary Plasti-City had emerged from Synthetica's chemical plants and molding presses.[58]

The industry's transformation over two decades owed success to a convergence of new materials, new manufacturing processes, and new consumer-oriented marketing techniques. Baekeland and his managers, most notably Allan Brown, had worked hard to educate manufacturers, and then consumers, to perceive plastic's worth. Once the message got through, however, some manufacturers demanded more than plastic could deliver. Sometimes requirements of a particular application stimulated development of a new material. What might be called "designer materials" based on polymers built to order at least in theory began to replace exploitation of fortuitous discoveries. The history of Plaskon, a urea formaldehyde molding compound announced in 1931, demonstrates this responsiveness of plastic to outside cultural demands and also summarizes the complex interaction of scientific research, technological innovation, and marketing strategy that typified the industry during its years of emergence.[59]

Plaskon owed its development to specific needs of the Toledo Scale Company, which had been making commercial weighing scales in Toledo, Ohio, since 1901. After twenty-five years under the leadership of founder Henry Theobald, Jr., the company gained a new president in 1926. Hubert D. Bennett, a 1917 graduate of Williams College, had worked in advertising and as director of sales for Studebaker in Brooklyn before arriving in Toledo. Among the problems confronting Bennett was a cast-iron grocery counter scale too heavy for all but the burliest salesmen to carry (Figure 4–9). Grocers complained that its weight kept them from rearranging their stores—a prime strategy of machine-age retailing. Before 1923 the scale had weighed in at seventy pounds and was a popular seller. In that year the company switched from lacquer to porcelain enamel to obtain a more durable finish. Because porcelain had to be fused to iron at 750°C, the iron parts had to be thickened and ribbed to prevent their warping under the heat, and the scale's weight had jumped to 163 pounds. Bennett, willing to take risks to rejuvenate the company, attacked

4–9. Harold Van Doren (left), Hubert D. Bennett, with the Toledo Scale Company's heavy cast-iron grocery scale

this problem of marketing and consumer demand by simultaneously seeking a new metal and a new coating. As an associate observed at the time, he was "interested in perfecting entirely new ideas, so he can have them patented."[60]

Bennett's first move came in December 1928, when he engaged designer Geddes to work on several scales, including the cylinder scale for grocers. The two had met over a tennis net in Toledo when Geddes was visiting a cousin who had known Bennett at Williams; later Mrs. Stanley Resor, wife of the president of the J. Walter Thompson advertising agency, brought them together again in hopes that Bennett would employ Geddes. They got on so well that within six weeks of signing the initial contract Bennett was asking Geddes to design a new factory in which to make the scale. Work progressed on both fronts, with sheet steel the material of choice for the scale. Geddes's draftsmen produced a series of striking charcoal renderings reminiscent of grain elevators and other machine-age forms—one of which brought considerable publicity when it appeared in *Fortune*.[61] In the meantime Bennett initiated an investigation at the Mellon Institute into the possibility of developing a lightweight durable coating that could be used on metal in place of porcelain enamel. When preliminary research indicated water-white urea formaldehyde as a potential resin for such a coating, Bennett funded an industrial fellowship to investigate it systematically, and Arthur M. Howald formally set to work on the problem in January 1929.[62] Although Bennett intended the two efforts to converge on a new scale, the situation became more complex.

Each of the lines of inquiry Bennett set in motion began to lead in its own way to the radical notion of abandoning metal entirely in favor of using plastic for the new scale's housing. As Geddes told the story some years later, he became fascinated by a round glass window through which shoppers could watch the operating mechanism of the old cast-iron scale. Why not make the whole thing transparent, he thought, and if glass was too fragile, then one of those new plastics might work. It is a sign of the mingled ignorance and wonder surrounding plastic that Geddes was surprised to learn that a large transparent Bakelite molding was out of the question.[63] His enthusiasm for plastic ironically cost him his job. Bennett found the idea of a plastic scale so fascinating that he abandoned Geddes's sheet-metal plans to focus full attention on new materials even if it meant a delay in replacing the old cast-iron scale. When he discovered Bakelite was the only plastic strong enough for his purposes, and that it could be had only in black or dark brown, he had ten scales in black enamel installed in groceries around the country to test shoppers' reactions to weighing food on black. Not only did they prefer the "sanitary appearance of the white machines," the study revealed; many of them "refused to buy food weighed over the black scales."[64] This rejection might have sent him back to Geddes's sheet-metal designs. But then Howald at the Mellon Institute suggested using urea formaldehyde not as a coating for metal but as a plastic molding compound indistinguishable from Bakelite except in its crucial capacity for accepting any desired color—including white.

After solving problems of water absorption and fading in sunlight, Howald's work

yielded Plaskon, a cellulose-filled urea formaldehyde molding compound produced
in hundred-pound lots in a pilot plant at Mellon in 1930. Recognizing potential uses
beyond scale housings, Bennett formed the Toledo Synthetic Products Company
under the presidency of James L. Rodgers, Jr., another 1917 Williams graduate.
Commercial production began in April 1931 at a plant in Toledo under Howald's
supervision. From the start they promoted Plaskon as a material especially suited
for consumer goods. Rodgers observed that Plaskon, removing any need "for cheap
imitation of woods and marbles," would initiate a new type of design in plastic
"relying rather on the intrinsic beauty of the material itself." An arty promotional
booklet of 1934, illustrated with luminous tinted photographs of Plaskon objects
printed on black, was called simply *Plaskon Molded Color*.[65] By then Bennett had
hired a new design consultant, Harold Van Doren, both to promote Plaskon with
visually effective prototypes and to design the housing for a new grocery counter
scale of molded Plaskon. Another graduate of Bennett's class at Williams, Van Doren
had studied at the Art Students League and in Paris, and had worked as an editor
before taking a position in 1927 as assistant director of the Minneapolis Institute of
Art, where he remained until 1930 when he set up shop as a designer at Toledo.[66]

Early Plaskon moldings designed by Van Doren and his partner John Gordon
Rideout introduced the material to manufacturers. Among them were lids for cosmetic jars, closures for tubes, small clock cases, a salt shaker resembling a Buck
Rogers observatory, and a small biscuit cutter given away by the millions in boxes
of Bisquick (Plate 5). Designing a one-piece molded case for the Air-King radio of
1933 gave Van Doren his first experience with a large housing (Figure 4–7). A foot
high, nine inches wide, and nearly eight inches deep, the molding taxed manufacturing processes but ultimately proved technically feasible and visually attractive to
consumers. Popular response to this modernistic radio perhaps suggested to Van
Doren and Bennett that a Plaskon scale should "look as different from the scale made
of cast iron as the modern automobile does from the buggy it replaced."[67] The old
cast-iron scale was an ungainly, top-heavy profusion of visually complex parts. By
contrast Van Doren's new design was simplicity itself (Figure 4–10). According to

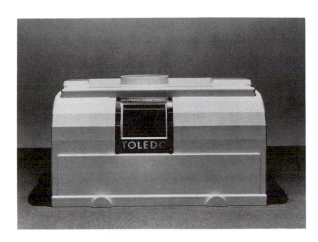

4–10. ONE-PIECE HOUSING FOR
TOLEDO SCALE COMPANY'S SENTINEL
GROCERY SCALE, DESIGNED BY HAROLD
VAN DOREN AND MOLDED BY GENERAL
ELECTRIC COMPANY FROM PLASKON
UREA FORMALDEHYDE, 1935

121

his presentation rendering, a low rectangular case enclosed both the works and the revolving cylinder from which weights were read; the weighing platform discreetly hugged the top. Receding bevels rounded the upper half of the molding, provided visual interest, made it easier to read the cylinder from a high angle—and strengthened the molding. About as wide as the old scale at eighteen inches, Van Doren's new design was less than half as deep and less than half as high at about a foot. Its smooth horizontal enclosure of the mechanism within a single volume embodied the public's passion for streamlining. But there was a big problem. Only after the design was completed, and the mechanism was substantially reengineered as well, did Bennett discover that no one could mold such a large piece—that in fact, if successful, the scale housing would be "the largest plastic molding ever made in the United States."[68]

By then, having funded development of a new plastic to reach the goal of a lighter, less bulky scale, Bennett refused to give up. He entrusted manufacture of the scale housing to General Electric's custom molding department. The job required constructing the largest hydraulic molding press then in existence, a task carried out by the French Oil Mill Machinery Company of Piqua, Ohio. When installed in GE's plant at Fort Wayne, Indiana, the new press stood twenty-two feet high, weighed eighty-nine thousand pounds, and exerted fifteen tons of pressure. The mold was cut from a seven-ton block of steel. These oft-repeated statistics suggested the awe with which mechanical engineers regarded the accomplishment; a drawing of the press used in technical journals depicted it from a point near ground level as a behemoth towering over a gnomish figure in laboratory garb.[69] Despite this image of mastery, the molding process proved difficult. Some pieces warped when removed from the mold. Many revealed cosmetically unacceptable flow lines. Others lost their polished surfaces when pulled away from the mold. Dust, not a factor in smaller moldings, contaminated the smooth, white surfaces. The press room had to be air conditioned and elevated pressure maintained to keep air flowing out rather than in. The very size of the equipment brought unexpected delays when it had to be repaired. Despite all these problems, the momentum established by Bennett prevailed. In August 1935 Toledo Scale began taking orders for its Duplex model, later known as the Sentinel.

The new scale more than met Bennett's initial goal. With a total weight of fifty-five pounds (the case itself weighing about eight pounds), the Sentinel could be lifted easily by salesmen and grocers. Its smooth lines radiated efficiency. Sales increased by 300 percent in six months as grocers discarded cast-iron reminders of the general store era. Success helped offset development costs of some $500,000 and the use of more expensive materials. While cast iron had cost only four cents a pound, Plaskon cost thirty-five cents, and the aluminum used for most working parts cost forty cents; Toledo Scale thus paid three times as much for the materials in a Sentinel. Offsetting that increase were considerable reductions in assembly and shipping costs. Less than a third as heavy as the old scale, the Sentinel incurred lower freight charges and was packed in a cardboard carton rather than a wooden crate. Beyond such corporate concerns, the success of Bennett's experiment convinced other manufacturers to abandon traditional materials for molded plastic. Soon after the Sentinel's introduction, an engineer at General Electric predicted that "business machines, radios, electric apparatus, vending machines, recording instruments," and food-handling equipment would soon sport plastic housings. Toledo Scale's experience made it possible to take plastic seriously as a material.[70]

Bennett and his associates recognized that more than size distinguished their accomplishment from earlier uses of plastic. Faced with a specific need for a new material light in weight and color, they had followed a complex trail of development without regard for expense or traditional wisdom. In the process they had created not only a new material, Plaskon, but also a method for molding it and a visual style appropriate for its presentation to consumers. Applied chemical research, production engineering, and the marketing strategy of industrial design came together to offer a solution that none alone could have provided. No longer a novelty, plastic was emerging during the 1930s as a staple essential to other industries. No one was ready to take it for granted, however. The many journal articles devoted to Plaskon and the Sentinel scale testified to a continuing, even intensified regard for plastic as something of a miracle substance. The quick shifts from cast iron to sheet metal to plastic offered a paradigm for other industries and products. But only gradually did the full extent of plastic's subversive effect on the material landscape—its potential for malleability—become apparent. For the time being Depression-weary Americans wanted stability—a utopian desire reflected in the streamlined forms of the New York World's Fair of 1939. Only superficially dynamic, its rounded buildings actually promised an ideal state of static closure. The styling of molded radios, clocks, telephones, and grocery scales echoed this vision of stability. Their durable material substance also promised a permanence uncorrupted by rot or corrosion. But the process by which plastic was proliferating in the material environment promised anything but stability.

As plastic became a matter of everyday experience rather than abstract conjecture, expressions of plastic utopianism tended toward extravagance. Slosson and other promoters in the early 1920s had remained satisfied with projecting a democracy of material abundance to be gained by exploiting nature's waste products. At their most optimistic they had conceived of artificial substances as improvements on nature's imperfect efforts. By the end of the 1930s, however, there were "no limits" to the potential for "creating new things that Nature forgot," or so *National Geographic* asserted. The average citizen was already experiencing a "rising flood of plastics" of "unlimited uses," as an article in the mass-circulation *American Magazine* reported. Before long, according to a breathless survey of synthetics published by the more refined *Arts & Decoration*, everybody would "hear the hum of airplanes on plastic wings, live in plastic houses, and ride in Bodies by Beetle." Not to be outdone, *Popular Mechanics* announced that "the American of tomorrow . . . clothed in plastics from head to foot . . . will live in a plastics house, drive a plastics auto and fly in a plastics airplane."[71]

Most of these treatments of plastic utopianism ignored stability by suggesting an utter lack of limits. The science director of the New York World's Fair claimed that "possible varieties [of plastic] are almost endless." Already plastic could be had in "*any* color and texture . . . cheaply molded into *any* form." Not everyone found the prospect pleasant to contemplate. As early as 1936, the author of *Fortune*'s first survey of plastic recoiled from Catalin cast phenolic because "nothing else can match it in potential gaudiness." The reader was asked to "shudder for the future, because . . . there is no law to restrain designers from doing again in all the colors of Joseph's coat what they [formerly] did in simple chromium." Four years later, when *Fortune* had shifted its tone to enthusiastic celebration of the plastic landscape's surrealistic derangements, the magazine observed that synthetic materials "transmit[ted] a fever contagion to the American imagination" by the way they could be "pressed and squeezed and rolled and sawed and drawn and cast and carved into the very image of a new world." Under the effervescent influence of new materials, as the fair's science director conceived it, even architecture would yield to "mobility and quick change." Any notion of building for "stability and permanence" was "a relic of the social outlook of previous centuries."[72]

This new celebration of endless transformation reached an extreme in the effusions of Waldemar Kaempffert, a science editor at the *New York Times* who envisioned the typical house of a hundred years in the future as containing nothing "not molded, pressed, or fashioned out of an appropriate synthetic." Housecleaning would be "reduced to a matter of washing everything that can be reached with a hose." Dirty plastic dishes would be dissolved in hot water and washed down the drain. Dirty clothes of synthetic fabrics would be thrown away. "In the synthetic world of tomorrow," he explained, "it costs more to wash and iron" than "to buy new ones." Such were the advantages of an age when science would "assume complete control over matter, and free men from their ancient dependence on animals and plants and the crust of the earth."[73] Overlooking the irony that Kaempffert was addressing the liberation of women, not men, one can find in his absurd article an accurate burlesque of the feeling of impermanence that plastic was soon to evoke— not in a century but in a decade or two. The image of melting or consuming one's possessions in a hot desire to possess and consume ever more goods suggests a basic transformation occurring in public attitudes toward plastic as it assumed a significant place in a shifting, ever more artificial landscape. Formerly conceived as a means of creating permanent abundance from waste materials, plastic became a means of creating an abundance of ephemeral waste. A thermoset world melted into thermoplasticity.

NYLON: DOMESTICATING

A NEW SYNTHETIC

*P*lastic did not truly emerge as a major presence in American material life until after the Second World War. Old-timers who had experienced the industry's earlier growth regarded the war years as a great divide separating marginal status from genuine acceptance. When the armed services faced shortages of aluminum, brass, or rubber, they called on the plastic industry for substitutes often superior to the originals. Military needs stimulated new materials and applications, as well as construction of new processing machines and factories to house them. Soldiers, sailors, and fliers returned to civilian life with an appreciation for synthetics that had served well when lives depended on it. On the other hand, people who remained behind often retained memories of makeshift goods, of inferior ersatz endured while quality plastics went off to war along with steel, brass, and tin. An undercurrent of distrust continued into the late 1940s and early 1950s. Although the industry had to confront this negative image, wartime experience pointed toward phenomenal success. "It was a whole new ball game after the end of the war," recalled Gordon Kline of the National Bureau of Standards. With federally subsidized plants converting from military to civilian production, the industry was "set to take off on a tremendous surge."[1]

Statistics revealed the extent of this expansion. In 1939, before the war, American companies produced 213 million pounds of synthetic resins. Two years later, during the first year of mobilization, output doubled to 428 million. By 1945 annual production nearly doubled again to 818 million pounds, and it reached 2.4 billion pounds in 1951. During the same period, the annual value of molded products rose by a factor of six from $47 million to $303 million. However, these figures did not reveal the full extent of the industry's transformation. Journalists often observed that plastic's impressive progress was as much qualitative as quantitative. According to the *New York Times* in 1941, such defense applications as acrylic bomber enclosures demonstrated a shift "from novelties to more important uses." No longer "amusing conversation pieces," plastics were "full-bodied headline news." Confirming that statement, a headline in *Life* declared that "War Makes Gimcrack Industry into Sober Producer of Prime Materials." According to *Harper's* in 1942, plastics had moved from lighthearted aesthetic indulgence into tough functional seriousness. No longer substitutes, they were "materials in their own right," preferred when they could get the job done better than anything else. As *Time* declared in identical words, plastics were "passing the stage of being chiefly substitute materials" and instead were becoming "major materials in their own right." Allowing for wartime propaganda, it was still accurate to refer to "Plastic—The 'Stand-In' That Became a Star."[2]

Wartime discussions of plastic touched on a significant change in the way chemists synthesized new materials. New plastics were tough and strong, well adapted to a bewildering variety of specifications, because chemists had learned to design at the molecular level. Rather than seeking uses for every gummy resin deposited in a test tube, they were endowing each plastic with qualities precisely defined in advance. Or so ran a frequent argument. General Electric's claim that chemistry offered "materials with qualities . . . built-in and controlled" was mild compared to other statements. According to *Harper's*, "man" [*sic*] could now "make a list of the properties he would like to find embodied in a new material" and then "custom-build that material as he never could before in all history." Even more enthusiastic, *Life* described "the variety of plastic materials [as] unlimited" because "use of heat, pressure and catalysis" enabled chemists to "juggle the atoms of these elements into an infinity of molecular patterns . . . each chemically tailored to a particular function." Not to be outdone, *Newsweek* predicted "a new industrial era" founded on materials "tailor-made to fit the finished products." Already "molecule engineers" could "draw blueprints of the kind of new molecules . . . they need for a given purpose."[3]

This theme of impending material transformation was not surprising in popular magazines engaged in morale building on the home front, but it also emerged in comments aimed at molders and fabricators—insiders who could recognize exaggeration if anyone could. When the levelheaded Williams Haynes addressed an SPI convention in May 1942, he flattered the group's collective sense of realism by admitting he was "not going to prophesy that you are coming out of this war the veritable lords of modern industry . . . masters of The Plastic Age." Such opinions would have reflected "the moonbeam popular science . . . of the Sunday papers." All the same, Haynes told members gathered at Hot Springs, Virginia, to discuss the war effort, that they stood "on the brink of a great revolution . . . in which man-made materials—not substitutes or makeshifts, but new materials, better and cheaper and more adaptable to our use, will play the historic role of bronze and gunpowder and steel."[4]

That revolution seemed already under way a few years later, in 1947, when *Collier's* published a postwar survey of the "Plastic Age" written by a journalist named Ruth Carson. With her enthusiasm tempered by a certain perplexity, she expressed both an appreciation of the everyday convenience of new materials and a distrust of their almost sinister derivation through scientific processes based on a questionable meddling with nature. Carson's ambivalence revealed itself in an uneasy, possibly unconscious balancing of positive and negative connotations. Echoing the faith of wartime promoters in the precise control to be obtained through molecular engineering, she declared that "each plastic is tailor-made by the chemist to meet special needs." On the other hand, an event as ordinary as a supposedly unbreakable melamine bowl cracking on a hard floor dispelled much of the "magic, dream-world talk." Despite such disillusion, it was true that "in hundreds of ways" new materials were making life "easier, safer and more varied." But as plastics gradually spread "all around us, many of them [remained] unseen"—a possibly discomfiting thought, offering no guarantee to ordinary individuals of personal control over them. After these hints of distrust, which a breezy style obscured, Carson's article took an unusual turn. She abandoned formulas of popular journalism to indulge an unsettling personal reaction. Despite obvious benefits, plastics remained fundamentally distinct from other materials—unnatural, even alien. Although they were derived from natural materials, "the witchery the chemist performs turns them first into something unearthly, that gives you the creeps. You feel, when you go into a chemical plant where plastics are made, that maybe man has something quite unruly by the tail." This expression of fear, even revulsion, disturbed the surface of an article otherwise similar to other celebrations of plastic in women's magazines of the late 1940s. The discordant tone and unexpected violence of Carson's remarks validated them as a sincere emotional eruption.[5]

Her reactions owed something to the debate over atomic energy, whose potential for transforming everyday life in a beneficial manner was overwhelmed by its capacity for destruction.[6] As physicists received the public attention that chemists had worked for since 1918, plastic shared in an undercurrent of distrust of science that was Hiroshima's legacy. Ten or fifteen years earlier, Depression austerity had infused the simplest of solutions with the power to convince. The streamlined curves of a bright Catalin radio appeared emblematic of a future whose technical experts would engineer an efficient society. By the time Carson expressed her doubts about the "Plastic Age," the situation was not so simple. The atom bomb had proven the fallibility of scientists. Plastic was corrupted, however tenuously, by its scientific origin. In its purest, most visible plastic-as-plastic manifestations, its reputation suffered from a cultural fallout that even a journalist found she had few words to express. Consumers in the postwar era demanded that plastic goods be domesticated, rendered more traditionally homelike, cast into forms that suggested, however inaccurately or dishonestly, continuity with the past. Domesticating plastic offered the consumer a means of taking control of new synthetic materials whose entire processing, from chemical creation through surface styling, otherwise revealed a colonizing of scientific and technical values upon the material realm of everyday life. If atomic energy, in both its creative and destructive aspects, remained an abstract concept not open to individual comprehension or active control, such was not the case with plastic. Physically present, malleable, capable of assuming any forms demanded by affluent consumers, plastic offered itself as an analogue to other processes not open to individual control. By taming plastic, by making it assume the outlines of a supposedly predictable everyday domestic life whose continuous transformation no one quite yet perceived, postwar Americans could alleviate their fears of less tangible scientific advances.

Commercial uses of plastic in consumer goods won mass approval by neutralizing the stigma of science that surrounded synthetics and by humanizing the emotional associations that defined them. This domestication reflected a symbiotic relationship between manufacturers who accommodated popular demands and consumers who appropriated new materials to their own devices. The process was most clearly exemplified in its earliest manifestation, Du Pont's development and commercial introduction of the first synthetic fiber—nylon. The most important result of a program of pure research launched by the chemical company in 1927, nylon was promoted from the moment of its public announcement in 1938 as a triumph of laboratory science. Whatever the reality, the new fiber appeared more than any other synthetic as a "tailor-made" polymer with a precise combination of desirable qualities. When Du Pont began introducing nylon in the form of women's stockings, however, the company found that its scientifically oriented promotion provoked misunderstanding, even active distrust. Consumers quickly devised their own agenda for the new synthetic. Although not exactly a plastic, at least not when used as a fiber, nylon offered an almost pure case study of the domestication process to which Americans submitted plastics after the war was won.

PURITY HALL

Nylon owed its inception to an unusual gamble proposed by Charles M. A. Stine to Du Pont's governing Executive Committee in December 1926. Arguing that routine applied research was "facing a shortage of its principal raw materials" (i.e., creative ideas), Stine requested funding to set up a program of "pure science or fundamental research work" for the purpose of "discovering new scientific facts." Such a program would attract bright young chemists to the company, develop ideas that could be sold to other companies, and earn the prestige of scientific publication. If lucky, it just might yield discoveries ripe for commercial development. When Stine submitted a detailed proposal in March 1927, one area in which he hoped to engage researchers was polymerization, of which he observed that little was "known about the actual mechanism of the change which takes place, so that the methods used are based almost solely on experience." After the Executive Committee allocated $250,000 to initiate Stine's project, he hired Wallace Hume Carothers to direct the polymer research. Carothers was a thirty-one-year-old Harvard instructor with a doctorate from the University of Illinois whose reluctance to teach overcame his reluctance to abandon the freedom of academic life. He was soon installed at Purity Hall, as skeptics referred to the new lab building constructed at Du Pont's Experimental Station just outside Wilmington. As it turned out, the association was phenomenally successful, with basic research undertaken by Carothers leading directly to nylon and to neoprene synthetic rubber.[7]

More of a theoretician than a laboratory practitioner, Carothers excelled at directing Purity Hall assistants into experiments testing his hypotheses. Before arriving at Du Pont, he had determined to explore polymerization along lines recently suggested by theoretical work of Hermann Staudinger, a German organic chemist. Although earlier researchers had succeeded in altering natural organic polymers like cellulose and rubber, and Baekeland and others had managed to synthesize formerly nonexistent organic polymers, no one knew quite how polymers were constructed. Some, like cellulose, silk, and protein, were thought to be quite large and complex. Before 1920 common opinion held that complex organic polymers followed nebulous principles distinct from the simple molecular bonds of inorganic chemistry. Colloid chemistry, whose concepts suggested heterogeneous particles agglomerated in suspension by vaguely understood processes, offered the most promising model. Or it did until 1920, when Staudinger initiated debate among German chemists by declaring that individual molecules of organic polymers formed in almost inconceivably long chains of thousands of atoms bonded according to the simple principles that governed inorganic molecules. At that time Staudinger had little more than logical assertion to support his concept of long-chained molecules. Six years later, at a key symposium, his theory of "macromolecules" received its first confirmation when Herman F. Mark, a Viennese trained in Berlin, announced the results of structural analysis of long-chained cellulose molecules carried out using the new technique of X-ray crystallography. Although Staudinger's academic opponents continued to grouse that Mark's results proved nothing, the reality principle of commerce triumphed when IG Farben hired Mark in 1927 to begin systematically exploring synthetic organic polymers along the lines of Staudinger's theory. Mark remained with IG Farben, directing profitable development of polystyrene, polyvinyl chloride, and polymethyl methacrylate, until 1932, when the impending Nazi rise to power led the company to dismiss him because of his Jewish ancestry. He returned to Vienna and eventually fled to Canada and then to the United States, serving as a consultant to Du Pont after Carothers's untimely suicide in 1937. After the war Mark became one of America's leading polymer chemists.[8]

When Carothers joined Du Pont in 1928, he knew nothing of the work at IG Farben, only then getting under way. His approach differed fundamentally from that of Mark, whose analytical training prompted him to work backward from promising resinous gunks by analyzing their components and then determining precisely how similar commercially viable polymers might be made. Carothers, on the other hand, with his insistence on pure research and an overriding interest in theory, hoped to

129

prove conclusively the validity of Staudinger's long-chained-molecule theory by constructing some examples from scratch. As he wrote soon after accepting the Du Pont position, "I have been hoping that it might be possible to tackle this problem from the synthetic side . . . to build up some very large molecules by simple and definite reactions in such a way that there could be no doubt about their structures." Although the idea seemed even to him "no doubt a little fantastic," that is exactly what he proceeded to do.[9] The plan of attack Carothers followed at Purity Hall appeared simple only in hindsight: to select a basic molecule whose structure allowed it to be enlarged at either end by means of a simple reaction and then to continue that reaction indefinitely. He put his team to work reacting acids with alcohols to yield esters, with water given off. They succeeded in building a series of somewhat long-chained polyesters but were unable to exceed molecular weights of about six thousand because some of the water entered into a reverse reaction leading to a state of equilibrium. With the help of Julian W. Hill, his most important associate, Carothers devised a molecular still in which polymerization occurred in a vacuum at 200°C, with water condensed on a cold surface and then removed. Although the process took nearly two weeks, the still enabled construction of linear condensation polymers with molecular weights of more than twelve thousand, so long that he referred to them as "superpolymers" in general and "superpolyesters" in particular. As Hill later phrased it, Carothers had "finally laid to rest the ghost . . . that polymers were mysterious aggregates of small entities rather than true molecules."[10] The chemist had done more than experimentally confirm Staudinger's hypothesis, however. He also extended polymeric theory with a series of articles in the *Journal of the American Chemical Society*.[11] If pure science was all Du Pont had in mind, Purity Hall was an unconditional success.

Even Carothers realized the company hoped for more than scientific prestige, and in fact his work quickly promised commercial applications. In May 1930, in a letter to his former professor Roger Adams, he mentioned "looking forward to your next visit" because "we will have quite a lot of things in my division to show you." These unspecified "things" included an elastic mass later commercialized as neoprene and a polyester filament that suggested the possibility of a synthetic fiber to compete with silk and rayon. Hill had taken the first practical step toward nylon when he dipped a rod into a beaker containing a long-chained polyester and discovered it could be "drawn taffy-like from its viscous melt" into a thin, brittle filament. If he pulled it after it had hardened, an amazing change occurred. The filament stretched to four times its original length, locked into that position, and became quite flexible and strong. Years later the excitement of this "cold drawing" remained as Hill described how when "suddenly any further pulling simply cuts your fingers, you have the sensation of virtually feeling the molecules lock into place as they fall into parallel array and the hydrogen bonds take hold." No doubt Adams experienced this sensation on his next visit to the Experimental Station.[12]

Despite excitement provoked by Hill's discovery of strong cold-drawn filaments, the path to a marketable synthetic fiber proved uncertain. Various polyesters obtained by the processes of Carothers and Hill exhibited too low a melting point to be viable as textile fibers that had to survive laundering and ironing. Switching from esters to amides in search of a useful cognate reaction, they distilled a variety of long-chained amide polymers (or polyamides) that also reacted positively to cold drawing. Although these were abandoned because melting points proved too high for practical spinning, a return to polyamide research later yielded nylon. In the meantime Carothers lost interest because he had completed his research agenda for linear condensation polymers and hoped to find a new research problem. Nor was there much outside interest. No matter how promising the work seemed at Purity Hall, it made little impact on the larger corporation. More than six months after Hill's discovery, a member of the Executive Committee unwittingly inquired "whether we had ever carried on any research with the idea of finding an entirely new material with which to make rayon," perhaps with "some of the new resins . . . we have been developing recently."[13]

Already, however, a practical approach was beginning to dominate Du Pont's research program. In June 1930, only three months after discovery of cold drawing, Elmer K. Bolton, an expert on applied dyestuff chemistry, took over the Chemical Department. From the start Bolton was "strongly opposed" to Purity Hall because he did not believe "fundamental research could be administered logically as part of an industrial organization."[14] After assuming responsibility for the Experimental Station, he moved to exert control over publication of Carothers's research, to direct the organic section toward applied goals, and to protect the company's interest in its researchers' work.[15] As part of this shift, in July 1931 the company filed a broad patent application on Carothers's behalf covering his work on "linear condensation polymers." It described "preparation of high molecular weight linear superpolymers" from ethylene glycol and succinic acid and claimed "the formal possibility . . . of infinite length." More to the point from a commercial perspective, it also described production of "pliable, strong, and elastic" fibers of potential use for "artificial silk, artificial hair bristles, threads, filaments, yarns, strips, films, bands, and the like," with secondary application as molding compounds, protective coatings, and paint additives. Most of these predictions proved accurate; at the time, however, they smacked of blue-sky dreaming.[16]

The patent application established Du Pont's priority and freed Carothers to publicize his work. Together he and Hill wrote a paper on "Artificial Fibres from Synthetic Linear Condensation Superpolymers" for delivery in September 1931 at a meeting of the American Chemical Society at Buffalo. The title itself, with "artificial fibres" taking pride of place over "superpolymers," offered evidence of the company's shift toward practical application. Although the two authors noted their work's theoretical origin, the paper's intellectual frame placed it in the plastic utopian tradition of Slosson. A long introduction referred to the nineteenth-century history of chemical synthesis, in particular to dyestuff pioneers who learned to synthesize exact imitations of vegetable dyes "more cheaply and in a purer state from dead matter" than they could be taken from nature. Not "content . . . with imitating nature," early organic chemists learned "to draw up an architect's sketch of the kind of a molecule that would have to be constructed to obtain a particular colour" and then "proceeded to synthesize such a molecule." Introducing their own work, Carothers and Hill explained they had mastered a similar but more difficult process for constructing "giant molecules" similar to those of silk and cellulose. Although these natural substances were too complex to reproduce exactly, the two chemists had learned to make "closely-analogous compounds . . . by rational methods." In fact, by introducing "minor variations in the structure of the product molecules," they could "obtain, within certain limits, any desired set of properties." Those people who depended on "cellulose provided by Nature" for rayon had to suffer "limitations that the inherent properties of this material impose." By contrast, the "synthetic method" made it possible "to obtain fibres having any desired properties." Carothers and Hill were claiming they could construct an artificial silk superior to the natural substance. Almost as an afterthought, however, they confessed to "no intention of immediately exploiting the new fibres commercially" because, as these self-proclaimed molecular architects admitted, they still had "to overcome many practical difficulties."[17]

All the same, the revelation attracted attention in an age that celebrated marvels conjured up by wonder-working laboratory chemists. The following morning, the *New York Times* announced that Du Pont was developing a synthetic fiber "at least as good as silk" from a mixture of antifreeze and castor oil. Although the article described the fiber as too expensive to be put on the market, a careless reader might have skimmed over that qualifier to the forceful declarations that it not only exhibited "good strength, pliability and high lustre" but also recovered its shape after being stretched—a boon to women tired of the baggy knees of rayon stockings, the only marginally fashionable alternative to silk.[18] But with nothing solid, no real goods, to accompany the announcement, it sank into that nebulous realm of half-remembered scientific promises of material abundance, each lending credence to the others, from which some Americans drew their continuing faith in the future.

Before vanishing, however, this premature report of artificial silk generated its share of rough humor, a verbal thumbing of the nose at scientists in white lab coats who, when they did not seem superhumanly dedicated or inhumanly sinister, appeared as absentminded professors abstracted from reality. In this vein the *Literary Digest* picked up an item on artificial silk from the *Detroit News*. "Man, after experimenting for years," so it ran, "has finally discovered that by an ingenious mixture of castor-oil, ethylene, glycol, carbon, hydrogen, and oxygen, he can make a silk fiber almost as good, and not more than three times as expensive as the one a Chinese worm has been manufacturing for centuries." Carothers, having proven his theoretical point, might have been content to leave it at that. Du Pont, a going business concern, could not afford to. As soon as a reinvestigation of polyamides suggested the possibility of overcoming those "practical difficulties" to which Carothers and Hill had referred, the company set up a crash program to develop a synthetic fiber with as little delay as possible "between the test tube and the counter."[19] But the marketing side of the company's program lacked the military precision of the chemical side. Du Pont had no clear idea of the need to remove ambiguous laboratory associations from the new material before offering it as a personal, even intimate addition to the lives of everyday people. Fortunately for the company, people found nylon so desirable, such an improvement over natural fibers, that they were willing to domesticate it themselves.

BETTER LIVING THROUGH CHEMISTRY

A few weeks before Carothers and Hill first publicly described the research that eventually made Du Pont a topic of conversation from coast to coast, President Lammot du Pont defended his company's lack of concern for public opinion. In August 1931, refusing an invitation to participate in the Century of Progress Exposition at Chicago, he observed that chemical companies made "raw materials for others." Unlike General Motors, whose products were "sold to the individual consumer all over the country," the company had no need to impress the public. His opposition to promoting corporate image soon cracked, however, as du Pont personally came under attack in 1934 for wartime munitions profiteering. A Senate committee spent three days grilling Lammot and his brothers about alleged overcharges for military powder and explosives from 1915 through 1918, when the company's annual profits mushroomed from $5 million to about $60 million. Committee chair Gerald P. Nye maintained that Du Pont and other "merchants of death" posed a serious threat to world peace. With the charge widely reported, anonymity was no longer a viable corporate option.[20]

Confronted by a serious crisis, the company turned to America's best-known adman, Bruce Barton, famous for a book portraying Jesus as a consummate business executive. Barton maintained that Du Pont had no choice but to accept "vocal and aggressive . . . hostility toward war" as an attribute of many "voters and customers" even if it derived from "hysteria" or "national decadence." The company was "living dangerously—like a man in an atmosphere of plague"—because people thought of Du Pont as "the powder people" even though 98 percent of its business came from "peace-time" products. Citing a survey that found 80 percent of the population well disposed toward General Motors and General Electric, while only 20 percent had anything good to say about Du Pont, Barton advocated spending $650,000 to change the company's image. It was inevitable, with "so wonderful a laboratory, working so diligently," that Du Pont would become "more and more a maker of products which go to the public under your own name." Rayon, one of the company's few consumer products, had proven difficult to sell because people doubted that the "powder people" knew anything "about style." The future would be different, however, once his agency, Batten, Barton, Durstine & Osborn, had created "a vast constituency which *knows* that du Pont knows about style, and about many other important phases of everyday living."[21]

The resulting image campaign represented chemistry's contribution to civilization in the array of superior material goods it offered modern consumers. The campaign's overall theme probably originated in reaction to a brutal two-page "merchants of death" cartoon appearing in the *Forum* in July 1934. On the left-hand page an endlessly repeating list of twenty munitions manufacturers, from Du Pont to Krupp and Mitsui, formed a patterned backdrop; on the right appeared a similar backdrop of repeating names of munitions—machine guns, poison gas, dirigibles, gunpowder, and so on. Superimposed on the left was a huge black anthropomorphic cannon barrel with arms and legs. "We are the armament makers," ran the caption. "This," with the figure gesturing toward a skull and crossbones superimposed on the right, "is our design for living."[22] Du Pont's response to this hard-hitting attack came in October 1935 with adoption of a slogan devised by BBD&O for promoting the company's own design for living: "Better Things for Better Living . . . through Chemistry." The message reached a large audience through a new network radio program, *The Cavalcade of America*, which advanced Du Pont's patriotism by dramatizing episodes from the nation's past, while its advertising celebrated present and future chemical wonders flowing from Du Pont laboratories.[23]

This new agenda shaped the Wonder World of Chemistry exhibit mounted by Du Pont at the Texas Centennial Exposition less than a year later (Figure 5–1). Although

5–1. Cotton and wood display at Du Pont's Wonder World of Chemistry, Texas Centennial Exposition, Dallas, 1936

the regional fair at Dallas paled by comparison to the Century of Progress, the newly sensitized company seized the opportunity to tell its story. Photographs in a souvenir booklet handed out to visitors revealed women in cotton-print dresses and men in shirtsleeves and straw hats escaping the Texas sun as they learned about products made by Du Pont chemists from basic raw materials, especially from cotton, wood, turpentine, molasses, and vegetable oils, all agricultural substances indicating a "partnership between farming and chemistry." Streamlined railings swept around displays of plastics and artificial rubber, antifreeze and cleaning solutions, refrigerants and insecticides, dyes and paints, all clearly keyed to the raw materials from which they originated. Munitions dwindled to the vanishing point, obliquely suggested by a case of Remington hunting rifles and a small display of blasting powder. Compressed into a few words for the souvenir booklet, the exhibit revealed "how du Pont chemists take Nature's raw materials and convert them into articles we all know and enjoy today."[24]

During the six months of the Texas Centennial more than 1.5 million people experienced the Wonder World of Chemistry. To survey reactions, the company collected written reports from the Southern Methodist University students hired as lecturers and demonstrators. According to comments reported from the floor, the array of peacetime products overcame the "merchants of death" issue for most visitors. "I get this kind of lesson from your exhibit," one man concluded; "du Pont doesn't need another war." A Catholic priest who "had got the idea du Pont was a munition manufacturer" from the newspapers declared the exhibit "worth seeing" as a corrective. Most visitors accepted the utopian gospel of wonder-working chemistry conveyed at Dallas. "Now everything is synthetic," enthused a woman who found it "wonderful how du Pont is improving on nature." The section on cotton attracted attention for its colorful plastic objects of celluloid and cellulose acetate, more fascinating than antifreeze or insecticide. "This display carries itself," reported a guide, "because the public enjoys hanging over the railing and delving into the secrets of plastics," especially when stimulated by "the bright colors of the phenolyn tubes and the toilet articles." Other responses to cellulosic plastics ranged from the star-struck utopian ("You mean you make all of these from cotton?") to the rationalist ("I'm glad to see how my dresser set is made"). Some visitors remained skeptical. "Isn't Pyralin highly explosive?" asked one woman, while another had the gall to point out the emperor's new clothes by inquiring whether Pyralin was not in fact "the same as celluloid"—hardly a twentieth-century age-of-synthesis miracle-material.[25]

When a lecturer observed that "people are keen on products that are new and out of the ordinary," he identified the reason for Du Pont's public relations success at the Texas Centennial. For some, however, the exhibit suggested the human race was moving too far too fast. These doubters came mostly from the ranks of the elderly or the less sophisticated. "You don't mean to tell me," one woman began incredulously, "[that] Du Pont claims he can produce a product superior to nature, do you?" More accepting of the claim but skeptical of its worth, an elderly Frenchman declared himself "suspicious of all synthetic things" because, as he put it, "I don't see how they could be as good as the real thing." Among this minority of doubters a paranoid tendency prevailed. A lecturer observed how "odd" it was that "the public always speaks of du Pont as 'he'—like 'Public Enemy No. 1,' if one is to judge by the intonation in some voices." As citizens of an agricultural state with a populist heritage, Texans distrusted big business and international connections. Even so, the undertone of anti–Du Pont sentiment also stemmed from emotional responses to the company's line of work. Few people could argue about making "better things for better living." But rather than integrating those "better things" into a vision of "better living," the company emphasized too much that they came into being "through chemistry," a science whose mystifying processes few people could understand. Among those who explored the Wonder World of Chemistry, many must have entertained thoughts similar to those of an old man who responded simply, "Man is getting *too* smart!"[26]

Taken together, the promotional rhetoric at Dallas and the public reaction to it exemplified the cultural context into which Du Pont introduced nylon two years later. But if mixed messages and uncertain responses defined the first meeting of nylon with its public, indicating a company that had not mastered its own newly engineered image, the perfecting of nylon as a synthetic fiber indicated a company firmly in control of a complex development process. Although Carothers was identified in the public mind as the inventor of nylon largely because journalists noticed the series of patents issued in his name, the development process involved well-coordinated efforts of dozens of people whose work continued long after his death. Du Pont executives may have been uncertain how to present their products to ordinary consumers unfamiliar with chemistry and sometimes skeptical of its values, but they had learned how to manage research and development.

FROM COAL, WATER, AND AIR

For several years the difficulty of handling the superpolymers described by Carothers and Hill in 1931 prevented work on synthetic fibers. Late in 1933 Bolton convinced Carothers to resume work on the problem. As Bolton told the Executive Committee, "an entirely new textile fiber" was "one of the most important speculative problems facing the chemist today."[27] Within a few months Carothers decided to return to the polyamides. In May 1934 an assistant succeeded in drawing a viable fiber from a four-gram batch of amide polymer he had spent two months preparing under Carothers's supervision. From that point on, development accelerated. A survey of more than eighty polyamides yielded five candidates for commercialization. On February 28, 1935, a Du Pont chemist for the first time drew a fiber from a batch of polymer 6-6, so called because its constituent molecules, hexamethylenediamine and adipic acid, each had six carbon atoms. Although Carothers preferred polymer 5-10, whose low melting point would have made spinning easy to accomplish, Bolton chose to develop 6-6 despite a troublesome melting point of 263°C because its components were more easily derived from the raw material benzene.

With Carothers increasingly distracted by psychological depression, Bolton of the Chemical Department and Ernest K. Gladding of a reorganized Rayon Department assumed responsibility for coordinating all the chemical and mechanical processes and subprocesses necessary to the manufacture of a practical synthetic fiber—from raw chemicals and polymerization through spinning and knitting to finished garments. Early on they decided to perfect a fiber for use in women's hosiery, and to capture the high-end market for so-called "full-fashioned" silk hose. Full-fashioned stockings were knitted flat, tapered to the ankle by dropping stitches, and then joined into a tube by a seam up the back, rather than being cheaply knitted in a continuous tube. Although silk made up less than 3 percent of U.S. fiber consumption in 1930, American women annually bought an average of eight pairs of silk stockings apiece and earned about $70 million for Japanese exporters.[28] Breaking into this market required perfecting a host of discrete physical operations—melt spinning, pretwisting, draw twisting for strength, uptwisting for elasticity, shrinkage removal, sizing to protect filaments during further processing, spooling, knitting, preboarding of stockings to prevent wrinkles, dyeing, and so on. Many procedures were new to the textile industry. Du Pont never actually manufactured nylon stockings for sale but definitely discovered how to do it. The entire process from laboratory to sales counter cost the company more than four million dollars, an investment it handily recouped.[29]

As technical development moved toward its goal, those involved worried about how to introduce a new synthetic fiber to the American public. It had to be set apart as something innovative, something distinctive, something superior. The central problem surfaced during a wear-test conducted among Du Pont office workers in the summer of 1937. Although most of the women declared the stockings "better than silk," their high luster suggested they were made from an improved rayon.[30] To most women, especially those purchasing full-fashioned stockings, rayon exuded second-rate cheapness. Unwisely first placed on the market as "artificial silk," rayon was still not as elastic as silk despite years of laboratory work. No matter how superior to silk, the new fiber would suffer if presented under the imitative banner of rayon. To introduce it as a substitute for anything was to court disaster. Crucial to its success was its naming—the synthesis of a new word for a new substance.

The issue of naming the synthetic fiber that researchers were calling Rayon 66 (for lack of anything better) first surfaced in January 1936. B. M. May of the Du Pont Rayon Company questioned "whether it is wise or unwise to denominate it a rayon." He suggested that "to call it something else might be better" because it was superior to anything "comprehended under the word 'rayon.'" Within the next few months colleagues suggested such acronyms as the humorous Duparooh ("Du Pont pulls a rabbit out of hat") and the multivalent Duparon ("Du Pont pulls a rabbit out [of] nitrogen/nature/nozzle/naptha"). For a short time Duparon gained acceptance as an in-house designation written into lab notebooks over the crossed-out Rayon 66. Officially, however, the synthetic fiber remained Rayon 66 until October 4, 1937, when a directive ordered that the term be "completely and entirely discontinued" and replaced by the more neutral Fiber 66. By that point, Polymer 66 (the flake from which filament was extruded) was being produced in fifty-pound batches, several textile mills were secretly knitting stockings, wear-tests had been run, and Du Pont was ready to erect a pilot plant. It was time to seek a name in earnest.[31]

In December 1937 a call went out to upper-level executives throughout the company to conceive a name with which to present Fiber 66 to the world. A list of more than four hundred suggestions ranging from Adalon to Yarnamid revealed a variety of baptismal motives.[32] Dufil and Pontex continued the firm's tradition of extrapolating from "Du Pont." Carotheron and Wacara honored the fiber's inventor, recently deceased. Elasta, Longduray, and Supraglos highlighted the fiber's characteristics. Novasilk and Synthesilk dangerously suggested imitation, while Amida, Hexafil, and Polamid echoed the new fiber's chemistry. Whether signifying glamour (Extasil, Shimaron) or conveying chemical origins (Fibermide, Polycromide), most suggestions evoked specific imagery in fanciful polysyllables. Responding to the list, Vice President Walter S. Carpenter, Jr., offered variations on the company name with Duponese and Pontella; he also expressed a liking for Lustrol. President Lammot du Pont, on the other hand, avoided reference to the company and preferred Neosheen, Durasheen, and Delawear, the latter his own pun. Nylon, the eventual choice, did not appear on the list, though possible ancestors included Dulon, NuRay, Nurayon, and Nusilk. With the exception of Dulon they lacked the imageless neutrality that made nylon an ideal name for a fiber or a material of many applications in sometimes contradictory contexts.[33]

How the company abandoned such extravagant names as Duponese and Delawear in favor of the neutral generic nylon is unclear. Several improbable explanations surfaced later. One rumor, repeated as late as 1982, derived the name from New York (NY) and London (LON), wrongly claimed as sites of nylon's simultaneous discovery.[34] The most persistent rumor asserted that Du Pont intended nylon as an acronym for a contorted challenge to the Japanese silk monopoly: "Now You Lousy Old Nipponese!"[35] In March 1941 Ernest Gladding, who had become manager of the Nylon Division, offered Lammot du Pont a version of the derivation of nylon "in case you get put on the spot some time." Gladding claimed they had begun with Nuron (phonetically similar to NuRay and Duron from the original list but semantically distinctive because it spelled "no run" backward). Owing to conflicts with other trade names, Nuron became Nulon, which also posed a legal conflict. Changing U to I produced Nilon, which could be pronounced at least three different ways. Changing I to Y finally yielded "nylon"—and a semantic neutrality that allowed the word to refer to the material of such different objects as stockings, toothbrush bristles, and automotive bushings.[36] An in-house memo first revealed this choice in October 1938, about two weeks before Du Pont publicly announced the new synthetic fiber. The memo stated simply that "the word nylon has been adopted for the base material heretofore designated as '66.' " Since "nylon" would be used "in a generic sense," it would be "unnecessary to capitalize it, put quotation marks around it or otherwise to create any impression that it is anything but a common noun." And since no definition of this "common noun" existed, the memo provided one: "Nylon is the generic name for all materials defined scientifically as synthetic fiber-forming polymeric amides having a protein-like chemical structure; derivable from coal, air and water, or other substances, and characterized by extreme toughness and strength and the peculiar ability to be formed into fibers and into various shapes such as bristles, sheets, etc." [37]

The deliberations surrounding the naming of nylon suggested Du Pont had progressed far in sensitivity to public opinion since Lammot's refusal seven years earlier to exhibit at the Century of Progress Exposition. The public announcement of nylon and subsequent introduction of nylon stockings also indicated awareness of the need to proceed carefully when presenting a new synthetic, especially one that might be misinterpreted as an inferior substitute. As it turned out, however, the company's pride in its scientific accomplishments and the prevailing machine-age orientation of business culture during the 1930s blinded Du Pont to the need to humanize or domesticate its new product. While the company promoted nylon as an unprecedented synthetic material emblematic of the new world of wonder-working chemists, housewives and salesclerks whose enthusiasm for "nylons" quickly became apparent actually mistrusted Du Pont's rhetoric and reinterpreted the new material in an informal domestication process that left the company scrambling to keep up.

Du Pont went to great lengths to retain exclusive control over announcement of the new fiber. Somehow the company maintained secrecy through most of a development process involving dozens of chemists, engineers, and other employees. A chemist who carried a few bobbins of nylon yarn to a mill for the first knitting test in February 1937 later claimed that security measures (including collecting and weighing stray bits of yarn) rivaled those of the Manhattan Project. [38] Three months later, only a single on-site employee at a mill chosen for more extensive tests knew that the unmarked yarn came from Du Pont. [39] Despite such measures, textile industry insiders quickly suspected that something practical was about to emerge from the research that Carothers and Hill had described back in 1931. In July 1937, only five months after the tight security of the first knitting test, a trade journal predicted in an article entitled "Silk Is Done" that "one of the larger and most reliable organizations" was about to introduce a synthetic fiber with none of silk's defects and irregularities. [40]

The company itself began dropping hints of a major development. On December 29 a commercial on *The Cavalcade of America* described "Du Pont chemists . . . working to develop fibers . . . finer, softer, tougher or less expensive than those we have now." While the company was "confident" that textile research would "give us new and better things for 1938," it was impossible to say "exactly what they'll be."

In July 1938 the *New York Daily News Record* ran an article long regarded by corporate folklore as a scoop on nylon but probably based on a leak intended to build suspense for the synthetic's unveiling. Based on "reliable sources," it described an "entirely new synthetic yarn for women's hose" derived from a "new complex chemical compound" similar to "the new plastics." The reporter had seen stockings knit from the new fiber and quoted reports that "they just last and last." He concluded that "one cannot help but feel . . . there is truth to the du Pont slogan, 'Better Things for Better Living Through Chemistry.' " Business magazines looked into the rumors; within a matter of weeks several upbeat notices appeared. While *Business Week* promised the new fiber would "outwear silk," *The Wall Street Journal* reported it was "said to be so strong and elastic that it will solve the problem of manufacturing virtually run-proof hosiery." Such comments were dangerous because nylon could do no such thing.[41]

At this point Du Pont began losing control of pre-release publicity. The situation got out of hand after September 20, 1938, when the U.S. Patent Office issued three Carothers patents for which the company had submitted applications between 1935 and 1937, including No. 2,130,948, describing a synthetic fiber made from diamines and dibasic acid. Two days later the *New York Times* and the *Washington News* published stories based on the patent; other newspapers and magazines followed. For the most part reports were positive, asserting the new fiber's superiority to silk in strength and elasticity and predicting reduced economic dependence on Japan. No one at Du Pont could have objected to such advance notice. But reporters did not stop there. Almost all accounts compromised the favorable publicity by exploiting a bizarre bit of information that would have disgusted many readers. As the *Washington News* pointed out, basing its observation on an interpretation of the patent, "one of the ways to prepare the new synthetic silk fiber might be to make it out of human corpses" by using cadaverine, a foul-smelling chemical substance formed during putrefaction "after burial." Fortunately chemists did "not need to depend on death" because they could also obtain cadaverine "out of sticky black tar, formed as coal is heated." By that point in the article "sticky black tar" may have seemed the lesser of two evils. The *News* compounded the damage by mentioning that Carothers had taken his own life twenty days after applying for the patent—thus providing food for a ghoulish speculation that in theory the inventor could have been consumed by his own invention. Nearly identical phrasing appeared in *Science News Letter*, which went out to public schools around the country and provided fillers for small newspapers.[42]

Two years later, in 1940, a distraught woman confronted a lecturer at Du Pont's exhibit at the New York World's Fair with a schoolteacher friend's assertion "that your nylon hose was derived from . . . acids taken from corpses." The young lecturer and his supervisor could not convince the woman otherwise; she kept insisting that her friend "knows the person who du Pont employs to take the acids from the dead bodies."[43] The anecdote suggested that early references to cadaverine had a definite impact on the reputations of nylon, of Du Pont, and of chemistry in general. A month after the cadaverine stories appeared, long before any stockings were ready for sale, and thus somewhat prematurely, Du Pont christened its new fiber under auspices designed to lend it the cachet of machine-age utopianism. The announcement and subsequent publicity so frequently mentioned "coal, air, and water" as nylon's ingredients that the phrase suggested a mantra contrived to ward off cadaverine's macabre influence. That was exactly how the phrase was invoked two years later by the exhibit lecturer even though he remained ignorant of the source of the bizarre misinformation he was trying to neutralize.

Few commercial products have enjoyed the fanfare that accompanied nylon's announcement on October 27, 1938. Seeking a stage appropriate to their achievement, Du Pont executives chose the site of the impending New York World's Fair. Some four thousand people, mostly middle-class club women, had traveled by special train to attend the final session of the *Herald Tribune*'s annual Forum on Current Problems. Attracted by an opportunity to preview "the multicolored buildings of the fair," especially the Trylon and Perisphere already towering over the grounds, they shared the enthusiasm of little Shirley Temple, who exclaimed by long-distance telephone from Hollywood, "Gee, I wish I could be with you at the fair today." After crowding into an auditorium for talks on "how science is making life richer and easier," the audience listened as corporate speakers outlined the shape of things to come. Consolidated Edison promised electronically synthesized music broadcast into nighttime streets as bright as day. General Foods predicted whole meals cooked at home in a minute or two and served in their attractive packaging. Politeness in the face of such patent absurdities yielded to avid attention when Charles Stine veered from a general discussion of chemistry's wonders to announce nylon, "the first man-made organic textile fiber." Derived from "coal, water and air," it could be "fashioned into filaments as strong as steel, as fine as the spider's web." At that point, according to the *Herald Tribune*, "a woman on the platform applauded, and the whole audience, visualizing stockings strong as steel, joined in." No newspaper, whatever its size or market, could ignore the story. Nylon became a household word overnight, long before more than a handful of people had seen or touched the stuff. Du Pont, perhaps seeking to link nylon by the sound of its name to the utopian symbolism of the Trylon, had scored a public relations coup. But from the moment of nylon's announcement, the company had also lost control of its creation to the practical concerns of American consumers.[44]

The promise of run-proof stockings quickly engaged the entire nation from the White House on down. Five days after nylon's announcement, President Roosevelt's cabinet addressed its "vast and interesting economic possibilities" in a discussion described by one member as "the most interesting thing that came up" all day.[45] Other observers were less circumspect. "New Hosiery Held Strong as Steel," ran a *New York Times* headline the day after the Forum. "No More Runs" declared the *Orlando Sentinel*, because Du Pont had promised "stockings as hard as steel!" Similar stories and headlines appeared in dozens of newspapers across the country. Some focused on the practical benefits "for the poorest working girl, waitress, or stenographer," for whom flimsy but expensive silk stockings had become an "indispensable . . . social convention." Others adopted a humorous tone, like the columnist who joked that with a stocking "made of chemical steel fiber . . . as strong as steel," a "girl" would have "to get an acetylene torch . . . to get a run in it."[46] According to one executive, "the press had nylon doing everything but climbing trees." Recognizing the danger of such extravagant claims before nylon even reached the marketplace, Du Pont tried to get out of its "embarrassing position" by scaling back the announcement's original terms, especially the notion that nylon possessed the strength of steel.[47] But the damage was done. Even so, by giving nylon a domestic role that ordinary consumers could understand, even extreme claims of runless stockings proved more effective as publicity than futuristic predictions and references to chemical synthesis or machine-age alchemy. Eliminating one of life's minor irritations and saving some money in the process proved more compelling than molding a synthetic utopia. Although nylon did not go on sale across the country until May 1940, American women eagerly awaited indestructible stockings.

BATTLING FOR NYLONS

The public's first glimpses of nylon stockings came at the two world's fairs of 1939, the Golden Gate International Exposition, opening in February at San Francisco, and the New York World's Fair, which followed two months later. In both cases industrial designer Teague retained the Wonder World of Chemistry theme he had devised for the Texas Centennial. At San Francisco an entrance foyer effected a transition from the domestic themes of most of the Homes and Gardens Building to the chemical utopianism of Du Pont's self-contained space. Visitors first confronted "a large chemical tower made up of over-sized laboratory apparatus with a maze of bulbs and spiral tubing, inside of which liquids boil and churn and lights flash." Behind this Tower of Research appeared a mural whose images of "chemically-created materials" suggested "how the chemist takes products of mine, forest and field and converts them into products better suited to man's needs."[48] Nylon received no greater attention than three dozen other chemical products and processes. The nylon display, one of eight devoted to coal derivatives, offered the sight of disembodied mechanical hands extending from a wall, endlessly stretching and releasing a nylon stocking to demonstrate its strength and elasticity. Nothing suggested the fabric's sensuous touch or its potential stylishness. Not allowed to touch the material, visitors who came to see nylon had to be satisfied with this mechanistic display of physical characteristics. The display simply ignored expectations raised by enthusiastic journalists.

Although New York's Wonder World of Chemistry was larger than San Francisco's, nylon initially played an even smaller part. Du Pont constructed its own building for the World of Tomorrow, a nearly circular structure entered through an out-door Court of Chemistry defined by a concave semicircle of steel trusswork seventy feet high. In the center of the court stood another stylized Tower of Research complete with Lucite retorts, coiled tubing, and colored lights, rising a hundred feet. For the interior Teague abandoned the concept of placing displays around the perimeter of a large open space and devised a more complex layout. After viewing a sixty-foot illuminated Lucite mural depicting "better living," visitors proceeded down a one-way corridor whose flow lines carried them past exhibits devoted to chemical research and raw materials, processes of chemical control, and methods of chemical manufacturing. Displays of visual interest included an injection-molding machine turning out combs, a small-scale cellophane machine extruding a three-inch ribbon of film, a model of a "sink-and-float" plant for separating coal from slate, a transparent chamber in which clouds of flies were exterminated with insecticide, and, at the end, a circular room with five small stages on which marionettes acted out everyday uses of chemical products. Overwhelmed by these dramatic attractions, a static display of nylon stockings, toothbrush bristles, fishing leaders, and sewing thread lacked even the minor visual thrill of San Francisco's mechanical hands. At the World of Tomorrow, however, as at the Golden Gate Exposition, the "sheer magic" of nylon stockings graced the legs of each young woman who served Du Pont as lecturer or receptionist. That single human touch among the vast chemical panorama made all the difference.[49]

Fascination with nylon was the public's most frequent response to the exhibit. The manager had expected "a fair number of questions on munitions." Instead he found such thoughts submerged in "the great battle cry": "How soon can we get nylon hose?" Weekly reports by lecturers and supervisors supported his generalization that many people "came to this building specifically to see nylon." "Everyone has heard about nylon and that is all they show any interest in," a lecturer observed. "Once they have it all explained," he continued, "they are content to move on and go home and tell their friends they saw the stockings made of coal." A supervisor concluded that "at least 50 per cent of our questioners inquire about 'spun steel' from which we are making stockings," or "the coal stockings that will not run." Most visitors seemed less interested in the chemistry of nylon than with experiencing it in their lives as soon as possible. One lecturer recalled a group of women who "came to the platform and were ready to take money out of their purses to purchase a pair of nylon stockings"; another recalled a woman who "displayed a cotton-clad leg . . . and groaned when I told her that we expected NYLON hose to be on the market in about six months."[50]

Despite such displays of chemical achievement as the injection molding machine and the insect death chamber, people wanted nylon. As the 1939 season wore on, Du Pont's management responded to demand by altering the meticulously designed and controlled exhibit to give nylon a progressively more central role. Sometime in June, two months into the season, they decided to enliven the research section's dull nylon display by elevating a female lecturer to a platform seven feet above the floor for an impromptu fashion show. Here was no stylized refinery of coiled Lucite, nor a pair of mechanical hands endlessly flexing, but an attractive young woman, a girl next door displaying nylon-clad legs with a "judicious lift of skirt."[51] Giving lip service to an agenda of chemical utopianism, a male lecturer introduced her by observing she was "dressed by chemistry from head to toe," from her Lucite jewelry and rayon dress to the cellophane bows and Pyralin heels of her shoes—and, of course, her nylon stockings. Stepping down after the presentation, she was invariably surrounded by women asking when they could buy stockings and how much they would cost. As she confided in her weekly report, "many ladies . . . surreptitiously felt them, saying, 'I didn't think you were wearing any stockings . . . on the platform.'" A male colleague suggested to no avail that "word-of-mouth advertising we could secure by allowing people to feel the stocking would be incalculable." Despite the risqué element of the new display, a Catholic nun declared herself not at all shocked—"no indeed"—because a girl on a platform offered "the best way to show the stockings." Without quite knowing it, the manager who had transformed Du Pont's research section was responding to public demand that nylon be domesticated, that it be taken away from the molecule engineer and rendered meaningful to everyday life. One woman hinted at this by declaring she had "never imagined chemistry and du Pont contributed so much to the fashion picture." Another stated it more directly while talking with the young woman in her nylons who had come down from the platform to mingle with the crowd. "This was such a bewildering chemical display," she admitted, that she was "glad to see something she could understand." Even utopian aspirations inspired by nylon revealed a domestic slant suggested by two elderly women who rejoiced that "everyone will become religious all over again, thanks to du Pont," because nylon stockings would mean "less mending on Sunday."[52]

The new display proved so successful that in late August, even as the 1939 season was winding down, it became the centerpiece of the Wonder World of Chemistry. No longer one among many displays, a reconstituted Lady of Chemistry became the final attraction seen by visitors before exiting to the marionette theater. Rather than climbing a visible stairway to her platform, the glamorous "nylon girl" in an evening dress was effortlessly revealed by the action of a small turntable. "You would arrange yourself in it," one of the models recalled twenty-five years later, "then get whirled around to face the audience." A male lecturer lamented that "a lot of the zip" was gone from his own research section but described "the climactic effect" of the new show as "overpowering." Coming so late in the season, this reworking of the exhibit indicated a continuing attempt to keep up with public demand. Even so, the Lady of Chemistry seemed more abstract than her predecessor. The girl next door had become a mannequin absorbed into the synthetic realm of the materials adorning her. "I do not talk to as many people now," reported one of the women who played the part; "they seem to be awed by the evening dress and are afraid to approach me." After finishing a show and returning to the exhibit floor, she found that women were "willing to talk about their hosiery problems, but I had to collar them first." Although 85 percent of the women and 65 percent of the men who went through the Du Pont exhibit in September still mentioned nylon as the attraction that drew them, the spontaneous nylon mania had subsided. Maybe it yielded to an autumn mood at the fairgrounds, reinforced by reports of war in Europe—an event leading some visitors to hope that Du Pont had not forgotten how to make munitions. But it also seemed that the limiting of nylon to a Lady of Chemistry who was divorced from the life of the exhibit floor marked a victory in another kind of war, a partial retaking by abstract chemistry of ground previously won for domesticity. Or perhaps idle talk and flashy demonstrations now simply irritated women impatient to possess a pair of the new synthetic stockings.[53]

When the fair opened for its second season in the spring of 1940, the Wonder World of Chemistry had further reasserted its identity as an apotheosis of chemical utopianism. The Lady of Chemistry of the previous autumn, her title suggesting a tenuous link to chivalric traditions, had yielded her place at the heart of the exhibit to Miss Chemistry, whose prosaic title suggested machine-age visions already diminished in the eyes of a public confronting war. A Du Pont publicist explained that the display revealed "the beauty added to familiar things by scientific research, symbolized by a model garbed in the products of chemistry." Miss Chemistry herself seemed synthetic, "her emergence from a giant test tube" contrived "in such a way that it appears she has literally been 'created by chemistry.'" Interrupting the lecturers, impatient visitors blurted out their desire to see the "test tube girl." A small boy declared he "felt awfully funny when that lady came out of the test tube" because "it was so mysterious!"[54] Miss Chemistry embodied the era's technological ideology of artificially perfect surfaces and smooth, efficient control—both to be attained through plastics and other synthetics. But as a popular attraction she proved superfluous. Despite the brief subversive thrill of seeing a pretty girl displayed in the name of scientific education, or the honest bewilderment of a young innocent, Miss Chemistry lacked the immediacy of the girl next door proudly showing off her nylon stockings to enthusiasts unable to try them out for themselves.

In the meantime nylon stockings had actually gone on sale throughout the country. Despite shortages, many women had obtained at least a single pair by summertime, when the renovated Du Pont exhibit experienced its greatest attendance. Once a week from October 1939 through February 1940, shops in Wilmington received shipments of three grades of full-fashioned nylon stockings, priced like silk, for sale only to local residents who "rush[ed] in like so many hens at feed time," as *Time* cavalierly reported.[55] After Du Pont's yarn plant at Seaford, Delaware, reached full capacity around the first of the year, the company authorized several knitting mills to prepare shipments of stockings for the first national offering at selected stores on

May 15, 1940. With consumers anticipating run-proof stockings and a worried textile industry appalled at the thought of stockings that would last forever, Du Pont tried to dilute the impact of "unauthorized and exaggerated statements." "Strong as steel" became an equivocal statement that "on the average, nylon stockings will wear at least as long as other fine quality hose of equal weight and workmanship."[56] On the eve of the sale, a Du Pont chemist even appeared on *The Cavalcade of America* to explain patiently in response to questions posed by a "representative housewife" that "any hosiery made by standard methods of knitting *will run* when a thread is broken"—though nylon's "extra *strength* and *elasticity*" made it "less likely to develop runs."[57]

Despite the company's lingering desire to have it both ways, the initial offering proved phenomenally successful throughout the country. Shoving in line past display windows "trimmed mostly with test tubes, beakers, coal and water," hordes of shoppers surpassing those of any remembered Christmas emptied most stores of their initial supply of nylon stockings by noon—at a premium price of $1.15 a pair.[58] In the first flush of enthusiasm consumers declared them even better than expected. Quibbling over snags and runs vanished in the sheer pleasure of it all. "My legs felt different," one woman reported; "I felt like kicking them up in the air." Another recalled that "somehow, they made you feel you had gone to town on yourself, like spending fifty bucks on a hairdo, a facial, a bottle of perfume . . . when all you'd done was put on a new pair of stockings."[59] In the face of such tangible, sensuous reactions to nylon, the abstract Miss Chemistry held no particular interest for women who visited the Du Pont Wonder World of Chemistry in 1940.

As in 1939, however, nylon continued to stimulate the most questions and comments for Du Pont's lecturers. But the focus was shifting now that most female visitors had experienced nylon or knew someone who had. Some complained that nylon stockings snagged and developed runs more easily than silk. Several salesclerks who admitted no longer recommending nylon stockings explained that customers "were dissatisfied with them because they were not *perfect*."[60] Informed by the era's utopian bent and by Du Pont's own extravagant publicity, many people expected no less than perfection. Eventually most consumers reached an accommodation with nylon's failings. As summer approached and the new stockings became increasingly common, one lecturer found that most women were "getting away from believing they . . . are made of steel." Some people were no longer "expecting miracles."[61] Others who took seriously the myths of scientific infallibility and machine-age perfection found it hard to give up the promise of run-proof stockings. Several lecturers observed a ground swell of rumors to the effect that "nylon was made stronger at first," or that it "was called back . . . because it was too good," or that its strength was reduced "because it would have ruined the hosiery industry."[62]

Suspicion about Du Pont's economic motives seemed related to a basic undercurrent of distrust of synthetics as unnatural laboratory products—a response sometimes approaching physical revulsion. A litany of complaints filled one lecturer's weekly report. Women told him that nylon stockings "feel like snakes when wet . . . give you nylon poisoning . . . are dyed with special dyes that are harmful to the skin . . . will stop perspiration from coming through the leg pores . . . will melt in hot water . . . will dirty the leg since nylon is made from coal . . . feel metallic since nylon is mechanically made." His report might have seemed intended to confound his supervisor had not other lecturers mentioned similar comments.[63] Rumors also surfaced about cadaverine, about women "already getting cancer of the legs from the nylon," and about a woman who "had walked through the exhaust of a bus" and was shocked to discover "the fumes had taken the nylon hose right off her legs."[64] Some rumors were grounded in fact. Dye used at a particular knitting mill did cause allergic reactions severe enough to attract the attention of the American Medical Association. More alarming, scattered reports over the next decade confirmed that warm humid air polluted with acid could indeed "melt" and "shred" nylon stockings in a matter of seconds. "How disconcerting it would be," ran a ditty in the *New York Herald Tribune*, "When trigly clad by chemistry, / In fabrics made from coal and air, / To find they were no longer there, / But changed again, returned from whence / They came—back to their elements!"[65] A subject for sexist humor at the expense of women who placed their dignity in the conjuring hands of latter-day alchemists, melting nylons also offered a frightening image of science out of control, corrupting the most personal items of everyday life. But that perception remained only an undercurrent. If Miss Chemistry seemed not so enthralling in 1940, it was not because her degree of abstraction seemed threatening but because women had already made nylon their own. From another perspective her bland exterior offered reassurance that ordinary people could absorb the products of science, new synthetics and plastics, without jeopardizing accustomed ways of living. When American women accepted nylon's embrace, as Williams Haynes perceived two years later, they also gave up "the unpleasant connotations that have so long clung to the idea of a chemical substitute." Synthetic materials had become part of the American way of life.[66]

Reconciliation of consumers and chemists appeared certain as the 1940 season of the New York World's Fair wound to a close. By September few visitors doubted nylon's material quality or questioned its safety. They had learned to appreciate nylon, even to consider it a necessity, and feared they would lose it as the nation geared up for defense. Although there had never been enough stockings to satisfy demand, rumors spread that the federal government was causing the shortage by diverting nylon yarn for military parachutes. As one of Du Pont's lecturers put it, "everybody and his Aunt is asking about the Government taking over nylon."[67] The U.S. Army Air Corps was indeed experimenting with nylon parachutes, and rumors became so persistent—not only at the fair but in the textile industry—that in September the company officially stated that most of its nylon production went for civilian uses. Stockings absorbed 90 percent of the nylon produced by Du Pont in 1940 (not until February 1942, two months after Pearl Harbor, did the company convert to military production).[68] Consumers who complained of continuing shortages late in 1940 were simply targeting the military as the most obvious scapegoat. By doing so, they indicated the degree to which nylon, though not produced in sufficient quantity to satisfy demand, had become a material they did not want to live without.

Nylon's wartime success as an engineering material contributed to a high general opinion of synthetics. Not only did it prove superior to traditional materials for some military applications; it also enabled manufacture of some equipment otherwise impossible to make. With increased production at a second Du Pont plant, opened in Martinsville, Virginia, in November 1941, the company shipped 77.7 million pounds of nylon for military use during the war. Until the end of 1942, escape parachutes absorbed most of it. Although first adopted merely as a substitute for Japanese-controlled silk, supply officers soon realized it was superior to silk in strength, resilience, and resistance to mildew and salt water. As the supply of nylon increased, it was used for cargo parachutes and, beginning in 1943, for tire cord for bombers. Eventually the synthetic appeared in glider tow ropes, as an interlayer in self-sealing fuel tanks for aircraft, as a laminate providing the protective element in flak jackets, in molded machine-gun parts, and in rot-proof shoelaces, mosquito netting, hammocks, and clothing for the humid jungles of the Pacific theater.[69]

Nylon had aroused such strong interest during its brief period of civilian availability that it became a symbol for sacrifices and subterfuges of life on the home front. Collection drives targeted cast-off stockings to be melted down and re-extruded as parachute yarn. Newspapers ran photos of young women "'Taking Em Off' for Uncle Sam" and "Sending Their Nylons Off to War." In 1943 an advertisement for B. F. Goodrich explained, "We Borrowed Their 'Nylons' to Make Tires for the Navy." A month after V-E Day another Goodrich ad headlined a photograph of two huge bomber tires by describing them as a "Pair of nylons . . . Superfortress size." Anticipating the return to civilian production, the company informed consumers that "a Boeing superfortress lands on enough nylon to make 4,000 pairs of stockings."[70] Despite that hint of things to come, women were still doing without as they had throughout the war. Many of them had experimented with nylon-colored leg makeup and had painted seams down the backs of their legs, while others were taken in by worthless "nylonizing" rinses that promised to convert "ill-fitting, slow-drying rayons into sleek, flattering stockings that look, fit and wear like cherished nylons."[71] A hit song of 1943, "When the Nylons Bloom Again," summed up bittersweet nostalgia for a product that consumers had hardly gotten to know, a product that stood for a promised future of material abundance whose arrival was in doubt.[72]

Not everyone was willing to wait. In some circles the question "Where'd you get those nylons?" became an expression of suspicion. A Louisiana burglary victim bemoaned the loss of eighteen pairs of nylons. Police in Chicago ruled out robbery as a motive for murder because the perpetrator had left behind half a dozen pairs of nylons. Entrepreneurs sold black-market stockings knit from hoarded prewar yarn at twenty dollars a pair. A shipping agent who diverted a shipment of new yarn to a small textile mill produced a horde of stockings worth $100,000.[73] To confuse matters, defective parachute fabric and other scrap nylon legally reached consumers as slips, blouses, shirts, socks, and sometimes, in the case of substandard yarn, even as stockings—all advertised as nylon. Du Pont again became the object of rumors of war profiteering. To counteract negative publicity the company launched an advertising campaign emphasizing that current production went "entirely for critical military purposes."[74]

The uproar that greeted the return of nylon stockings at the end of the war demonstrated that even in their absence they remained an element of everyday life. Nylons—or rather their lack—so directly symbolized wartime sacrifices of Americans on the home front, especially those of women, that they rushed to acquire stockings as tangible evidence of the nation's return to normality. Only eight days after Japan's surrender, in August 1945, Du Pont announced its reconversion to production of yarn for nylon stockings.[75] The first of the so-called nylon riots, as journalists referred to them, occurred late in September when small shipments of stockings went on sale in a limited number of stores, all besieged by mobs of people who had learned of the offering by word of mouth. The riots continued through the middle of 1946, as long as the shortage remained severe, wherever and whenever nylons went on sale (Figure 5–2).

5–2. A SCENE FROM THE NYLON RIOTS, 1945

Most reporters who covered the nylon riots cast them in mock-heroic or at least pseudo-military terms, as if American women had been forced to fight the final battles of the war so they could earn the right to settle down and enjoy the fruits of the more bitter victories their men had won. "Nylon Sale and No Casualties," ran a headline in Cortland, New York. "Women Jam Store, Fight For Nylons," ran another in Columbia, South Carolina, where "elbows began to jab ribs, and the ladies and sprinkling of men became a shoving, pushing lot with but one thought in mind." At Indianapolis, where hundreds of women lined up two hours before G. C. Murphy's opened to contend for three hundred pairs at $2.25 a pair, the newspaper heralded a noteworthy oddity—"Lone Man Spearheads Push For First Nylons Here." By the end of the year the riots were more severe. In November twelve hundred women were dispersed by police from the sidewalk in front of a dress shop in Chicago. "Women Risk Life and Limb in Bitter Battle for Nylons," ran a headline in Augusta, Georgia, the following month. "Clad in raincoats, bearing dripping umbrellas, the crowd surged into the store, knocking over piles of dishes, bedroom slippers and other items in their eagerness," and many of those who succeeded sat outside in the pouring rain "on the bumpers of cars to put on their purchases." It continued on into January 1946. "Lady Raiders Take Nylon Beachhead," reported a newspaper in Los Angeles. "Women Win Battle For Nylons Here," ran a headline in Portsmouth, Virginia. At Pittsburgh, later in the year, the mayor arranged an evening sale in response to a petition from four hundred working women who complained they could not shop during regular hours. Forty thousand people lined up for sixteen blocks to compete for thirteen thousand pairs. As the newspaper reported, "A good old fashioned hair-pulling, face-scratching fight broke out in the line shortly before midnight." When it was all over a policeman moaned, "I hope . . . I hope I never see another woman." By then, complaining that nylon had "received more free advertising" than any other commodity "in the history of [the] newspaper business," a trade journal for newspaper editors had long since declared, "let's have no more free publicity on nylons!"[76]

Du Pont did not emerge unscathed from the nylon riots. As the shortage wore on into 1946, people began to suspect the company of engineering the situation for its own benefit. Holding sole manufacturing rights under patent law to a substance in universal demand, the company appeared greedy, even unpatriotic, in its presumed artificial limiting of supply. It turned out Du Pont's plants were actually operating at full capacity, as a letter from a public relations officer to the *Christian Science Monitor*'s science editor documented in some detail.[77] But mere assertion did not satisfy consumers or their advocates. Declaring the hosiery industry "virtually at the mercy of one producer—du Pont," a retailer from Paterson, New Jersey, took out a full-page advertisement in *Knit Goods Weekly* urging colleagues to convince fifteen million consumers to protest to their congressmen.[78] Possibly prodded by someone at Du Pont, the head of the compliance division of the Civilian Production Administration tried to defuse the situation by blaming the shortage on "piggish" housewives with "leisure to stand in line and buy." He argued that by March 1946 production of stockings had reached thirty million pairs a month, enough for everyone. But greedy housewives hoarded them—he personally knew a woman with thirty-three pairs—and there were "a lot of working women" who had none.[79] Insulting explanations did not prevent newspapers like the *Hattiesburg American* from denouncing Du Pont as "the only company which knows the magic nylon secret." Declaring that "News Is All Bad on the Nylon Front," the *American* suggested that "mournful news" about nylon shortages "should cause women's pages from coast to coast to be bordered in black."[80] The situation evaporated as production increased and nylons reached everyone who wanted them. Although consumers generally expressed satisfaction, even with stylish new ultra-sheer stockings designed to wear out more quickly, competitors in the textile industry remained restive. Five years later Du Pont yielded to a threatened antitrust suit. Despite a clever argument that nylon faced competition from silk, cotton, wool, and rayon, the company licensed nylon to the Chemstrand Corporation in 1951 and eventually enrolled other licensees.[81]

Licensing of nylon marked the signing of a peace treaty for a war that had long since ended. By making nylon technology available to other companies, Du Pont gave up de facto rights to the name nylon. The material for the first time legally entered the realm of the generic, the everyday, the universal. It was a substance whose success proved it belonged to everyone. Women who protested Du Pont's monopoly back in 1946, as the nylon riots were drawing to a close, had already recognized that fact and had been asserting their rights. The nylon riots represented a struggle for personal control over a material that had emerged from the alien realm of industrial chemistry only to be literally stripped away and expropriated for the alien realm of military technology. Journalists who used tongue-in-cheek military imagery to describe the nylon riots accomplished several things. Most obviously, they were enjoying the luxury of casting a frivolous subject in terms only recently reserved for events of life-and-death significance to the whole nation. In addition, they were acknowledging women's right to take possession of a reward for their sacrifices on the home front. Less consciously, they also revealed an ambivalence about the forcefulness, even violence, of people, mostly women, who had lived in limbo "for the duration" and were now bent on getting on with life. Nylon, to the extent that women valued it for intimate garments associated with sexual attractiveness, was entering the service of human reproduction and thus of life itself—a far cry from a few years back when newspapers announced it could be made of cadaverine taken from corpses. Cleansed of inhuman laboratory associations and recaptured from the military realm of death, nylon was fully domesticated. Parallel tendencies characterized postwar applications of other synthetics and plastics. Streamlined plastic-as-plastic forms of the 1930s yielded in the early 1950s to imitations of natural or traditional materials. As *House Beautiful* and *Better Homes and Gardens* informed nylon-riot veterans who were moving into new suburban ranch houses, plastic did not have to be "shiny, sleek, and a little too strange-looking for the living room." It could also be "homey, chintzy, and comfortable." American women could enjoy "plastics" that "don't look it."[82] As plastic moved from utopian projection to postwar proliferation, its uses tended to embody traditional values and suggested a distrust of cold, inhuman science and technology. Domestication of nylon exemplified one pole of that unresolved tension.

GROWING PAINS: THE

CONVERSION TO POSTWAR

*A*lternately desired and despised, plastic never enjoyed the near universal acclaim of its upscale cousin nylon. The latter had it easy, arriving on the material scene as a synthetic substance christened with its own name and clearly superior to the natural silk it replaced. People first encountered nylon in a single high-fashion form that seriously engaged half the population. Apprehensions about its chemical origins affected only a minority of consumers who tried it. Although the case of nylon offered a clear paradigm of domesticating a new synthetic, like all such examples it obscured messy reality. While wartime growth had brought economic respectability to the plastic industry, the cultural perception of plastic remained ambiguous. Its domestication occurred unevenly, by fits and starts. For one thing, the category of plastic included so many different materials with such a variety of appearances, textures, characteristics, and uses. A host of new military applications—and sometimes inappropriate home front substitutions—compounded the problem. Predictions of postwar test-tube miracles clashed with the reality of lipstick tubes or mixing bowls crudely pressed from resin-impregnated paper. Throughout the 1940s people remained uncertain why some materials qualified as plastic but not others. Lack of clarity about the word's definition extended from corporate boardrooms to artifacts of popular culture. Plastic Man, a crime-fighting comic book superhero who debuted in August 1941, seemed more elastic than plastic as he bounced and stretched through adventures that left dazed gangsters jabbering about "a-a man of *rubber*! . . . a Plastic Man!!"[1] If comic books insisted on presenting plastic as a glorified chewing gum, then the public could be forgiven its uncertainty. Even an advertisement for General Electric, "Making Magic with Plastics," added to the confusion with a comic strip biography of chemist Winton Patnode in which the bespectacled scientist stretched a rubbery substance that a later generation would have identified as Silly Putty.[2]

Uncertainty whether to consider plastic a miracle material or a third-rate substitute was so common that it inspired a cartoon released by Walt Disney Studios late in 1944. "The Plastics Inventor" featured a hapless Donald Duck molding a plastic airplane at home according to step-by-step instructions broadcast during "The Plastic Hour," a radio program whose announcer launched into a breezy *Popular Mechanics* parody and ended with a running commentary on Donald's eventual disaster. "Good morning, junior inventors," the announcer addresses his listeners before telling them they are "going to bake an airplane out of junk." After making sure he has a "good hot fire under [his] pot for cooking the plastic," Donald tips a mountain of household junk into a big iron kettle. A cast-off washboard and an old pair of boots quickly melt into a golden batter, which the duck pours "on the griddle" (an airplane-shaped mold) to "bake." In the meantime, to the accompaniment of "some music played with all-plastic instruments," Donald rolls out the excess dough, uses cookie cutters to make engine parts, and shoves them into an oven. A layer of plastic dough slapped round his head and baked in a hair dryer yields a flyer's helmet. "See how light it is; it floats like a feather," the announcer intones as Donald removes the fuselage from the mold and assembles his airplane. It is now "the moment we have been waiting for, the first flight in our plastic plane." With his trusty radio beside him, commentary pattering on, Donald exuberantly soars and loops through the sky until an unfamiliar voice breaks in: "Tell me, professor, does the plane have any faults?" Only one, the daring duck learns: "It melts in water . . . so don't get it wet and avoid all rain clouds." As a thunderstorm materializes, the plane disintegrates into a gooey mess and Donald's helmet unwinds behind, twisting into Wagnerian braids. "If you just keep flying plastic planes," the gleeful announcer puns, "you'll be in the dough." Finally, as Donald appears headed for catastrophe, the viscous melt forms a parachute and carries him to an undignified landing. "Did your plane fall apart?" asks the announcer. "That's funny, you know it always happens that way to me, too"—to which Donald responds by seizing a handy watering can and melting down the plastic radio that has propelled him from innocent enthusiasm to complete fiasco.[3]

The Disney cartoon embodied a sardonic commentary so well-informed that it suggested public awareness of the promises and disappointments of plastic utopianism. When Donald melted down a heap of discarded junk to make a plastic batter, the scene evoked two decades of promises, stretching from Bakelite to nylon, that new materials would come from cheap waste substances. The cartoon might have shown kitchens, cars, or entire houses molded of plastic, but to choose an airplane tapped into wartime obsession with aviation and into a prominent strand of plastic utopianism. The public was fascinated by acrylic domes and bubble tops and by airplanes made of cheap plywood laminated with plastic-resin glues, of which How-ard Hughes's vast "Spruce Goose" offered a momentarily breathtaking example. Back in 1939, *Scientific American* had promoted a fuselage of laminated phenolic resin as a means of realizing a "long-standing dream" of "the Ford of the skyways."[4] But such visions melted away as quickly as Donald Duck's plastic plane when exposed to the reality of shoddy home-front plastics. By portraying plastic as a so-called miracle material that dissolved in contact with water, Disney animators relied on their au-dience's familiarity with similar catastrophes—with plastic sink strainers that melted in hot water or buttons that became greasy blobs at the dry cleaners.[5] Even the experts admitted problems with plastic, the cartoon implied, and not always without a certain glee at the plight of people who took all the miracle talk too seri-ously. On a level that transcended particular materials, "The Plastics Inventor" of-fered a cautionary tale on the folly of blindly trusting scientific experts. At the same time it suggested that any attempt to "do-it-yourself" was doomed.

MAVERICKS, ENTREPRENEURS, AND DO-IT-YOURSELFERS

Oddly enough, considering the esoteric knowledge implied by the phrase "modern alchemy," plastic utopianism of the 1940s did seem to offer the ordinary citizen a role in shaping the future. At times, as if reacting against control by experts, those who promoted plastic invoked an individualist "do-it-yourself" spirit that might be interpreted as the male version of domestication. That spirit certainly pervaded the most striking prewar expression of plastic utopianism, which began with Henry Ford's search for a soybean plastic and ended with an experimental auto body fab-ricated of plastic panels.[6] Always a champion of the American farmer, Ford hoped to use agricultural products as raw materials and often claimed it would soon "be possible to grow most of an automobile."[7] Sometime in the mid-1930s he converted a glass plant at River Rouge to the manufacture of a casein-type protein plastic with soybean meal as a base. The typical Ford car of 1936 contained ten to fifteen pounds of this soybean plastic in window frames, steering wheel, gearshift knob, horn but-ton, accelerator pedal, and electrical parts. The company had arrived at this point without benefit of the chemical establishment after Ford decided to go it alone with the help of a bright young man whose diligence impressed him. Robert A. Boyer was a son of the manager of the Wayside Inn, a historic site in Massachusetts owned and restored by Ford. Around 1930, at the age of twenty, Boyer abandoned a Dartmouth education at Ford's insistence to direct the soybean project. Boyer's experimental plastic autos of the early 1940s attracted wide press coverage and seemed as eccen-tric as the airplane of Disney's "Plastics Inventor."

Early in November 1940 Ford invited a group of reporters to witness a demon-stration of the plastic automobile's unique potential. The brittle-looking septuage-narian picked up an ax and swung it with all his might into the trunk panel of a

6–1. Henry Ford swinging an ax into a phenolic plastic trunk panel, 1940

custom-built car while a photographer recorded the event (Figure 6–1). Rather than crumpling and losing its paint like an ordinary sheet-metal body panel, the back of Ford's custom car rebounded into shape as good as new because it was made of a tough phenolic plastic with its black color embedded throughout. As a *Time* reporter declared, "fenders of this Buck Rogers material . . . withdraw from minor collisions . . . like unhurried rubber balls." Ford himself, pointing in triumph to the undamaged trunk lid, announced that he would soon be "mass-producing plastic-bodied automobiles."[8]

The first and only one appeared nine months later in August 1941. Before a crowd of ten thousand people at an annual Dearborn Day celebration, Ford and Boyer unveiled an experimental car with a body of fourteen plastic panels attached to a tube-steel frame. The panels were compression-molded from hemp and ramie fibers embedded in a matrix of urea formaldehyde with wood and soybean flours. The car weighed only two thousand pounds, a third less than a standard sheet-metal sedan, and had acrylic windows. Always a champion of plastic, *Time* described "the new Ford," which Boyer promised to put in limited production by 1943, as "the first gun in a technological revolution that may begin when the other guns are stilled."[9] But Ford's plastic car was never put to the test, as mobilization intervened. All the same, the jaunty cream-colored plastic Ford provoked newspaper commentary around the country. Some editors took the announcement straight. The *Saginaw News*, for example, reported that people would "soon be riding around in plastic car bodies"; the *Decatur Herald-Review* declared, "Here is something an America on wheels has been waiting for." Other editors took it with a grain of salt—or a dash of vinegar. The *St. Louis Globe-Democrat* described the new car as "part salad and part automobile," while the *Cleveland Press* thought Ford should "strengthen his plastic by adding spinach."[10] But even skeptical comments were embedded in laudatory prose indicating faith in Ford's ability to transform automotive transportation with a new plastic technology that came not from Du Pont or Union Carbide with their highly trained chemists but from a small team of enthusiastic young men formed in the mold of Henry Ford, their naive independent creativity freeing them to devise unprecedented solutions. As Boyer proudly told the *New York Times*, "We were dumb enough to try anything in those days. . . . I was 20 and oldest."[11]

Almost as if the specter of technocratic control haunted a nation dedicated to ridding the world of totalitarianism, promotion of plastic during and immediately after the Second World War emphasized not only the cornucopia of wonders ordinary citizens could enjoy but also the industry's openness to pioneering mavericks. Major chemical companies continued to consolidate their hold on the supply of synthetic resins, but molding and fabricating remained labor-intensive, craft-oriented activities carried out by small companies. Individual entrepreneurs who took advantage of wartime needs were often quickly propelled to business success beyond their wildest dreams. Dave Swedlow, for example, whose small company in southern California had fabricated transparent acrylic novelties before the war, saw business boom in response to the demand for aircraft enclosures (Figure 6–2). As a Du Pont manual admitted, shaping acrylic sheet over wooden forms and by blow molding was "not an exact science"; there were no "exact procedures." Working with the National Bureau of Standards by trial-and-error methods, Swedlow devised a

6–2. Blow molding of Lucite polymethyl methacrylate gun-sighting dome for B-29 Superfortress

process of multiaxial stretching (similar to cold drawing of nylon) that toughened sheets of Plexiglas or Lucite by reorienting their molecular chains. Because the process demanded intuitive awareness on the part of skilled workers, it was plagued by "many, many quality problems all through the war years." The craft-oriented approach of the Swedlow Aeroplastics Corporation, small enough to avoid the pitfalls of mass-production psychology, proved successful where a larger operation might have failed.[12]

Other new plastics with military applications attracted the attention of people willing to take risks. Although Kline complained that the term "plastic plywood" generated too much nonsense about the "so-called plastic airplane," plywood laminated with urea formaldehyde resin proved so superior to older varieties that it overcame plywood's shoddy reputation and constituted a new material. Compressed at high pressure by inflated rubber bladders against compound-curved forms, resin-bonded plywood saw wartime service in the fuselage and wings of training planes and gliders, and in form-fitting seats for fighter pilots. Many of those who perfected techniques for shaping resin-bonded plywood enjoyed postwar success making casual furniture, small boats, and prefabricated housing systems.[13] Wartime needs also stimulated development of the plastic material later popularly known as fiberglas. More accurately referred to as glass-reinforced polyester, it consisted of a tough thermosetting resin reinforced by spun glass fibers chopped up or woven into a mat. Because polyester resin gave off no water or gases during its formative reaction, the material could be "laid up" by hand in complexly curved wooden or plaster molds and then allowed to cure under low pressure (or, more slowly, under atmospheric pressure). Used for enclosing airborne radar in durable, lightweight, aerodynamic housings called radomes that did not interfere with transmissions, fiberglas required little equipment or experience. When the war ended, dozens of small companies started molding small boats of glass-reinforced polyester, one of the first a nine-foot dinghy marketed by Carl N. Beetle at New Bedford in 1947. Success quickly transformed the Beetle Boat Company into the Carl N. Beetle Plastics Corporation.[14]

Other wartime innovations could also be exploited with minimal investment. Injection molding continued to spread, especially with availability of polyethylene, a thermoplastic used in wartime for insulating electric wiring. Developed by ICI during the late 1930s, polyethylene was produced in the United States by Du Pont and Union Carbide using different processes. Most of Du Pont's immediate postwar output went to Earl S. Tupper, a custom molder from Massachusetts who in collaboration with Du Pont engineers began experimenting with injection molding of household items from the soft, waxy, flexible material. In May 1946 a Du Pont advertisement described a polyethylene tumbler molded by Tupper Plastics as "light, tasteless, odorless, good-looking." *Time* was soon lauding Tupper for producing seven million polyethylene nesting cups for the American Thermos Bottle Company.[15]

For every Tupper who entered the folklore of consumer culture, there were dozens of other "do-it-yourself" entrepreneurs. Growth of the plastic industry often seemed emblematic of national expansion itself, as in Frank Capra's film *It's a Won-derful Life*, released in 1946. While the reluctant hero George Bailey (played by James Stewart) remains in his hometown out of a sense of duty to a family-run savings-and-loan, his high-school buddy succeeds in the outside world as a plastics entrepreneur. Bailey turns down an invitation to become a partner in soybean plastic and later watches from the sideline as his friend prospers during the war by fabricating acrylic bubbles for airplanes. At the end of the film it is the buddy who rescues Bailey's collapsing bank—and the citizens who had put their trust in it. Although Capra's sympathies lay with small-town life, the progress of the outside world as exemplified by the expanding plastic industry clearly stood for other positive values.

To some veterans returning from the war, "plastics" radiated an aura similar to that suggested by Capra's film. The business appeared easy to get into and impossible to fail at. An engineering manual on plastic product design observed that life was "never dull for the plastics engineer" because he received "secret communications from hundreds of private inventors who believe that everything should be made of plastics." In 1949 an intelligent popularization, *Meet the Plastics*, opened a chapter on "Opportunities in the Plastics Industry" by quoting a string of sentiments typical of those heard at the end of the war. "I want to get into plastics!" went the refrain. "A person ought to go places in plastics." And finally, echoing a line that Capra's hero might have spoken, "I have a friend who got into the plastics right after the war—He's sitting pretty now." The book's author concluded that "the postwar years could hardly have been more favorable for the entry of newcomers." Military applications found their way into civilian products. Starved consumer markets absorbed all the "worthwhile" products presented to them. Even shoddy goods "produced by persons who were trying their wings" sold readily "on the non-competitive market of the postwar period." Less enthusiastic, *Modern Plastics's* editor complained about "hundreds of fledgling entrepreneurs" who thought all they had to do to get rich was "set up a press in an old garage and start turning out finished products." His editorial was reacting to a congressional push to allocate scarce presses and raw materials to returning veterans even if they had no molding experience. Fearful that consumers would reject all plastics after experiencing tacky vinyls and warped moldings, he insisted that plastic molding and fabricating was "no business for a novice."[16]

Despite efforts of experienced hands like Boonton's Scribner and SPI's Cruse to regulate the industry, it did attract a host of inexperienced newcomers who saw in it the future of America. Many "do-it-yourselfers" prospered. Louis H. Barnett, for example, founded Loma Plastics at Fort Worth in 1948 with five employees and a small rented injection press in a metal building. His first year he cleared a thousand dollars selling fishing lures; by 1956 Loma was a major supplier of such housewares as the first plastic garbage can, with a capacity of twenty-two gallons, the largest object yet produced from injection-molded polyethylene.[17] Even so, many postwar newcomers quickly went out of business after littering the landscape with shoddy products. A sense of unprecedented expansion, compromised at times by needless waste, ran through the plastic industry as through the larger culture itself. In the case of plastic these opposing attitudes ran all the way back to the forced situations of wartime, when well-engineered military materials coexisted uneasily with make-shift civilian expedients. Beneath plastic's gleaming utopian surface was a gritty reality. The conflict proved nearly impossible for industry leaders to resolve.

PLANNING FOR POSTWAR

If one believed *Modern Plastics* and SPI, then plastics had won the war. After the journal lost an editor to the army in 1943, it joked he was about to discover "first-hand just how the phenolic helmet liner, the vinyl raincoat, the ethyl cellulose canteen, the urea buttons, the cellulose acetate butyrate scabbard . . . the cellulose acetate gas mask parts, [and] the acrylic goggles" performed in action.[18] Some did not work well at all. The canteen had to be scrapped because it could not be heated to thaw frozen contents. For the most part, however, plastic war materiel from nylon shoelaces to atom bomb parts of Bakelite served their purposes. Addressing an SPI conference at Los Angeles in 1943, Kline read a laundry list of items of government issue for which procurement officers had approved plastic. They ranged from combs and razors, through molded bayonet handles, binocular cases, and mortar shell fuses, to resinous linings for jerry cans and stiffeners for fabric skins of carrier-based airplanes. A book of wartime case studies published in 1946 by the General Electric Plastics Division portrayed its military role extending across the material spectrum from the most insignificant—nameplates and instruction plates of phenolic laminate—to the most essential—lightweight rocket launcher tubes of rolled paper impregnated with phenolic resin. Even some of the bugles that awakened recruits were injection-molded from a cellulosic plastic in "conventional olive drab" to avoid wasting precious brass.[19]

Wartime requirements clearly accelerated demand for synthetic materials. Just as important, forced expansion led manufacturers to greater understanding of plastic as a category encompassing a dozen types with hundreds of variants, each unique in performance owing to differences in chemicals, plasticizers and catalysts, types of fillers, and processing techniques. Keeping up with the industry's growth required intensive gathering and disseminating of information. To procure technically sophisticated equipment on a scale suitable for global warfare, the federal government had to require identical resins and interchangeable moldings from many different material suppliers and processors. That goal in turn required commonly accepted chemical definitions, physical specifications, and methods of testing for a confusing, ever-increasing variety of plastics. Cooperation came hard to an industry whose short tradition included dozens of patent suits and whose chemists and molders jealously guarded proprietary formulas and processes.

Among the organizations facilitating the flow of information was Kline's Organic Plastics Section at the National Bureau of Standards, which grew from two individuals before the war to a staff of forty in 1945. Mobilized in July 1941 at the request of the director of the Materials Group of the Army Air Corps, who was frustrated by the slow progress of the plastics committee of the American Society for Testing Materials, Kline and his section set to work. As he later recalled, they were soon "doing all of the major testing for all of the military services," mostly of items "prepared for them by the various plastics molders." Still technical editor of *Modern Plastics*, he continued to compile bibliographies of new resins and processing techniques; he also facilitated communication between chemists and engineers across company lines. He extended his role as a conduit of information by serving on a team of technical experts sent to inspect German plants and interrogate German chemists in the spring of 1945. On the scene before V-E Day, the energetic Kline had set out by autobahn for a Bakelite plant at Hamburg when a front-line sentry stopped his jeep to warn him the city was still in enemy hands. Later, at a polyethylene plant captured intact near the Austrian border, he and three associates recovered a complete set of IG Farben plastics reports. Smuggled home in a duffel bag to avoid their disappearance into Army Intelligence, the reports yielded a series of articles for *Modern Plastics* describing such German innovations as polyurethane foam and extruded vinyl pipe.[20]

Just as important was the work of SPI, which expanded under war's impetus. Formerly distrusted by many executives for its tentative evolution beyond a golfing society, SPI became the industry's liaison with government and the source for mass-media information about plastic. After noting "the gravity of the present world situation," the board of directors in January 1941 appointed a Plastics Defense Committee "to deal with technical plastics problems" facing the government. In announcing this step, the SPI newsletter emphasized the "tremendous responsibility" and "rare opportunity" of the occasion. As a "young, virile industry unhampered by ancient ideas and stodgy procedure," the makers of plastic could offer "new materials . . . , new and revolutionary productive methods, great productive capacity, and above all, ingenuity," as the country prepared "to ward off a military catastrophe." As initially appointed, the Plastics Defense Committee included twenty-seven members almost equally divided among material suppliers and processors, along with two representatives from manufacturers of molding presses and one from *Modern Plastics*. Despite this apparent balance of power, a smaller executive Special Defense Committee of four revealed a definite shift toward material suppliers over processors. Long before Pearl Harbor, SPI established smaller committees to address such issues as standards and specifications, and in March 1941 Cruse, one of its board members, was sent to Washington as full-time liaison with the Office of Production Management; his resignation as editor of *Modern Plastics* to become executive secretary of SPI came seven months later. Even as the war provided entrepreneurial opportunity for molders and fabricators, it stimulated an opposing trend toward consolidation, not only by strengthening the influence of chemical companies but also by placing greater power in the hands of SPI. It was no accident that the war years witnessed the formation of the Society of Plastics Engineers (SPE), a rival

group representing the interests of individual engineers, molders, and sales representatives.[21]

Orchestrating various elements of the plastic industry in support of the war effort
proved noncontroversial if not easy. Motivated after Pearl Harbor by a tide of patriotic outrage, companies large and small won *E* pennants for their contributions. But
SPI was not nearly so successful in controlling publicity devoted to plastic wonders
to be enjoyed at war's end. The outbreak of hostilities may have dampened world-
of-tomorrow utopianism, but remnants lingered like a nagging hangover. Manufacturers of consumer goods who had converted to military production found wartime
advertising limited to celebrations of patriotism that soon became repetitive. With
no goods for a home front weary of rationing, advertising shifted to images of attractive new products people could enjoy when the war ended. Such propaganda
boosted civilian morale but yielded a corrosive disillusion if carried on too long. The
theme of an early example, a General Electric ad from September 1942, exemplified
the genre. Under an image of a pigtailed seven-year-old All-American girl, the text
suggested that readers were "in the presence of royalty." Someday she would reign
over her own "kingdom," a house unlike any yet seen, "made of inexpensive materials that you haven't even heard of yet." Wearing plastic shoes instead of glass slippers, enjoying the labor of electric servants, and flying a plane "as readily as you
would drive a car," she would inhabit a "fairyland" made possible by "new materials
like plastics, new developments like television, new sciences like electronics." To
make sure it would all come true, "today's job is fighting for that better world."
Another advertisement portrayed a kitchen mixer to be molded of plastic and exclaimed, "Here's another design for modern living that's just waiting for America to
win the right to enjoy it."[22]

Envisioning a postwar world brought industrial designers into the picture. Although they worked on military projects, they also planned for reconversion to consumer goods—a business fad in 1943. The custom molder in the novel *Cash McCall*
devised a "Post-War Program" for expansion and welcomed his daughter's decision
to study industrial design because she "would be a real help to him when Suffolk
moved into consumer goods after the war."[23] Sometimes hired by material suppliers
or custom molders, sometimes by end-use manufacturers, designers created images
of postwar products for use in advertisements and other promotions. In this prototype work they enjoyed the luxury of ignoring practical considerations because none
of their designs could be realized. Even so, the prospect of having to deliver at the
end of the war reined in some of the extravagance that had characterized the world-
of-tomorrow era. The general tone projected by designers appeared in a *Modern
Plastics* advertisement from May 1942. When the "smoke of conflict" cleared, both
warriors and workers would "return to find quite a different world" where "old,
familiar things will have been made more beautiful, more useful, more durable and
far, far more desirable because of PLASTICS." Or, as the advertisement headlined its
theme, "When the Minute Man returns to his Plow . . . it will have *Plastic* handles!"[24]

Few designers were as active as Carl Sundberg and Montgomery Ferar in promot-

ing postwar plastic. They continued on retainer with Kurz-Kasch to provide renderings of postwar products for use in advertising. In February 1943 they offered a
"Preview of the Century *un*Limited," a drawing of a "swift, sleek streamliner" with
acrylic "engineer's turret" and "windows extending to the roof." Their design was
only a sketch, shown on a drawing board surrounded by compass, pencil, and rule.
Although the train was not yet "engineered or built," Kurz-Kasch presented it as "a
sound design by qualified designers." Among other advertising sketches by Sundberg & Ferar was a so-called "Supercar" with wraparound "full-vision plastic windows" and a teardrop body made possible by "new [but unspecified] moulding techniques." Apart from this association with Kurz-Kasch, the two designers called for
"post-war planning and designing" as early as August 1942 in a *Modern Plastics* article illustrated with renderings of a molded "Victory" refrigerator, a molded chair
with transparent acrylic legs, and a shark-nosed bus with a "plastic turret" for the
driver. As peace approached, they appeared to repent of their predictions. The May
1945 issue of *Modern Plastics* contained Sundberg's more sober assessment of "The
Realities of the Future." He admitted that prophets inspired by "rumors and reports"
filtering through a "screen of military secrecy" had misled the public with "promises
of every type of super-super product." Designers had offered superficial pictures
"sensational and good to look at." Although critical of himself along with the rest of
a guilty profession, he accompanied the article with lush color renderings of Flash
Gordon bubble-domed cars and boats to be made of glass-reinforced resins (Figure 6–3), and renderings of children's vehicles whose extravagant molded shapes
suggested the Big Wheel trikes of thirty years later. Prediction was too much fun to
resist even as the day of reckoning approached.[25]

6–3. CARL SUNDBERG AND MONTGOMERY FERAR, PROJECTED POSTWAR ALL-PLASTIC AUTO TRAILER/CAMPER/MOTOR-
BOAT SHELL, 1945

A series of sophisticated advertisements for Durez Plastics & Chemicals appeared in *Fortune* as well as *Modern Plastics*, seeking to convince manufacturers to specify plastic when converting from military to consumer products. Echoing the Bakelite series of the 1930s, Durez commissioned several designers to envision postwar developments. Renderings or photographs of models offered a convincing verisimilitude. Photographs of individual designers extended the cult of design. Prototypes included a vacuum cleaner by Dreyfuss, a cylindrical resin-bonded plywood refrigerator shell by Egmont Arens, a sewing machine by Muller-Munk, a sculptural speaker-phone by Deskey (cordless, push-button, and equipped with a microfilm directory), and a kitchen mixer by Brooks Stevens. Even a mock-up of a vacation cottage clad in decorative striated plywood (designed by Deskey) suggested "almost endless" possibilities—that "the navy's new mosquito boat fleets" would yield to "Durez-bonded plywood pleasure boats," that "army barracks" would "give way to vast economical housing projects," and that peacetime aviation would "become one of plywood's best customers" (presumably with a plastic plane in every garage).[26]

Most designers' portrayals of postwar plastic revealed a degree of realism in their concentration on specific products different only in material from those available before the war. But restraint disappeared as their ideas reached newspapers and magazines intended for a mass audience. Popular accounts of plastic plunged ahead with predictions equaling anything from the New York World's Fair. In August 1943, for example, a *New York Times* reporter enthused about life "In a Plastic World." Her article described the minds of leading scientists as "filled with the wonders that will emerge from war-bound laboratories for use by ordinary civilians when peace comes." Designers' drafting boards were "covered with plans for such amazing and unfamiliar things as plastic automobiles and transparent refrigerators and lights that can shine around a corner." Even gardening would become "futuristic" with hoses made of "pliant, indestructible plastic." In the following year *Scientific American* said of an approaching "Plastics Tomorrow" that no other innovation offered "such promise for rebuilding our war-torn industrial economy." Even "serious suggestions" indicated that "our houses, our automobiles, our ships, may soon be drawn cheap, complete, and finished from the wonderful grab bag of plastics." Although the magazine cautioned against regarding plastic as "a miraculous answer to universal needs," such second thoughts did not occur to editors of *Popular Mechanics*, who projected "a plastic postwar world" based on a faith that "man really began to mold the world to his own design when he invented the plastics."[27]

This wartime burst of utopianism triggered anxiety among those whose business it was to mold plastic's image. Earl Lougee, who had established the virtual identity of new materials and modern design while editing *Modern Plastics* in the mid-1930s, continued to promote their connection until his death in 1945. At a West Coast SPI meeting in the spring of 1944 he suggested that the "rather fantastic" predictions of the popular press would predispose people to appreciate plastics "for what they really are" once postwar production got under way. Several months later he amplified his thoughts in *Art & Industry*, a British journal widely circulated among American design advocates. According to Lougee the war created an "ideal situation for the advancement of design" because prewar inventories of consumer goods were used up, dies and molds for those products were melted down for scrap, and everything could be made anew. He advised designers and manufacturers to take advantage of wartime advances in plastics when planning new products. He also warned that designers held the power to "establish plastics in the mind of the consumer, either as desirable merchandise of good taste, or worthless trinkets for the immature." Lougee seemed in little doubt of a positive outcome as he presented blue-sky renderings of plastic products envisioned by American designers. Plastic was in good hands with industrial designers.[28]

But not everyone shared Lougee's faith that well-designed products would swing public opinion toward plastic no matter how extravagant its promotion. Cruse of SPI was so pessimistic that he began a campaign to neutralize plastic utopianism. In a statement to the *New York Times* in August 1943, within a week of the paper's enthusiastic portrayal of life "in a plastic world," he complained about the negative impact of "too many Sunday supplement features" portraying plastic as a "'miracle whip' material with which anything can be done." He revealed that an SPI postwar planning committee was concentrating on a "deglamorization" program to provide accurate information about plastic to manufacturers, engineers, and the general public so that postwar reconversion would find no one anticipating the impossible. At about the same time, Cruse informed SPI members of steps to deglamorize plastic, to clarify its actual potential, and to eliminate misapplications that left impressions of cheapness and shoddiness. SPI had hired a full-time public relations expert. It was also working with the National Retail Drygoods Association to prepare a simply worded handbook of information for retailers. Most important, the society was planning a national plastics exposition. As Cruse looked ahead to canceling of military contracts and return to a competitive civilian economy, he predicted a "promising future" only "if plastics are correctly applied, if misapplications are avoided, [and] if the industry is prevented from becoming a wayward glamour girl." It was necessary as *Modern Plastics* later maintained to offer an antidote to the vision "nebulously" held by the "average man" of "a Wellsian world manipulated by plastic buttons and transported in plastic airplanes."[29]

Deglamorization emerged as a form of damage control. As long as plastic had entered everyday life in a few applications like celluloid dresser sets or Bakelite radio cabinets, things known for quality, for toughness, sometimes for machine-age design, then utopian extrapolation seemed plausible. But once people began to experience flawed plastic goods, then utopian visions became untenable, even laughable. A certain democracy of frustration united everyone who had suffered wartime shortages. Tolerance for shoddy or nonexistent goods became lower with each passing year. A single encounter with a so-called miracle material that broke or melted away could lead a person to be "scornful of all plastics," as *Modern Plastics* warned in

1944. In that same year, custom molder Scribner, then president of SPI, complained about a no-win situation. On the one hand, millions of people thought "practically everything from automobiles to bathtubs" would be made from plastic after the war. On the other hand, thousands more who could not distinguish between "good and bad plastics" already thought of anything plastic as "no good."[30] For the moment there seemed no middle ground; it was either a utopian miracle-stuff or a third-rate substitute.

To some degree the plastic industry's contribution to the war effort was responsible for the situation it faced as peace approached. Mobilization accelerated such developments as extrusion and injection-molding of polyethylene, low-pressure molding of glass-reinforced polyester, and vacuum-forming of acrylic sheet. Material suppliers devised dozens of new formulations of older plastics like phenol formaldehyde and cellulose acetate. What molders accomplished with these new resins would have been inconceivable a decade earlier. But few civilians experienced tangible results of this innovative energy because "all the best plastics went to war" along with the traditional materials for which less reliable plastics substituted on the home front.[31] In fact some chemists and engineers worked hard to perfect *inferior* molding compounds for civilian goods using high proportions of cornstalk or wheat straw as fillers—with the bare minimum of resin needed to "Keep Em Molding," a slogan that yielded an acronym, KEM, by which the Makalot Corporation proudly referred to its own shoddy but strategically essential Bakelite substitute.[32]

By the time the troops came home, the people they left behind were tired of substitutes. They were fed up with shower heads of cellulose acetate that softened in hot water, with laminated products that separated when wet or stressed, with small moldings so devoid of resin they shattered when dropped.[33] They were beginning to think of plastic "as ersatz—something to be worried along with until more common materials are once more attainable." When the war ended they wanted "genuine" materials, not "artificial." A Fifth Avenue shop reported that its customers preferred "to wait for hard-to-get leather luggage rather than accept similar long-wearing, good-looking plastic styles."[34] To make matters worse, many plastic products sold at war's end came from poor-quality scrap rejected for military use and cheaply processed by fly-by-nighters who rushed in to capitalize on the industry's predicted expansion while possessing only rudimentary knowledge of manufacturing techniques and material limits. Consumers expressed disdain for tacky raincoats and shower curtains made from improperly plasticized vinyl sheeting (Figure 6–4). "They smell, they sweat, the print comes off and they get brittle," an editorial in *Modern Plastics* reported; enough rejected sheeting existed in storage "to curtain the whole world." Public disgust was so widespread that one fabricator even dropped the word plastic and referred to it as "flexible synthetic."[35]

New plastics were too new, too different from the celluloid and Bakelite of childhood, to be accepted without question as evidence of progress. Stuff that crazed, cracked, broke, discolored, or exuded a greasy residue argued against the industry's platitude that "there is no such thing as a 'good' plastic or a 'bad' plastic" because "it

6-4. POLYVINYL CHLORIDE RAINCOAT

depends entirely on the use." As Cruse and Scribner recognized, the war had proven to be a mixed blessing. Mobilization brought new resins, new processing techniques, expansion of plants, and an ability to engineer materials "tailor-made" for demanding specifications. It also spread thousands of defective products distinguished from earlier novelties only by the fact that there promised to be so many more of them. The raw materials of Synthetica may have helped win the war, but machine-age abstractions had lost their attraction. Two years after the war ended, Charles Breskin applauded the fact that "this 'Plastics World' idea is gradually being abandoned." The notion that "the chemist dumps coal into a hopper and, by some miraculous transmutation, extracts plastics at the other end of the machine" was exposed as a wild exaggeration.[36] As even plastics inventor Donald Duck discovered when his airplane melted away, things—especially plastic things—were never that simple. New materials had to prove their desirability to the postwar generation.

DAMP-CLOTH UTOPIANISM

Fears about growing pains, whether realized as a tacky shower curtain or an army veteran molding polystyrene in his garage, receded briefly in April 1946 when SPI put on the first National Plastics Exposition in New York. "Nothing can stop plastics," declared president Ronald Kinnear as he opened the show by cutting through a Vinylite film sealing the entrance. Impatient visitors jammed into the hall and made their way past some two hundred displays, many of World's Fair quality. Attendance the first day totaled twenty thousand and would have been higher had fire regulations not forced repeated closing of the doors. Three days later, when the general public was first admitted, people lined up four abreast for several blocks. Altogether eighty-seven thousand went through the exposition during a six-day run—a total claimed by SPI as larger than "any other industrial show has ever attracted." Exhibitors reported tremendous enthusiasm. "The public are hungry for something new and evidently plastics are the answer," ran one comment; according to another, "the public are certainly steamed up on plastics." So were businessmen. "Our exhibit is paying off many times over," one man claimed; another pronounced the exposition "the first show where we really got our money's worth" because customers were "showing a terrific interest . . . the only question is when can you deliver."[37]

Postwar euphoria stimulated some of the enthusiasm but could not claim all the credit. The show symbolized, as its organizers claimed, "a great industry that has come of age," no longer dependent on "unimportant if interesting gadgets" but instead offering "products, useful and beautiful, that have never before existed." They hoped displays of materials, machines, methods, and products would give manufacturers new solutions to old problems and suggest new possibilities. A public that had heard "many glamorous tales about plastics" would receive "a wealth of information . . . no less glamorous because it is factual."[38] The exposition did indeed suggest the coming proliferation of plastic. Using an encyclopedic "from/to" construction running all the way back to Hyatt, a journalist reported "hundreds of applications ranging from lightweight luggage and translucent walls to textile spools and switchboard panels." Surveying the field, *Modern Plastics* found one-piece boats, fishing rods, skis, baby strollers, and luggage, all made from glass-reinforced polyester. Polyvinyl chloride appeared in everything "from transparent hat boxes to saddles and drapery material." Polystyrene could be found in toy trains, battery cases, wall tile, and push buttons for radios. Even nylon, the miracle material par excellence, appeared in an unfamiliar guise as a tough new molding compound.[39] Many of these products were prototypes. Even so, gathered together in overwhelming diversity, they suggested the outlines of the plastic world of the next half century. The people who crammed into the first National Plastics Exposition experienced a preview of things they would soon be able to enjoy after two decades of deprivation.

Aside from a burst of publicity, the expo had little effect on a public that knew plastic only in everyday life where it often performed poorly. Almost everyone concerned about plastic's image problem in the late 1940s concluded that it stemmed from misinformation or lack of communication. There were few problems with older plastics. Consumers knew all about them. Celluloid was a tough substance for traditional applications. Bakelite seemed indestructible after forty years of experience. New materials like polystyrene, polyethylene, and polyvinyl chloride suffered by comparison. As thermoplastics they lacked Bakelite's hardness and were sensitive in varying degrees to heat, chemicals, water, and sunlight. Even as new plastics reached the marketplace in finished goods, material suppliers and processors continued to experiment with plasticizers, stabilizers, catalysts, retardants, pigments, and other chemical ingredients, and to tinker with molding techniques.[40] Not only were these new materials untested and sometimes defective; they also went out to people who had no experience with them, no sense of their limitations, and who had, to the extent that Bakelite served as an archetype, a faulty concept of plastic as a substance so tough, durable, and chemically inert as to be practically indestructible. As *Modern Plastics* complained, few people knew "that cellulose nitrate and urea buttons are as different as iron and aluminum; that cellulose acetate and polystyrene are as unlike as oak and pine; that Saran and nylon are as far apart as cotton and wool."[41] This ignorance extended from manufacturers who assembled products from molded parts to wholesale buyers, retailers, and the general public. Producers, distributors, and consumers at all levels lacked accurate information about strengths and weaknesses of new plastics. A sense of crisis dominated the second National Plastics Exposition, held at Chicago in May 1947 under a banner dedicating the industry to "The Right Plastic for the Right Purpose." The key to success was building faith in trustworthy materials.[42]

Although the issue of ignorance and information achieved crisis status in 1947, it was not a new problem. Toward the end of 1945 marketing consultant Alfred Auerbach conducted an informal survey revealing complete ignorance about plastic among New York salesclerks. While purchasing housewares and toys ranging in price from ten cents to $3.50, his investigators garnered an abundance of misinformation. A melamine bowl (similar to those used on navy ships) could not be boiled without melting, for example, and a Lucite jewelry box could be cleaned with steel wool. Behind the scenes Auerbach discovered that professional buyers knew little more than the clerks. While some considered plastics "the biggest thing after the war," others dismissed them as "ersatz products . . . on the way out" and insisted no one would buy them "if he can get anything else." Even those who professed neutrality complained that "there are so many kinds of plastics," or that "everyone is so uninformed about plastics."[43]

For several years SPI addressed the crisis of confidence by means of "informative labeling," a term emerging from a talk delivered by Scribner in May 1944. Warning that many people "condemned all plastics because of an unfortunate experience with [a single] one," he noted that misapplications of plastic occurred when manufacturers or consumers mistook one type for another. Scribner suggested a simple labeling system to acquaint consumers with differences among plastics and to provide minimal information about use and care of a specific product. His system elevated the word "plastic" to a kind of super-generic. Simplified chemical names such as "acrylic" or "styrene" became standard generics that all manufacturers could use.

Individual trade names added little or no information about the material itself. A typical label would have described a product as made of "Bakelite phenolic plastic" or "Du Pont Lucite acrylic plastic."[44]

Scribner's comments were well received. The SPI board of directors took up the subject at its next meeting in October 1944; within a year an Informative Labeling Committee had prepared a small pamphlet for molders and fabricators. As a committee member declared in 1947, however, informative labeling was "not only a hot potato, but a tough nut."[45] For one thing, most suppliers of basic materials did not want them known by generic names any company could use. More to the point, not many molders adopted the committee's recommendations because they thought of themselves not as makers of consumer goods but as suppliers of parts to end-use manufacturers. To make matters worse, SPI had no way to impose a labeling policy on manufacturers or retailers. From 1947 to 1951 the society debated proposals for an SPI mark of approval to be awarded to products whose labels provided adequate information or, more radically, whose use of plastic conformed to an objective code of standards. Such ideas found little favor among realists who wanted to avoid policing an entire industry or endorsing individual products. Although SPI's deliberations on labeling yielded no policy, the issues reached most people who made, molded, or distributed plastic. Even manufacturers of novelties realized it was in their best interest to avoid fiascos like drink coasters that melted and charred the furniture when people used them as ash trays.[46] A cheap salt-and-pepper set distributed by Carvanite Products of Los Angeles included a printed label stating that these "modern new seasoners" were "made of a Thermo Plastic Carvanite product" that would provide "perfect satisfaction" if properly used. The label ended with a warning—"Do Not, however, wash them in boiling water or subject them to an undue amount of heat."[47] A member of the SPI committee claimed in 1950 that such ad hoc efforts had worked. Just as the industry had outlived "the period when plastics were generally supposed to be miracle materials that were better than everything," so had it outlived the "period of 'mis-application' when plastics were damned up and down as shoddy substitutes." At last the public considered plastic "dependable . . . when properly applied to products that are well designed and manufactured."[48] Not all of Scribner's program was realized, however. Except in a few isolated cases, labeling did not make other types of plastic "as common a household word as 'nylon.' "[49] People still thought of plastic as plastic—for whatever that was worth.

In addition to promoting informative labeling, SPI intervened on specific occasions to bolster plastic's reputation. During the Christmas season of 1948, for example, hundreds of consumers returned defective tree lights whose bulbs were hooded with thermoplastic birds, Santas, and bells that had softened, become deformed, or charred. Here was a case that threatened to contaminate all plastics with the epithet "shoddy." By cooperating with Underwriters' Laboratories in discovering the causes of the failures, SPI strengthened its own reputation for objectivity. The society then worked directly with material suppliers and molders to solve the problems and assisted the Noma Electric Company with a presentation intended to reassure a national convention of fire marshals.[50]

If pushed to it, SPI could come out swinging, especially when rival industries threatened plastic's reputation. That happened in 1951, when the Vitrified China Association issued a report "purporting," as *Modern Plastics* put it, "to tell about the horrible things that will happen if we eat from melamine dishes." The report portrayed melamine dishes as so soft that ordinary scratches could harbor bacteria, so water repellant that they could not be properly cleaned, and so chemically unstable that they released formaldehyde (or "embalming fluid") when subjected to hot water. "Soon no doubt," *Modern Plastics* sarcastically observed, "the citizens of this country will have an epidemic of plasticitis and the undertaker won't even be necessary because they will already be embalmed."[51] At issue in the Vitrified China attack were large contracts for melamine dishes awarded by the military services as they expanded to fight the Korean War. SPI responded with its own printed report, of which seventeen thousand copies went out to military procurement officers and to managers of institutional food services. Although the issue died, it was not the last time SPI had to fend off an attack by a rival trade association.[52]

For the most part SPI exerted influence through more gradual means. In 1947 the society routinely handled a thousand requests for information each month, most from businessmen but some from the general public. Far more important than informative labeling in reaching the public was SPI's involvement with mass-circulation periodicals. The society responded quickly to derogatory or even mildly harmful references. When *Life* published a breezy "modern miracles" item, for example, a member of the Public Relations Committee took time from his busy schedule as an executive at American Cyanamid to visit the magazine and offer assistance with future articles.[53] Not content merely to react, the society also actively collaborated with magazine editors to present plastic to the general public, and in particular to young women shaping postwar households. In such articles plastic came across as the perfect substance for molding a world modern and up to date but also comfortable and conservative—certainly not a streamlined version of "things to come." SPI's public relations experts registered a grassroots distrust of artificial materials and responded by engineering a postwar image of plastic domesticated.

As plastic utopians had predicted, synthetic materials did invade the American home after the war, but in the guise of traditional materials. Vinyl floor tiles and upholstery masqueraded as ceramic and leather. Wood-grained Formica protected table tops, desks, and wall surfaces. Rayon draperies and lamp shades of cellulose acetate appeared strange only on close examination. In the words of Mary Roche, a *New York Times* interior design columnist, the "house of plastics was one of those outlandish things we heard a good bit about a few years ago but never expected to see." But now "you might look right at it without even recognizing it" because designers were "blending synthetic products with other materials that do not come out

6–5. ALL-PLASTIC POSTWAR LIVING ROOM IN TRADITIONAL STYLE, 1949

of a test tube." Traditional appearance at less cost and with less maintenance became primary concerns. For those threatened by artificially precise machine-age surfaces reminiscent of the assembly line or the chemical plant, who sought a reassuring domestic realm as escape from an increasingly uncertain modernity, it was refreshing to learn that you could indeed "put plastics together in a comfortable, normal house." To be modern one need not live in an inhuman "facsimile of a Statler cocktail bar" (Figure 6–5).[54]

SPI, collaborating with the staff of *House Beautiful*, stood at the center of an effort to convince American women of exactly that. Several editors of the magazine had visited the second National Plastics Expo in 1947. Impressed by the displays and convinced of plastic's centrality to postwar life, they decided to dedicate an entire issue to plastic and sought SPI's assistance. Public Relations Committee chair John Sasso, who had written on plastic product design, agreed to serve as consultant for the special issue, made available SPI's mailing list, guided *House Beautiful*'s requests for information, coerced assistance from companies, suggested topics, and revised copy for accuracy. The special issue was published in October 1947 with fifty pages devoted to "Plastics . . . A Way to a Better More Carefree Life." In effect it served as a well-illustrated consumer handbook complete with information about specific materials and suppliers. An introduction promised to reveal "The Truth About Plastics." Neither "miracles" nor "junk," synthetic materials would "improve your life a thousandfold IF you know what to expect of them." To get the most from them, people had to "forget the dream world stuff you've heard about plastics and learn what they really are and what they can really do for you." The crux of the matter came in an article that defined "This New Era of Easy Upkeep" by illustrating three homemaker's nightmares—a child scribbling with lipstick on a wall, a lamp shade covered with dust, ink being spilled on an upholstered chair—and their happy resolutions. Although it was "all very well for the chemists to talk of their molecules and polymers, and for the engineers to carry on about extrusions and injections," there was "only one good reason why you, personally, should be interested in plastics." That reason was "*damp-cloth cleaning*"—a concept "about which you'll read constantly in the pages that follow," readers were told. And they did, ad nauseam. Despite articles suggesting that plastic did not always have to look like traditional materials, or that people had to use the right plastics in the right situations, the big draw was damp-cloth cleaning. An all-plastic kitchen featured "plastics you can wipe clean with a stroke, plastics you can spill almost anything on." To a housewife who aspired to keep her kitchen as spotless as those in the magazines, damp-cloth cleaning ushered in "practically the millennium."[55]

A former editor of *Modern Plastics* was not exaggerating when he recalled how "we used to write stories for many years—blah, blah, blah, and you can wipe it clean with a damp cloth—all the stories ended like that." Everyone who discussed new postwar materials celebrated damp-cloth cleaning. Sometimes soap entered the picture, as when *Life* featured a room whose walls, floor, and contents could be renewed "with swipe of soapy cloth." More often plain water sufficed. According to *Better Homes and Gardens*, neither dogs, kids, nor even one's husband could damage a house furnished with plastic because "you can clean [it] with a damp cloth." The same magazine later attested that "cleaning can be as easy as the swish of a damp cloth." Even *Fortune* praised Formica because "maintenance consists of a casual swipe with a damp cloth." This spate of swiping and swishing by which plastic promised to transform everyday life contrasted remarkably with the visionary plastic utopianism of earlier decades. As plastic made its way further into the domestic landscape, in laminated walls and dinette tops, in vinyl floors and moldings, in polyethylene waste baskets and polystyrene wall clocks, expectations became more mundane (Plate 11). The period might half-seriously be referred to as an era of damp-cloth utopianism.[56]

Over the next few years SPI continued to help journalists. In 1953 the society collaborated with *McCall's* in spreading the gospel to American women. Six months in preparation, the promotion focused on an "all-plastic" living room and patio exhibited by *McCall's* at the National Home Furnishings Show in New York. Photographs in the magazine revealed typically warm, bright "American modern" interiors. Captions identified plastic in the patio's corrugated fiberglas roof, the fabric webbing of porch furniture, a garden hose, laminated table tops, polystyrene wall tiles, lamp shades, a folding door, toys, frames of sunglasses, a pair of knitting needles, even metallicized threads in pillow covers—all readily available to consumers. Plastics were "everything a woman could ask for . . . glamorous, practical and luxurious." It was time to "jump into an exciting new never-never land . . . where fabrics can be washed and don't need ironing, where tables can't be marred, where spots on rugs and upholstery are swished away with a damp cloth . . . and where moths can't get a mouthful to eat." After this string of damp-cloth encomiums, the text addressed the issue of the right plastic for the right use with a set of performance questions to be asked of each plastic product before purchase.[57] *McCall's* also arranged newspaper stories in New York and Chicago, coordinated plastic exhibits in sixteen department stores, distributed promotional kits to other department stores around the country, and mailed out eight thousand reprints of the article.

The magazine also distributed an eleven-page booklet to anyone requesting further information—and to twelve thousand home economics teachers. If the brief article in *McCall's* served as a teaser, then *Plastics: Everything a Woman Could Ask For* delivered the goods. The goal was to make consumers knowledgeable about synthetic materials and capable of distinguishing among them so they would be less likely to blame all new materials for failures generated by ignorance. "To ask one plastic to behave like another," the text insisted, "is to ask wood to be like iron, or chiffon to turn into tweed." Soon, with a bit of training and experience, "feminine fingers, schooled in the touch of silk and wool and crystal, will tap and tell it's polystyrene." Or "if it's squeezy and opalescent, as in a spray bottle, it's likely to be polyethylene." Beyond intuition into the realm of intellect, "feminine tongues, which don't trip on words like carbohydrate or hydromatic, soon will be just as casual with all the basic and trade terms given to plastics." Even chemical "tongue-twisters," not to mention trade names like Boltaflex, Styron, and Velon, would soon be "part of your daily vocabulary." As the text emphasized, "the greatest asset of plastics is their difference, so you can select the right one for the job you want done." Following this condescending introduction, the booklet provided objective summaries of nine major types, including characteristics, major uses, trade names, and suppliers. Only the last page reverted to damp-cloth utopianism, ironically implying that even that dreamworld was illusory. "Don't get too ambitious about scouring, sterilizing, ironing, disinfecting," the text warned, because "the smooth, non-porous surfaces of plastics are easy to keep clean with just soap and water, or even a damp cloth . . . so just swoosh and smile." Some people who scrubbed too hard had already discovered just how destructible some plastics really were.[58]

As always, SPI's goal was protecting plastic's reputation from the greed or ignorance of a few processors or retailers. To a certain extent the society remained locked in the past in trying to influence public opinion. Worrying about the image of plastic made sense in 1945 when unfamiliar new materials confronted wary consumers. By the mid-1950s, however, no one was ignorant of plastic because it surrounded everyone. Sidney Gross, who joined *Modern Plastics* in 1952 and became editor in 1968, recalled that he had "agitated a lot" over the years to get SPI to quit trying to convince people "that plastic is not bad." It was a waste of money because plastic's image—good or bad—did not really matter. The key to plastic's success, as he saw it, was always "selling the manufacturer." Once plastic products filled the stores, people had no choice but to consume what they were offered. Most of the time, Gross maintained, after the industry had solved postwar quality problems, plastic objects did work better. Things made of plastic were better designed and lasted longer. People intuitively recognized that fact even if they retained an intellectual notion that plastic was bad or shoddy. In short, nothing succeeded like success.[59]

Often plastic did offer a significant improvement on whatever it replaced. A sleepy householder had to watch only once in disbelief as a polyethylene juice pitcher bounced off the kitchen floor to begin accepting plastic in a practical way no matter how strong the conceptual disdain for it. A Formica counter top was indeed easier to clean than a wooden counter that absorbed stains when its surface became worn. Even plastic toys, despite the brittle polystyrene items that broke on Christmas morning, proved superior in many ways. A toy soldier of molded polyethylene could not scratch the furniture or gouge a sibling as readily as an old-fashioned lead soldier. Most people who expressed negative attitudes about plastic used it anyway without thinking about it, either because a particular use had proven itself or because an inexpensive trouble-free alternative no longer existed. As *House Beautiful* observed in 1955, "The news is not that plastics exist, but [that] they have already been so assimilated into our lives." The average person was "conditioned to plastics."[60] They had penetrated so far into the material fabric of everyday life that their presence could not be denied no matter how many people considered them second-rate substitutes or a sad commentary on modern times.

By the late 1950s people muttered such complaints only privately, when a piece of plastic failed or the dime store at the new shopping center no longer carried a favorite item of wood or metal. Like the sprawling new developments of tract housing, plastic was expanding. The phenomenal growth predicted by promoters like Haynes during the grim early months of the Second World War actually materialized. Owing to the war itself, which forced maturity on a new industry, and to the postwar demand for consumer goods and construction materials for veterans and their young families, plastic was enjoying a success that no complaints about shoddiness could overcome. Annual resin production reached 3.6 billion pounds in 1955, more than four times the total for 1945.[61] Ever more visible in the material stuff of the boom years of the American century—in Corvette autos and fiberglas boats, in vinyl siding and extruded window frames, in casings of kitchen gadgets, in toys and sports equipment, in nylon carpeting and polyester clothes, and in the Naugahyde and Formica that gave so many things their surface textures—plastic became an integral element of an expansive era, one of its most revealing elements.

Late in 1959, SPI again collaborated on a major article with a women's magazine, this time *Good Housekeeping*. While damp-cloth cleaning remained a talking point, promoters had given up teaching ordinary consumers to distinguish among a plethora of types. No longer were people supposed to recognize different members of a "family" *McCall's* had condescendingly introduced as "Polly and Vinny Who?" Such details were beyond the capacity of most consumers. It was enough to remind them of the material itself, conceptually a single substance despite a diversity of manifestations. Plastic enabled ordinary people to enjoy a host of products either too costly or simply inconceivable in traditional materials. No one recalled Slosson's prediction of thirty years earlier, that synthetic chemistry would democratize the stuff of everyday life, but it was coming true. By the time *Good Housekeeping's* article appeared early in 1960, almost eight years before the opening scene of *The Graduate* turned the word into an expression of contempt for postwar America's materialism, it had become possible—and it was not yet a sinister absurdity—to celebrate "A New Way of Life in One Word: Plastics."[62]

A NEW ROCOCO

A few days before the nuclear explosions over Hiroshima and Nagasaki ended the Second World War, a gathering of marketing experts listened to a plea for creation and release of a far different explosive force. J. W. McCoy, a Du Pont vice president, addressed a workshop on peacetime conversion. Business would be good for some time owing to "a great backlog of unfilled wants." Satisfying desires for cars, washing machines, radios, and other consumer goods would create "an upward spiral of productivity, raising the standard of living, increasing the national income, making more jobs." But McCoy feared that "a satisfied people is a stagnant people." After satisfying immediate desires, merchandisers would have "to see to it that Americans are never satisfied."[63] This goal fueled an explosion as capable as the atom bomb of transforming the culture's conceptual parameters. An ever-expanding proliferation of consumer goods created an inflationary culture that invested ever more of its psychological well-being in acquiring material things but paradoxically considered those things of such low value as to encourage their displacement, their disposal, their quick and total consumption. Plastic became the material of choice for this never-ending expansion. The material's virtues were limitless. It was inexpensive because it was derived from an endless supply of petroleum. It was less solid or intractable than wood or steel. It was free of traditional preconceptions regarding its use and could be molded into any shape a restless drive for novelty might conceive. It was, finally, so lightweight and in some forms so insubstantial as to be discarded without a second thought. Plastic not only offered a perfect medium for this material proliferation. It conceptually embodied it and stimulated it.[64]

Several decades later, when social and cultural analysts looked back at the postwar period from a time of shortages, energy crises, and perceived limits to growth, they often cited plastic as crucial to the economy of an era that had ended. Recognition of plastic's significance came not only with hindsight, however. During the late 1940s, before the inflation of things really got under way, promoters began using language that implied a radical discontinuity in the fabric of everyday life:

"Doubling—Tripling—Expanding: That's Plastics." So ran the title of an article in a 1947 special issue of Monsanto's company magazine. Some observers conceived of plastic as a multiplier or extender of familiar benefits of mass production. "The Horn of Plenty Is Mechanized," announced the title of another Monsanto article. Referring explicitly to a high-capacity injection molding machine that "spews out the [molded] articles faster than we can tell you about it," the title also suggested plastic's wider social role. Along the same lines, a General Electric booklet celebrated plastic molding in 1946 as "publishing in three dimensions" and compared it "to printing from movable type—the invention that did more than any other to bring about modern civilization."[65]

These mechanical analogies paled next to biological conceits suggesting plastic was taking on a life of its own. Another Monsanto article, for example, focused on "Something Nature Could Not Supply." While the theme extended back to Baekeland, the rhetoric was subtly transformed as the author reported the "staggering" fact that with plastic "the laboratory was actually creating *new material*." This image hinted at the spontaneous generation of matter, a continuous fermentation of stuff that had to be processed and used to prevent its filling the world. Another executive explained in 1946 that the new field seemed "like something alive" because "no one can tell from one moment to another what is being made from these materials." Developing his organic conceit, he declared that "plastics constitute a sort of beneficent growth within the body of industry." From benign growth to malignant tumor was not so far. About fifteen years later Norman Mailer, then emerging as a social critic, first began railing about a central obsession—a vision of plastic spreading through America "like the metastases of cancer cells." Whether inclined to celebrate or to deplore, most people who generalized about plastic thought of it as transforming the material basis of everyday life.[66]

During the 1950s, dominance in the plastic industry shifted from thermosets to such thermoplastics as polyethylene that contributed to a flood of new uses—garbage pails, squeeze bottles, hula hoops—lighter, more flexible, less permanent than objects made from thermosets. The chemical feedstocks of the industry also changed. After a serious shortage of benzene, which was obtained as a by-product of the coking of coal, material suppliers turned to petroleum and natural gas for raw materials.[67] By a happy coincidence, the shift from coal to petroleum, from a dense solid to a viscous liquid, symbolized the shift from eternally permanent thermosets to infinitely shape-shifting thermoplastics. An affluent postwar America described by historian David M. Potter as the "people of plenty" had no trouble consuming an ever-increasing quantity of plastic that materially and visually contributed to a culture that was shapeless, ever-changing, impermanent, even ephemeral.[68]

The plastic cornucopia evoked by publicists often seemed more of a phantasmagoria. Although women's magazines reassured readers by explaining how plastic could emulate traditional materials, most consumers did not really need more than a comforting nod to the past. And just because the world-of-tomorrow aesthetic was

177

6–6. Toy polystyrene replicas of Russel Wright's American Modern dinnerware, 1950s

obscurely threatening—its cool hues and sweeping curves stimulating little emotional involvement—did not mean people would reject a more easygoing modernism. The new suburban middle class in fact embraced a warm, colorful, playful modernism because it promised ease and comfort, a trouble-free existence, the very virtues embodied in the plastic promoters' damp-cloth utopianism. No one recognized this more than designer Russel Wright, who with his wife, Mary, celebrated the new informality in *A Guide to Easier Living* in 1951. The soft rounded biomorphic forms and warm muted colors of Wright's popular American Modern dinnerware marked the emergence of a new postwar aesthetic later evocatively referred to as "populuxe." Although Wright's dishes were ceramic rather than plastic, it was the latter's malleability that inspired the look—too eclectic, too shifting, too commercially insistent on superficial novelty to be called a style—and that enabled its full expression into the 1950s and beyond (Figure 6–6).[69]

By 1957, when *Life* opened a series of articles on technology by exploring its impact on everyday life, that process of expression was well under way. The magazine posited a "new world" in which "man makes a multitude of new things and finds a multitude of new ways to make them" using "materials whose nature he could not even have guessed at a few decades ago." *Life* admitted uneasily that "even the familiar items have undergone such outlandish transformations that he [i.e., "man"] can hardly recognize them." Much earlier, during the mid-1940s, some commentators had expressed similar doubts while contemplating the uncertain outlines of the postwar world. Lougee, for example, warned that if a product were badly designed, then "the speed with which plastics molding will turn it out will flood our lives with ugly commodities." And John Gloag, a British design critic, feared plastic would enable overly imaginative designers to transgress far beyond simple imitation. Lamenting that it was so "easy to lose self-control in the matter of imagination when these synthetic materials are discussed," Gloag described fascination with plastic as "not so much a state of mind as a state of intoxication" that could "lead to the widespread misuse of plastics." Most appalling was an attitude he summarized in the statement, "These materials can do anything—let's do everything." He predicted that consumers starved first by the asceticism of the modern movement and then by the austerity of wartime would embrace "an orgy of ornament." Because plastic so easily took on any form or color, without being limited by constraints of traditional materials or craft techniques, it seemed likely postwar designers would "create a new rococo period" marked by extravagance, excess, vulgarity.[70]

By everyone's account that is exactly what happened. The postwar explosion of amoeboid, boomerang, and saucer shapes, all easily molded of plastic, certainly justified Gloag's prediction. Two generations of design critics addressed the issue. Whether they criticized Americans for splurging on "borax and chrome" or became self-consciously involved in "learning from Las Vegas," they confronted a vigorous popular culture that owed much of its exuberant surrealism to plastic. Echoing Gloag's prediction, a journalist later observed that postwar Americans "went on a baroque bender" by embracing products "in a lurid rainbow of colors and a steadily changing array of styles." As this proliferation went on, "commonplace objects took extraordinary form, and the novel and exotic quickly turned commonplace" (Plate 12). An architectural historian described "a spirit of originality uncontrolled by discipline." Referring specifically to the design of coffee shops, which relied on back-lit acrylic signs, on large fiberglas moldings, on Formica and vinyl, he characterized it as an "architecture of superabundance," of "extravagant gestures . . . lack of inhibition . . . a sense of exhilaration."[71]

Another historian proposed similar themes but perhaps erred when he suggested that the aesthetic of the postwar era rested on an assumption "that the world could be brought to perfection and all experience controlled through modern design."[72] The utopian impulse of the 1930s had indeed striven for static perfection, but that of the 1950s abandoned perfection in pursuit of an ever-improving flux. Even so, as new products poured from chemical plants and molding machines, they did convey to consumers a comforting sense of control. The very extravagance of artificial forms depended on synthetic materials that yielded easily to human manipulation. The more outlandish an object's form became, the more artificial and thus totally controlled it seemed—more so than ever before in history. But the process itself was another matter entirely. Although it too seemed under control, that perception proved an illusion. Without anyone realizing it, Thomas Hine has observed, "the very nature of *things* had changed" as people adjusted to "a disposable world."[73] In so doing, they lost control of the process in two crucial but contradictory ways. Neither became clear until much later. In the first place, as material things became attenuated, images or simulations of things assumed a cultural significance nearly equal to that of things themselves. At the same time, however, the flood of attenuated things swelled to such proportions that no matter how disposable each individual thing might be, taken together they threatened to swamp landfills and discharge chemical waste into the air and water on which life depended. Plastic offered an unprecedented degree of control over individual things—allowing extravagance

179

of form to coexist with precisely engineered function. But plastic also accelerated larger processes that society recognized as out of control only long after it had become dependent on the comfort and convenience of plastic.

No one predicted any of that during the plastic industry's expansion after the Second World War. Those who entered the business only knew they had found a good thing, a rising trend they hoped would carry them with it. As a founder of Union Products in Leominster recalled forty years later, it was obvious in 1946 that "virtually nothing was made of plastic and *anything* could be"—including the pink flamingos for which his company became best known.[74] That versatility explained why plastic became indispensable during the postwar era. Writing in 1947, product designer Donald A. Wallance praised "the magic of industrial chemistry" for the ease with which it could create either "a polyethylene bowl whose delicate, translucent walls yield to the slightest finger pressure" or a "tough, laminated plastic armor plate which stops a 45-caliber bullet." Among all the materials available for human manipulation, only plastic offered "such an extraordinary range of possibilities."[75] Polyethylene bowls were more common than laminated body armor, however. Used most visibly for domestic objects, plastic itself became domesticated. Wallance made his remarks on the occasion of an exhibition of plastic consumer products at the Walker Art Center in Minneapolis. Ranging from crystalline acrylic salad sets to Russel Wright's first Melmac dishes, from a vinyl garden hose to a baby stroller molded in one form-fitting piece of phenolic reinforced with rope fiber, the objects on display came mostly from the domestic side of life. Most characteristic of all was a set of bowls marketed as Tupperware, a line of products that served to domesticate plastic more than anything else since nylon (Figure 6–7).

Injection-molded in light, delicate colors from translucent polyethylene, Tupperware was sold from about 1950 onward only in private homes at parties given for friends and acquaintances by hostesses who hoped to earn bonus prizes. While Tupperware was one of the few inexpensive mass-marketed products to gain admittance to the permanent design collection of the Museum of Modern Art, the Tupperware party entered middlebrow folklore—and highbrow scholarship—as one of the most common symbols of the time. A former editor of *Modern Plastics* once said that Tupperware "gave plastics a very good name because it was terrific stuff."[76] It was lightweight, tough, flexible, unbreakable, and equipped by its inventor Earl Tupper with a patented air-tight seal that made it perfect for covered refrigerator dishes, though the line quickly expanded to include everything from mixing bowls to cocktail shakers and ice-cube trays. Versatile and nearly perfect unless accidentally set down on a hot stove burner, Tupperware had by 1949 "taken the country by storm," or so *Modern Plastics* claimed.[77] In a number of ways Tupperware epitomized plastic's development during the 1950s. For one thing, Tupperware was no mere imitation. It substituted for older materials but also offered qualities previously unattainable. Before Tupperware, people made do with storage dishes and water jugs of

6–7. Tupperware as presented by
the Museum of Modern Art

ceramic and glass. Heavy and breakable, they sweated condensation and became
slippery; the lids fell off and broke. Lightweight, unbreakable, and waxy-textured
for a good grip, Tupperware solved all those problems while also introducing the
crucial innovation of the air-tight seal. That it looked different, frankly new, and
unprecedented simply reinforced its superiority to anything that preceded it. Some
people may have worried about a slight chemical odor, faintly suggestive of ozone,
given off when an empty Tupperware container was opened, but others no doubt
considered that as proof of the material's status as a product of the most up-to-date
science. Both thermoplastic and a petroleum derivative, Tupperware was one of
those attenuated products that verged on the disposable. Even its method of selling,
which made it less a product than a vehicle of sociability, suggested impermanence.
It proved so functional, however, that people saved their Tupperware and used it for
years, as they later did with less sturdy margarine tubs actually designed to be
discarded.

Tupperware's popularity signaled overall acceptance of plastic. Its domestic convenience reflected aspirations for a casual life of leisure identical to those in the
promotions of the damp-cloth utopians. At the same time, however, Tupperware's
acknowledgment of plastic as plastic, as a material with its own unique appearance
and texture, indicated the beginning of a trend that continued through the 1950s
and into the 1960s and beyond. Forms taken by plastic increasingly expressed
American society's fluidity and mobility, its acceptance of change for its own sake,
its desire for impermanence, its urge finally to control all of life by transforming it
into the whimsical or fantastic play of entertainment. Even plastic's imitative potential, more convincing than ever before, often seemed less an opposing movement

181

6-8. Polyethylene juice
containers, 1963

than an especially outrageous instance of molding things to an arbitrary measure just because it could be done (Figure 6–8). During the last half of the twentieth century, as plastic became an ever more visible presence in the material world, its cultural significance was contained in the words of a skeptical John Gloag, who suggested that plastic should be understood as what happens "when the artificial becomes the real."[78]

DESIGN IN PLASTIC:

FROM DURABLE TO DISPOSABLE

*T*o dwell on plastic's malleability is to forget that it possesses little cultural value until it takes definite shape. Plastic's very presence can seem overwhelming. A visit to a molding plant reveals storage bins overflowing with millions of colored resin flakes. To watch them cascading into the hopper of an automatic molding machine recalls the flood-of-goods metaphor of twentieth-century discussions of production and consumption. But not until plastic emerges from a specific mold does it possess a definite form enabling it to satisfy functional demands, to convey cultural meanings, to interact with its environment in ways both intended and unforeseen. Each plastic object emerges from a design process involving social needs and desires, economic constraints, technical limitations, and marketing strategies, all coordinated with some degree of awareness. Any attempt to isolate plastic's impact on that complex process must remain imprecise and suggestive. Not even a designer or engineer responsible for a product's final form can distinguish its technically necessary elements from those that satisfy emotional or expressive needs. The problem is all the more difficult with plastic, which lacks intrinsic form or texture. A plastic imitation of wood or stone is no more arbitrary than an object whose unique form projects a high-tech aesthetic of artificiality. Each conveys multiple meanings, both tangible and intangible, even if purporting only to be useful.

Mid-twentieth-century design critics tended to judge plastic by strict either-or categories. On the one hand, "regressive" manufacturers imitated traditional materials with ever more convincing results, hoping to gain by reducing processing costs without alienating naive consumers who demanded clearly "domestic" surround-

7–1. Traditional wood interior in Uniroyal cast polystyrene, designed by Ving Smith, 1970

ings. On the other hand, "enlightened" consumers preferred artificial objects whose plastic forms suggested the outlines of a future technological environment. A split earlier defined by celluloid collars and Bakelite radios survived decades later as an almost schizoid aesthetic in two photographs provocatively situated on facing pages in an article on plastic design published by *Progressive Architecture* in 1970. On the side of tradition (Figure 7–1) appeared a wonderfully detailed room used by Uniroyal to demonstrate the beauty of its cast polystyrene by means of apparently hand-hewn ceiling beams, worm-eaten paneling, a large hand-carved chest, and built-in carved benches with cushions whose vinyl upholstery simulated leather. On the opposite side (Figure 7–2) appeared a space defined by warm, rough-textured, otherworldly organic curves sprayed from polyurethane foam to house a museum exhibit entitled "Plastic as Plastic," a space described as the first "completely abstract, totally synthetic environment" in the history of construction.[1] Thus continued a design controversy—falsely imitating the past versus honestly embodying the fu-

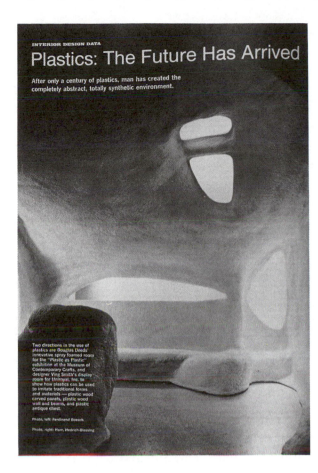

Plastics: The Future Has Arrived

After only a century of plastics, man has created the completely abstract, totally synthetic environment.

Two directions in the use of plastics are Douglas Deeds' innovative spray foamed room for the "Plastic as Plastic" exhibition at the Museum of Contemporary Crafts, and designer Ving Smith's display room for Uniroyal, Inc. to show how plastics can be used to imitate traditional forms and materials — plastic wood carved panels, plastic wood wall and beams, and plastic antique chest.

Photo, left: Ferdinand Boesch
Photo, right: Harr, Hedrich-Blessing

7–2. ROOM OF SPRAYED POLY-URETHANE FOAM DESIGNED BY DOUGLAS DEEDS FOR "PLASTIC AS PLASTIC" EXHIBITION AT MUSEUM OF CONTEMPORARY CRAFTS, NEW YORK, 1968

ture—that had engrossed machine-age observers from John Ruskin and Thorstein Veblen to Le Corbusier and Franklin Brill. With plastic's expansion, the issue became in theory compelling but in reality a moot point. Manufacturers would make and customers buy whatever they wanted.

To argue about imitative plastic versus plastic-as-plastic was to ignore a vast excluded middle. Few completely plastic environments existed, and even large pieces of furniture merged visually into promiscuous arrays of objects. More to the point, thousands of humble plastic products escaped either category. A green polyethylene wastebasket from the 1960s, its stipple-textured surface broken on one side by smooth raised images of stylized daisies, was neither imitative nor expressive of a unique aesthetic. It was simply itself—cheap, cheerful-looking, easy to clean, easy to discard when one tired of it. A British design critic attending a trade fair regarded huge injection molding machines in full operation as "masterpieces of technical ingenuity," but he was dismayed to find them spewing out "the nastiest rubbish ever conceived by man," evidence of "appalling abuse by indiscriminate manufacturers and users alike."[2] Such "rubbish" attracted contempt from critics who aspired to a

185

coherent aesthetic but exemplified many goods pouring from the plastic cornucopia of the 1950s and 1960s (Plate 13).

Few people reflected on plastic's presence among the stuff of everyday life. Only the visions of publicists and the nightmares of critics objectified a proliferation occurring across a wide spectrum of products. Useful, convenient, inexpensive, and so common as to be invisible, plastic objects attracted notice only when they broke or failed—an event whose occasional repetition damaged plastic's reputation but could not reverse the expansive trend. Although some artifacts did evoke modernity, most remained stylistically "rubbish," too anonymous and ephemeral to do more than add to a vague consciousness of an ever greater flow of plastic. New products and applications appeared almost daily. Few attracted more than passing notice; many were so quickly accepted it seemed they had always existed. To recall a series of these innovations risks artificially compressing time and creating an impression of overly rapid change. But such a reconstruction also reveals just how swiftly the material texture of everyday life did change at mid-century. Expansive trends were no less real because busy people remained oblivious or their descendants accepted the results as natural elements of the material world.

PLASTICVILLE, U.S.A.

Many older plastics gained new uses after the war. Bakelite continued to be used in molded handles for skillets and irons but also appeared in a series of one-piece compression-molded television cabinets introduced by Admiral from 1948 to 1950. The largest stood nearly three feet tall and weighed forty-one pounds without its ten-inch picture tube and speaker.[3] As it happened, molding technology could not keep up with demand for larger TV sets, but other prewar plastics enjoyed a steady stream of new applications. Rohm and Haas scraped by at war's end on sales of Plexiglas to veterans who went into business selling "poorly fabricated . . . geegaws and gimcracks." In a few years, however, as soon as chemists had mastered a non-fading red pigment for acrylic, the company was supplying carloads of molding powder to Detroit for auto taillight lenses whose baroque shapes transcended the limits of glass. Large back-lit thermoformed acrylic signs proved even more successful when Rohm and Haas learned to relax wartime quality standards for acrylic sheeting. First installed by oil companies at service stations, acrylic signs quickly conquered the neon jungle of the American night. As early as 1955 an architect described a "landscape of lettering" engulfing "whole sections of our environment" in "delightful fantasies of light, form and color."[4]

More evident in domestic situations was polyvinyl chloride, which protected children on rainy days, held water in their inflatable wading pools, and kept baby brother or sister's cotton diaper from leaking. Dolls cheaply molded from vinyl plastisols (polyvinyl chloride loaded with plasticizers) offered the advantage, as a publicist observed, that they "'feel' real as well as look real." A quest for a different kind of realism yielded Mattel's hard vinyl Barbie in 1959.[5] Vinyl also appeared in watch bands, garden hose, women's purses, and enjoyed a brief career as a material for draperies. It gained permanent acceptance as a resilient floor covering and notoriety as a "breathable" substitute for shoe leather. Vinyl proved ubiquitous as damp-cloth upholstery material—for cars (in turquoise, in pink, in fire-engine red), for living-room furniture (in Naugahyde imitations of the real thing), and for the seats of tubular chairs clustered around the chrome-banded dinette table of many 1950s kitchens.[6]

These dinettes with their melamine-laminated tops made the Formica Corporation so successful that it enjoyed "the enviable position of fearing that its trade name may become generic." By 1951 production of the popular tables—solid-colored, marble-swirled in synthetic hues, or flecked with gold—ran to more than a hundred thousand a month.[7] Formica, Micarta, and other brands of melamine-phenolic laminates in patterns simulating light woods surfaced classroom desks in thousands of new baby-boom schools. This innovation frustrated use of the jackknife to record adolescent angst but offered cool respite with a crisp chemical tang to the delinquent ordered to "put your head down on your desk." At home and at school, children and their parents ate from pastel-colored melamine dishes that were dishwasher-safe and practically unbreakable. The proprietary Boontonware line (Plate 14), which received *Consumer Reports*'s top rating in 1951, proved so successful that Scribner enjoyed the first real security of his career as he kept two of three Boonton plants busy molding plastic dishes.[8]

Postwar successes of these older plastics paled next to the market assaults of thermoplastics polystyrene and polyethylene. With a new high-impact formulation marking an improvement over the brittle stuff of broken toys, polystyrene appeared in Gillette razor-blade dispensers, heavy Crystalon bathroom tumblers, wall tiles, suction-cupped storage trays for the dashboard, lawn sprinklers, ice buckets, and such items as waste baskets and laundry baskets for which polyethylene soon became the preferred material. As molders became more confident with polystyrene and manufacturers more convinced of its durability, it appeared in transparent crisper drawers and other refrigerator parts. The Foster Grant Company of Leominster was so successful molding polystyrene that it bucked chemical-company consolidation by opening its own styrene monomer plant at Baton Rouge in 1954. Inexpensive injection molding enabled Foster Grant to market a line of trademark sunglasses in a changing array of ornate styles, including the harlequin designs of the mid-to-late 1950s. Polystyrene became the preferred material for molded radio cabinets. Because postwar radio tubes were smaller and their heat more easily dissipated, manufacturers could abandon slow compression molding of heat-resistant phenol or urea formaldehyde in favor of high-speed polystyrene injection molding. Capable of flowing smoothly into sharply angled molds, polystyrene facilitated a

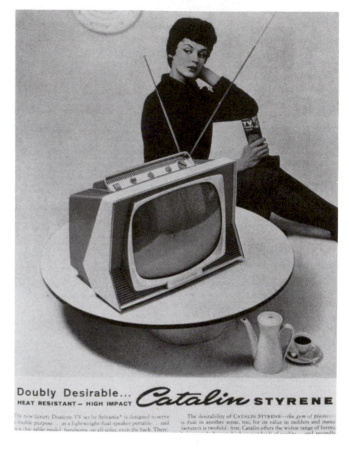

7–3. POLYSTYRENE
TELEVISION
CABINET, 1959

shift from bulbous streamlining to a flared look paralleling Detroit's jet-age tailfins. Once again plastic materials and processes contributed to an expressive visual style (Figure 7–3).[9]

Despite an initially poor reputation, polystyrene remained a major material for toys throughout the 1950s. Sometime in the late 1940s Fisher-Price's Buzzy Bee, a wooden pull toy with flat lithographed graphics, sprouted three-dimensional wings and a crown of injection-molded polystyrene nailed to its body. More than any other factor, a rapid shift to plastic propelled the company's success.[10] Polystyrene also took over model railroading when the Lionel and A. C. Gilbert companies turned from metal to plastic in search of lower manufacturing costs. At about the same time, American boys abandoned labor-intensive balsa-wood model-building in favor of mass-produced kits by Revell and Aurora with injection-molded polystyrene parts easily twisted from connecting sprues and glued together following step-by-step instructions. Plastic cars, ships, and fighter planes, their intricate realistic details a product of mold-making rather than model-making, seemed to take pride of place in the bedroom of every pre-teen male. Combining the two modeling trends, Bachmann Brothers of Philadelphia offered hobbyists snap-together polystyrene kits from which to assemble the houses, stores, diner, railroad station, and water tower of "Plasticville, U.S.A.," a model town scaled to O-gauge electric trains and populated with optional sets of "Plastic People" ("Do not accept a substitute," warned the company).[11]

Styrofoam differed radically from molded polystyrene and offered unprecedented qualities as a plastic. Made by pressurizing liquid polystyrene resin to obtain a light-weight cellular foam of hardened bubbles, it was used experimentally during the war for rafts and life preservers. Promoters at Dow Chemical Company hoped Styrofoam boards sliced from large blocks would provide superior building insulation. Other early uses included such novelties as "snowballs" for Christmas trees into which straight-pinned bangles and hanger eyes could be stuck with a distinctly synthetic crunch unlike any other sound. Decorators discovered Styrofoam as a cheap, easily carved base for retail displays and flower arrangements. By the early 1960s Styrofoam was common in picnic coolers, flower pots, flotation devices, disposable coffee cups, and packing materials. Ten years later it encompassed egg cartons, meat trays, and fast-food trays thermoformed from an extruded foam sheeting much lighter, with more densely packed cells, than the crude "snowballs" of 1950. As for high-impact molded polystyrene, less controversial than foam because less "disposable," it proved indispensable for casings of audio cartridges and cassettes—a use that absorbed sixty million pounds of resin in 1969.[12]

Polystyrene's earliest uses had echoed those of cellulose acetate or phenol formaldehyde. Polyethylene, on the other hand, with its tough flexibility, immediately appeared distinctive. "It was a polymer so unlike the polymers known at the time," a researcher at ICI declared, that at first "no one could envisage a use for it." By 1952, however, when Du Pont and ICI lost control of polyethylene in an antitrust judgment, it seemed the most promising of the new plastics. Construction of plants by eight other American chemical companies (among them Allied Chemical, Dow, Eastman Kodak, Monsanto, and Phillips Petroleum) soon left suppliers fearing overproduction and falling prices—a buyer's market for a plastic of which an optimistic molder observed that it was "almost difficult to mold a bad product." Polyethylene's price fell quickly from Du Pont's initial forty cents a pound to less than a dime—not at all surprising given that ethylene monomer went for less than two cents. Low cost resin stimulated entry into a host of new markets whose demands absorbed production and in turn stimulated speculative expansion. As an insider recalled, "profits depended on a constant flow of new applications." By 1958 polyethylene producers were churning out the industry's highest annual tonnage, 920 million pounds, and maintaining its highest growth rate, 30 percent. Polyethylene had become "the first of the mass plastics."[13]

Polyethylene changed the way people thought about plastic. The archetypal Bakelite had gained acceptance through durability. Polyethylene at first offered a variation on that long-standing theme. Tupperware containers delighted housewives and design critics by exhibiting resilient flexibility without sacrificing the durability people expected of high-quality plastics. But as someone observed in 1945, "The beauty of plastic articles is that they're cheap enough to replace rather than repair."[14] Polyethylene soon became an archetypal plastic of a different sort as endemic over-supply led producers to think not of durability but of disposability.

Packaging promised to consume a never-ending stream of plastic. Soft flexible polyethylene yielded the innovative squeeze bottle, first used in 1947 as a container for Stopette deodorant. The squeeze bottle was perfected by Jules Montenier, a lead-ing cosmetics chemist, in collaboration with engineers at the Plax Corporation who were developing a blow molding process for thermoplastic resins. By 1950 Plax was making a hundred thousand squeeze bottles a day and planning to triple its capacity to mold unbreakable containers for hand cream, suntan lotion, shaving cream, shampoo, and cologne. Polyethylene was so successful that a designer urged Procter & Gamble to use it for Crisco shortening. Because the blow-molded polyethylene bottle was several times as expensive as glass, manufacturers initially considered it a convenience item. But it fell rapidly in price, and cheap blow-molded thermoplastic bottles were soon used for detergent, bleach, milk, sauces, and condiments. Al-though promoters touted squeeze bottles and other plastic containers for the dura-bility that made them reusable, by 1956 they were celebrating disposability as "an important key to continuing volume." In the words of one commentator, plastic's future was "in the garbage can."[15]

Polyethylene appeared in other highly visible, equally ephemeral consumer goods, such as cheap toys. Young boys refought the war with realistic toy soldiers injection-molded of dark green polyethylene, or re-created the days of Fort Apache with cowboys and Indians molded with so little authenticity that a chief in full re-galia might appear as solid purple. Young girls played "dress-up" with stringless loops of pop beads, each molded with peg and socket to pop into flexible connection with another bead on either side. Offered at first with a pearly lacquer as novelty jewelry for adults, the beads soon appeared unpainted in gaudy popsicle colors and were taken up by children. Although pop beads consumed forty thousand pounds of resin each month in 1956, their economic and cultural resonance faded next to that of the hula hoop, one of the baby boom's first fads. Sold for about two dollars by Wham-O, a California molding company that later marketed frisbees, hula hoops were so popular they inspired "chiselers" and "jackals" to offer knock-offs for fifty cents, "scrap and junk, badly extruded with too-thin walls in dingy colors, and badly assembled." Twirling hoops absorbed about fifteen million pounds of a new high-density polyethylene (one-third of the total output for 1958), resin that otherwise would have served witness to the folly of overproduction.[16]

This new high-density or linear polyethylene was a marked improvement on the basic type developed during the Second World War. Polymerization of ethylene had remained difficult owing to extreme pressures required, up to 30,000 psi. A search for catalysts of low-pressure polymerization ended with several near simultaneous breakthroughs in 1953 and 1954. Desire for an easier process and lower production costs motivated the search, but it also yielded an improved product. An ordinary polyethylene molecule was "branched"; that is, its long chain of atoms possessed branches of irregular shape and length. Molecular length made the material strong; irregular meshing of the branched molecules made it flexible. The new "linear" polyethylene, by contrast, consisted of long unbranched single-chain molecules packed tightly together to yield a stronger material with greater dimensional rigidity and resistance to heat. Although Karl Ziegler of West Germany, an academic researcher, announced the discovery of linear polyethylene and began licensing his process, Du Pont chemists were the first to make the new material, and others at Phillips Petroleum and Standard Oil of Indiana devised slightly different processes. At about the same time, researchers at Du Pont, Phillips, Standard Oil, Hercules, and Montecatini of Italy all laid claim to another polymer even stronger than linear polyethylene but similar to it in derivation and structure. The Italian Giulio Natta long received credit for inventing polypropylene; in 1963 he and Ziegler shared the Nobel Prize in chemistry. Polypropylene entered commercial production in the United States in 1957; in 1983, after a quarter century of litigation, Phillips received a patent when the United States recognized J. Paul Hogan and Robert L. Banks as its inventors.[17]

Nearly as hard and glossy as polystyrene but without its brittleness, polypropylene was so light it floated in water and enabled manufacturers to stretch already cheap supplies of petrochemicals. Nearly as important to its acceptance, polypropylene retained the warm touch of linear polyethylene. While the latter was used for cosmetic bottles, pink lawn flamingos, dish pans, laundry baskets, and garbage cans, polypropylene was tough enough for milk crates, tackle boxes, tool kits, toilet seats, small appliance housings, one-piece chairs and tables, and luggage. One-shot injection-molded thermoplastics were approaching the durability of Bakelite or of fiberglas—without the expense of reinforcing fillers, fiber mats, preforms, or paper laminates, without the slowness of compression molding or the greater delay of fiberglas hand lay-up techniques. Polyethylene and polypropylene succeeded so well that thirty years later a veteran molder described them as the industry's "old workhorses."[18]

Many other plastics first appeared in consumer goods during the 1950s. Among them was ABS (a co-polymer of acrylonitrile, butadiene, and styrene), a rigid sheeting material for one-piece refrigerator liners and luggage.[19] Molded nylon, marketed by Du Pont as Zytel, replaced metals in gears and bearings of small automotive and appliance assemblies, and also gained visible exposure as roller-skate wheels.[20] For some uses nylon was eventually replaced by an even stronger Du Pont engineering plastic, Delrin acetal resin, a form of polymerized formaldehyde. Although it cost several times more per pound than the brass or zinc it replaced, its low density brought considerable savings. A one-piece instrument panel for the 1961 Plymouth Valiant, for example, contained only two pounds of Delrin compared with nine pounds of zinc in the 1960 model.[21]

At the opposite end of the synthetic spectrum Du Pont unsuccessfully imitated its original nylon strategy by introducing a new fiber, polyester, as a yarn for expensive men's suits. By allowing retailers to follow the precedent of nylon stockings and make extravagant claims about "miracle suits . . . that never need pressing," the company set up polyester fabric for a fall from which it did not recover until the mid-1960s when permanent-press pants became a necessity of middle-class life.[22] By then polyester was well established in other uses. As a resin it formed the thermosetting matrix in which fiberglas was embedded to make boats, surfboards, office chairs, and corrugated patio roofing. Extruded as high-tech Mylar film, polyester gave the "memory business" its first reliable magnetic recording tape, replacing cellulose acetate. In 1956 Luchow's Restaurant in New York marketed some of its popular dishes in frozen boil-in-the-bag Mylar pouches. Coated with a thin layer of reflective aluminum, Mylar film reached outer space as the inflatable *Echo I* satellite, a hundred feet in diameter; long before that, however, people had experienced it as a decorative metallic accent on interior car-door panels, often quilted or embossed, and as a metallicized yarn, woven among fibers more traditional in appearance, glittering from the upholstery of cars, airliners, and living room couches, where it provoked the destructive picking of bored children.[23]

New plastics and variations on the old appeared so frequently that promoters and business writers kept busy plugging new terms into a well-worn formula. Not every new plastic succeeded as quickly as publicity suggested, however. Some lived up to advance billing only after an uncertain development period lasting a decade or more. Not until 1954 did Americans begin exploiting the polyurethane foam that Kline's wartime report from Germany had mentioned as an excellent insulating material. In that year Monsanto and Bayer of Germany organized a subsidiary, the Mobay Chemical Company, to commercialize polyurethane. Glowing publicity for this "next great synthetic" noted the impending demise of latex foam rubber, but it was another five or six years before Mobay chemists formulated a polyurethane foam that survived heat and humidity without disintegrating.[24] Even more extreme was the case of Teflon, a polymer of tetrafluoroethylene accidentally discovered by Du Pont chemist Roy J. Plunkett in 1938 when a cylinder of gas in a refrigerant project yielded a waxy solid. The material remained stable at an unprecedented temperature of 315°C but proved too slippery to mold, fabricate, or bond using standard techniques. Eventually chemists learned to bond it to metal as a coating, and it was used at Oak Ridge during the war to protect equipment from corrosive materials during manufacture of uranium-235. Journalists frequently claimed in the early 1950s that housewives would soon enjoy "non-stick" Teflon cookware, but the promise was not realized until 1961. Delays in commercializing polyurethane and Teflon were unusual but not unique.[25]

Once in a while a plastic that garnered considerable publicity failed completely. Briefly in the 1950s the industry made much of a process developed at General Electric for raising polyethylene's heat resistance from 70 to 200°C by bombarding it with radiation. This so-called irradiated polyethylene disappeared after a few years—probably owing to development of high-density polyethylene.[26] Another loser with far more at stake was Corfam, a "breathable" substitute for shoe leather introduced by Du Pont with great fanfare in 1963 after ten years of research and development. Not a chemically new plastic, Corfam consisted of a mat of polyester fibers impregnated with porous polyurethane and surfaced with a tough, shiny polyurethane skin. It supposedly had a million pores per square inch. Although Du Pont orchestrated a sophisticated campaign to introduce Corfam at the high end of the market, it never caught on with consumers who confused it with cheap "nonbreathable" vinyl—"plastic shoes"—and who disliked its failure to conform with use to the shape of one's foot as leather does. Americans were not interested in a scuffproof "wipe and wear" leather substitute so perfect it never had to be polished and never took on the patina of old leather. Corfam lost $70 million before Du Pont gave up in 1971 and sold the process to Poland. Such spectacular failures remained rare. Most people accepted as normal a continuing transformation of the material world made possible by an accelerating flow of new plastic materials.[27]

DEPARTING FROM THE NORM

Popular media rhetoric about an emerging "plastic age" occasionally surfaced during the 1950s even though the mundane reality required no announcement.[28] Less intense than twenty years earlier when the industry required self-conscious definition, this publicity offered routine descriptions of new products instead of glowing utopian predictions. Industry leaders focused publicity efforts on specific problems—attacks by rival trade associations, or reports of faulty resin, poor engineering, or bad design. Although they abandoned the glamorous association of plastic and design of the 1930s, they continued to believe industrial design was crucial to their success. Not only did designers mold public confidence in plastic by selecting the right plastic for the right job. They also bore primary responsibility for giving form to malleable new materials—by recognizing, as a designer observed, "that you can't do a good design job [in plastic] without departing from the norm."[29] Jane Fiske Mitarachi, editor of the professional journal *Industrial Design*, took up this responsibility in a special issue coinciding with the seventh National Plastics Exposition in June 1956. During the postwar years, while plastic was evolving "from a promise into a major basic materials industry," industrial design was evolving "from a sales gimmick to a critical activity of top management." From a designer's perspective "the making of a synthetic material"—that is, of the basic molecule itself—could be considered "a creative act . . . *design* in its broadest sense." But the designer had to come to terms with the changes wrought by the chemist.[30]

Taking up this challenge in an essay on plastic and quality, Mitarachi wondered, "How can traditional satisfaction and prestige be designed into a family of untraditional materials?" She asked readers to contemplate an elegant set of sterling silver flatware with handles of matte-black Zytel nylon. Projecting an unconventional beauty, the silverware offered a sharp contrast to the appeal of most plastic objects, those whose imitative surfaces so often proclaimed, "See how much I look like the materials you like." That lame justification owed its logic to plastic's late arrival in a "marketplace of materials" with "distinct and interesting personalities." In defense, plastic had "retreated to a kind of visual sackcloth, seeking consolation in service-ability." Because various plastics remained "uncertain of their own personalities," they could not compete with "the sheen of oiled wood, the matte-like grain of ivory and ebony, the crisp polish of lacquer, the patina of leather, the fibrous texture of paper." The aesthetic appeal of traditional substances depended on three factors—a texture or depth derived from growth or manufacture (as in the grain of wood or the warp and woof of textiles); the picturesque irregularity of all natural things, organic or inorganic; and the effects of time, of "a lifetime of seeing them worked, and used, and worn, and mellowed." Plastics might simulate but could never emu-late these traditional markers of quality. As products of "invisible chemical miracles" they offered "no such intimacy with their character, to give us an intuitive sense of their beauty."

Writing at about the same time Roland Barthes was composing his meditation on plastic, Mitarachi confronted similar issues. She had little if any idea what to do about the problem. But she did believe industrial designers, whose work involved interpreting things to their users, would devise solutions. The problem "with such pliable stuff as plastic" was the temptation "to turn it into forms we already know." This "agreeable passivity" would never yield "real quality." For that she depended on designers to endow plastic with an "affirmative character" based on "inherent and untraditional" attributes. True quality could only be "designed into plastic by a frank exploitation of the things that make plastics unique." For Mitarachi, in the final analysis, quality was "as much a matter of visible effects as of physical performance." Throughout the 1950s and 1960s designers and architects struggled to unite the visual and the structural, the image and the object—in other words to develop a culturally expressive design vocabulary for plastic. The search proved to be difficult and uncertain. It was only too true, as designer Arthur J. Pulos observed at a confer-ence devoted to fiberglas, that it was "physiologically comforting to one who spends his day struggling with the future to be able to retire at night into the past." Few Americans at mid-century really wanted to live like the Jetsons. But for those who did, fiberglas proved to be the ideal material.[31]

CUSTOM CARS AND ORGANIC CHAIRS

With new polymers appearing faster than anyone could keep track, glass-reinforced polyester seemed an unlikely medium for a futuristic design aesthetic.[32] It lacked the public image of old materials like Bakelite and new ones like polyethylene. Even for insiders it had no clear identity. As if no one knew how to perceive the stuff, it was known by many names. Polyester indicated only the basic resin used to bind or laminate the glass fibers. People referred to it as low-pressure laminate (to distinguish it from high-pressure laminates like Formica), or as fiber-reinforced plastic (a category including Ford's phenolic auto body panels filled with hemp and ramie), or as glass-reinforced plastic, or as fiberglas-reinforced plastic. The latter term, incorporating an Owens-Corning trade name for glass fiber, yielded the confusing generic fiberglas as a word standing for a combination of two materials. Even so, no one ever forgot that fiberglas in that extended sense functioned not as glass but as plastic. When cars and houses were made from it, promoters extolled them as plastic—and so too did the public perceive them.

To compound the problem of uncertain identity, fiberglas lacked the high-tech prestige of nylon, Mylar, or even polyethylene. It was a primitive material processed by crude methods. While the plastic industry enthused about automated molding machines processing thermoplastic resin flake into identical products at unbelievable speeds, fiberglas remained labor-intensive. The simplest method of working fiberglas, evocatively referred to as "hand lay-up," required minimal equipment. Starting with a single-piece mold of wood, plaster, or even fiberglas itself, the operator lathered it with sticky polyester resin or "gunk," smoothed in a thin woven or matted layer of glass fiber, waited for the resin to cure or harden, usually at room temperature, and then repeated the process over and over until reaching the desired thickness. Although hand lay-up was a simple process, it was also slow, messy, and inexact. Molds must have seemed almost cost-free compared to no-tolerance machine-tooled hardened-steel molds used in compression or injection molding. But lay-up worked well only for one-of-a-kind or low-volume production. Even then, finished pieces had to be trimmed, ground smooth, and polished by hand. Not until the end of the 1950s did hand lay-up begin to yield to the more mechanized process of a gun spraying a self-adhesive mixture of resin and chopped fiberglas.[33]

Several other processes for molding fiberglas and polyester required greater investment in equipment but remained crude and labor-intensive. The "bag method" speeded the curing of large objects like boats by using a vacuum or compressed air to force a rubber bladder evenly against the material laid-up in a mold. The resulting products were stronger than those obtained by hand lay-up alone because applying pressure allowed a higher ratio of glass to resin and yielded a greater degree of reinforcement. A third method, using matched aluminum or steel dies to apply both heat and pressure to a preform or mat of resin and glass fiber, echoed the technique of phenolic compression molding. This "matched-die method" enabled molders of relatively small parts to increase production volume. Molders often converted secondhand compression presses for use with fiberglas; to the extent that phenolic molding already seemed out of date, so too did the analogous fiberglas process. Only one fiberglas technique, known as "filament winding," was at all innovative. In this process a continuous strand of fiberglas coated with polyester resin was mechanically and evenly wound onto a precisely shaped form like thread around a spool, and then, when proper thickness was reached, it was fused solid and cured by heat.

Used for cylindrical or tapered objects such as large pipes and storage tanks, the process yielded a strong, light, chemically resistant object that could not be duplicated in metal. Beyond this single high-tech innovation, useful only for specialty products, fiberglas molding remained a slow, messy business marked by guesswork, hand labor, and a sense of craftsmanship.

If it was ever true that a military veteran with nothing more than ambition could get into plastics, fiberglas made it seem easy. Two years after V-J Day, with fiberglas radomes for airplanes no longer needed, annual production of polyester resin had fallen by three-quarters to 1.5 million pounds, most of it shipped to California. Late in 1947 *Modern Plastics* reported "scores of very small companies" on the west coast molding fiberglas boats and aircraft parts. Often their premises consisted of "no more than a picket fence around a lot, with a tent set up for storage." For the time being the California sun, soon to offer an ideal environment for fiberglas surfboards, skateboards, dune buggies, and sports cars, contributed more directly to the material's exploitation by enabling molders to cure the resin outdoors. If enthusiasm grew in this supportive climate, so too did failure. Fiberglas molders gained a reputation for being "woefully inefficient, woefully sloppy, and woefully ignorant." As one recalled, "It always seemed that the guy who was your closest friend at one of the early convention meetings would be selling socks [when] the next one came along." By 1952 the responsible fiberglas molders (that is, the successful ones) worried about the allure of easy money for low-budget entrepreneurs. An SPI policy committee abandoned the phrase "low pressure" with its air of "gimcrackery and cheapness." Hiram McCann, editor of *Modern Plastics* and chair of the committee, argued for a public relations campaign to convince corporate customers that "reinforced plastics" had become "an industry of expensive tooling and professional equipment" capable of "big scale production." Nothing was so harmful to progress as the myth that "any guy with an empty garage or a rumpus room and a thousand bucks can hope to get into it and make a success of it."[34]

The simplicity of hand lay-up attracted enthusiasts whose involvement passed for investment but actually remained a hobby. Nothing fascinated them more than designing and shaping their own sports cars. The idea of a plastic car had engaged the *Popular Mechanics* mentality ever since Ford swung his ax into that rebounding body panel in 1941. There was something heartening about Ford contriving an eccentric soybean plastic with the help of a young self-trained chemist whose elevation to his post smacked of Horatio Alger. The story gave the whole idea of a plastic car a maverick status only reinforced by the ease of working with fiberglas. Although the first postwar fiberglas automobile was built in 1946 as a prototype for an established company, Graham-Paige, its designer, William B. Stout, was well known as an outsider for an earlier series of radical aerodynamic designs. As soon as he proved "it was physically *possible*" to make a fiberglas auto body, would-be molders throughout southern California laid in a supply of gunk and began laying up cars.[35]

Most experiments avoided the futurism of the ConVaircar of 1947, a "road-plane" prototype with a complete winged fuselage grafted to the top of a sedan. Instead molders focused on custom sports car bodies and sought speed, acceleration, and malleability of design. Some bodies were sold to be mounted on standard chassis, some came with fiberglas chassis, and still others came as kits to be assembled (not molded) at home. Bill Tritt, whose Glasspar Company went into business to mold plastic boats, completed his first fiberglas car in 1951. Known as the Brooks Boxer, it resembled an MG and was mounted on a ten-year-old Ford chassis. The Boxer garnered rave reviews that year at the Los Angeles Motorama along with three other unique fiberglas cars, including Jack Wills's Skorpion, a bulbous, boxy convertible available at $500 for mounting on a Crosley chassis. The rush was on as each small shop distinguished its own fantasy-arousing shell of polyester and fiberglas by endowing it with a unique sculptural form. The Woodhill Motor Company, an offshoot of Glasspar, soon offered the Wildfire in kit form as a body, or fully assembled and ready to roll. Low and sleek, it boasted a smooth front airscoop and a sensuously curving fender line flowing continuously from front to rear. Other competitors included the Vale Sports model, with a low-slung form defined by "aviation-type fins on the rear fenders," only one of many fiberglas design innovations later borrowed by Detroit stylists for mass production in sheet metal.[36]

Detroit capitulated to the fiberglas fad in 1953 when General Motors introduced the two-seat Chevrolet Corvette, less than a year in the making from designer's sketches to finished product (Figure 7–4). The company farmed out molding of the

7–4. Chevrolet Corvette and body shell of fiberglas-polyester, 1953

Corvette's forty-one plastic body parts to Robert S. Morrison of the Molded Fiber Glass Company at Ashtabula, Ohio, whose plant was equipped for matched-die work. The original Corvette assembly line at Flint, Michigan, accommodated six chassis at a time and finished three cars a day. After moving production to St. Louis, GM intended to build thirty cars a day—almost nothing compared with the seventy-seven hundred steel-bodied vehicles then rolling from the company's plants each day. Tooling for the Corvette cost only $500,000 instead of the $4.5 million involved in a standard steel automobile, and cost analysis determined that a fiberglas body remained economically advantageous up to the point where annual production exceeded fifteen thousand cars. Plastic was not likely to replace steel, except for sports cars, postal vans, camping trailers, and other specialty vehicles with low production runs. All the same, the Corvette, with its faired-in headlights, a virtually seamless surface, and the directness of a long flattened tubular body, brought the distinctive image of a plastic car from California to a national audience. But some people missed the point. Anyone who counted on fiberglas sports cars to glamorize the image of plastic must have cringed when *Fortune* quoted a GM executive as saying he didn't mind the fiberglas Corvette "just so long as they don't try to do it with any damned plastic."[37]

The fiberglas sports car enabled men to domesticate plastic in a manner distinct from that of women who domesticated it by insisting that it imitate or harmonize with traditional materials. Their husbands, at least those of the *Popular Mechanics* persuasion, accepted an artificial design aesthetic but sought to control it personally within a do-it-yourself framework. General Motors had made it easier to possess an unusual sports car, but the Corvette's very existence stimulated the weekend hobbyist's desire to create a unique, personally designed and crafted sports car. In 1954 *Motor Trend* published a "how-to" *Manual of Building Plastic Cars*, a seventy-five-cent paperback whose cover illustrated a muscular fellow hefting a full-length Corvette floor panel with one upstretched hand. According to the inside cover, "inexpensive car bodies of reinforced-glass plastics" were "being made by novice car designers in garages throughout the nation."[38]

When this guide to fiberglas hand lay-up appeared, Maurice Lannon's mimeographed treatise on *Polyester and Fiberglas*, written "for the hobbyist and small manufacturer," was already in its second edition. Affecting a breezy inspirational style, Lannon included a host of practical tips such as an exhortation to "get REAL rubber gloves, not plastic rubber" (which would dissolve in acetone solvent). He meticulously described the creation of his own custom auto body, designed by a no-nonsense linotype machinist from North Hollywood who had "never attempted to design or make a car body before." Because he "had a picture in his mind of what he wanted, he started right in on the mold" without fooling with sketches or models. It was that easy. Anyone could have a "light-weight, long lasting beautiful car body of [their] own design, worth many times what it cost"—a body "stronger than steel," impervious to California's salt air, and easily repaired at home with a bit of glass matting and a can of gunk. Here was a miracle material if ever there was one—fully domesticated by an American do-it-yourselfer with a taste for high-tech baroque. A simple process based on a crude technology enabled everyman to express a visually exciting plastic-as-plastic design aesthetic.[39]

That paradox also marked fiberglas as a good material for furniture of modern design. However much designers intellectualized about new materials and mass production, furniture making remained a craft. Even the mechanistic tube-steel Bauhaus artifacts of the 1920s were handmade. Thirty years later fiberglas attracted designers who found its qualities leading to an artificial aesthetic of smooth surfaces and flowing compound curves. At the same time fiberglas required the hands-on involvement of a craftsman or sculptor. The author of *Motor Trend*'s handbook spoke to these points when advising do-it-yourselfers to gain practice with smaller projects like a "hammock sun chair," a swooping curvilinear one-piece chaise lounge with sides flaring into armrests and headrest curving down to form a structural support. "We are not going to tell you how to build them," he explained, "for design and styles are so flexible with this material that they can be as original as you wish." More articulate about it, designer and architect Eero Saarinen maintained that the "compound shell of plastic" was the "most appropriate [form] for twentieth-century furniture." Design was no longer a "cubist" or "constructivist" project of assembling discrete parts, as in the early modern movement. Plastic made it a "sculptural" challenge. But Saarinen also believed plastic's unprecedented freedom of form entailed greater responsibility for the designer—an attitude of caution not found in *Motor Trend*'s easygoing optimist.[40]

Among the prominent designers of the postwar era, Saarinen and his colleague Charles Eames contributed more than anyone else to a design vocabulary fully exploiting plastic's material qualities. As the son of Finnish architect Eliel Saarinen, who came to America in 1923 and headed the Cranbrook Academy of Art near Detroit, Eero escaped the full impact of rationalist Bauhaus doctrines. He matured as a designer in a nurturing environment that promoted a craftsman's intimate knowledge of materials and recognized the validity of an expressive motive in their working. After studying sculpture in Paris and earning a Yale architecture degree, Saarinen returned to Cranbrook in 1936 to practice with his father and teach as a part-time instructor. He was also employed briefly by industrial designer Geddes, working on the extravagantly streamlined General Motors building at the New York World's Fair—an experience that carried Saarinen even farther from the dry purism of the International Style toward the expression of a fluid plasticity.[41]

When Eames arrived at Cranbrook in 1938, he and Saarinen quickly became friends and collaborators. While the latter tended toward the dramatic, Eames possessed an engineering mentality and sought to extend materials to the logical extremes of their structural characteristics. In 1941 they won first prize for "seating"

199

in a competition on "Organic Design in Home Furnishings" sponsored by the Museum of Modern Art. Their winning entries, an armchair and a side chair of laminated wood, suggested the directions they each later independently took with fiberglas furniture. Compressed in cast-iron molds, the wooden chairs' form-fitting shells possessed a complex three-dimensional curvature previously unknown in furniture construction. Only tapered wooden legs prevented each chair from being a one-piece sculptural whole. Although a thin layer of foam rubber was glued to each chair's seating area, and the entire shell, inside and out, was covered with cloth upholstery, the designs indicated clear innovation in a material—plywood—whose binder of synthetic resin was leading enthusiasts to promote it as a plastic. At that point glass-reinforced polyester hardly existed; working with plywood afforded experience that later translated into manipulation of the more versatile material.[42]

Of the two former collaborators, Saarinen first explored fiberglas as a structural material for furniture. Late in 1948 Knoll Associates began marketing Saarinen's "womb chair," a deep fiberglas shell with high back curving down into gently flared armrests. It floated on a structure of wire struts and contributed to the vogue for casual living by inviting one to curl up in it. As with the plywood chairs, Saarinen lined the shell with foam rubber, covered it with fabric, and even provided cushions. The womb chair became an icon illustrated by Norman Rockwell on a *Saturday Evening Post* cover as the refuge of an unshaven chain-smoking suburban dad in robe and slippers, reading the Sunday paper as wife and children paraded behind him to church. But Saarinen's chair went only halfway toward realizing plastic's design potential. While it demonstrated the structural use of fiberglas to support "organic" human posture, its upholstery prevented frank expression of plastic as plastic.[43]

The fiberglas shell chair, like so many other plastic artifacts, owed its final form to a substitution of plastic for a different material by someone already engaged in production—in this case Eames, who had continued to develop techniques for shaping plywood into complex curvilinear forms. In 1941 Eames had married Cranbrook student Ray Kaiser and moved to Los Angeles to go into partnership with her. During the war they manufactured form-fitting plywood splints and litters for the navy and graduated to plywood trainer aircraft and gliders. After military orders evaporated, the Eameses returned to molding plywood chairs and eventually licensed their manufacture to the Herman Miller Furniture Company, which had been promoting modern design since the 1930s. It was for Herman Miller in 1949 that Charles and Ray Eames developed a series of molded fiberglas chairs that became even greater icons than Saarinen's womb chair of design in plastic and of postwar material culture.

Despite their plywood furniture's commercial success, the Eameses were not satisfied they had found the best solution to the problem of the lightweight contour chair. The announcement in 1948 of a competition on "Low-Cost Furniture Design" at the Museum of Modern Art prompted Charles to experiment with other materials.

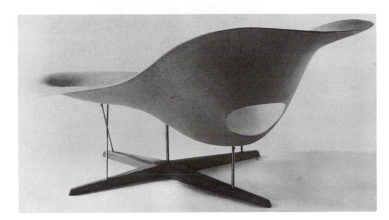

7–5. Experimental fiberglas-polyester shell chair, "La Chaise," designed and fabricated by Charles and Ray Eames

Frustrated by the hand labor involved with plywood, he proposed stamping the shells for an armchair and a side chair out of sheet metal and coating them with vinyl or neoprene. Along with these stamped-metal entries, the Eameses also submitted a design for a lounge chair of "large flowing forms," its strength and lightness achieved by means of "stressed skin construction"—that is, exterior layers of fiberglas plastic sandwiching a resilient core of foam rubber and Styrofoam. They had recently used war-surplus polyester and fiberglas to make translucent interior panels for their experimental Case Study House #8, sponsored by *Arts & Architecture* magazine. While that project employed plastic rationally as an engineering material, the free-form lounge chair enabled them to explore its surreal potential (Figure 7–5). Known as "La Chaise" because it could have accommodated one of Gaston Lachaise's gargantuan nude sculptures come to life, the lounge chair's asymmetrical stylized oyster shell floated on thin metal struts above a wooden base. In designing and crafting La Chaise, the Eameses revealed an understanding of plastic as a substance whose malleability liberated them to flow with vague cultural trends. As they admitted in a statement accompanying their entry, La Chaise did not "anticipate the variety of needs it is to fill" because those needs remained "indefinite" and their solutions "intuitive." The Eameses hoped La Chaise's sculptural form, an agent provocateur of the material world, would "suggest a freer adaptation of material to need and stimulate inquiry into what these needs may be." Among other names proposed for this free-form plastic chair, Gondola and Duchesse evoked historical resonance. Kangaroo was playfully biomorphic. And Psyche suggested the ambiguity of their project and of design for plastic in general. They confessed they did not know quite what they were doing. The vagueness of their language suggested the tenta-

7–6. Fiberglas-polyester armchair
designed by Charles and Ray
Eames, manufactured by Herman
Miller, Inc., from 1950 onward

tiveness of their attempt to design something to satisfy "a difficult-to-define need of the time."[44]

While La Chaise remained a prototype, the ease of working with reinforced poly-ester impressed the Eameses. Two years later, in 1950, when the Museum of Modern Art awarded their more conservative form-fitting armchair first prize in the "Low-Cost" competition, the actual chair shell was not made of sheet metal, as in their original proposal, but of fiberglas (Figure 7–6). Edgar Kaufmann, Jr., a champion of "good design," praised the winning chair for not disguising its material but instead exposing a delicate swirl of white fibers embedded in a matrix of colored resin. In Kaufmann's words, it was "the first one-piece plastic chair to feature the natural surface of its material, variegated and satiny."[45] Already being marketed by Herman Miller, the chair had a bucket seat curving outward to form narrow arms; it could be purchased with a variety of bases: four wooden or metal legs, several systems of wire struts, wooden rockers with wire supports, and a swivel device on wooden legs at drafting-table height. The designers had turned to fiberglas for the chair after stamped metal, the so-called material of mass production, proved ironically too ex-pensive. The shells were molded by the Zenith Plastics Company of Gardena, Cali-fornia, which got its start making radomes in the war. The molders used a matted preform of short strands of fiberglas and polyester resin formed by suction against a wire screen. Using the matched-die method, they molded each preform for three minutes at a temperature of 130°C and pressures ranging from 100 to 150 psi. Each shell had to be trimmed and buffed by hand with power tools before being joined to its base assembly (Figure 7–7).

As *Modern Plastics* observed, the resulting "polyester chair" was "light in weight, colorful, and long wearing." Its "pleasing surface grain and design" would never

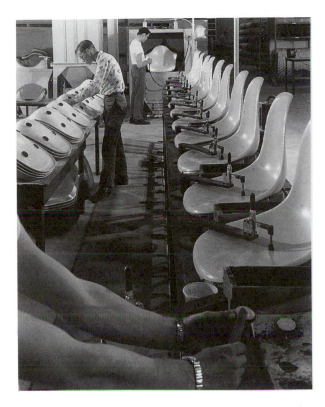

7-7. FASTENING LEG MOUNTS ON FIBERGLAS-POLYESTER SHELLS OF EAMES SIDE CHAIRS.

"chip, peel, or fade." Above all, its shape afforded a "maximum of body support and relaxation" and its surface was "always warm to the touch."[46] As it turned out, the fiberglas chair was often *too* warm to the touch, leaving people to steam in their own unevaporating humidity. For that reason as much as any skepticism about functional modernism, the plastic chair became associated in people's minds with schools, waiting rooms, and other places known for a certain nervous funk. Herman Miller promoted the armchair and its companion side chair for kitchens and family rooms, but their success was mostly institutional. That association reawakened some of the old prejudice about plastic as cheap or second-rate. Even so, the indestructibility of the Eameses' fiberglas chairs guaranteed them continuous production for forty years. Their success—and high visibility—inspired other structural uses of fiberglas and contributed even more than the plastic automobile to an artificial plastic-as-plastic design aesthetic.

Eero Saarinen eventually carried the idea of a formal plastic aesthetic farther than his friends Charles and Ray. While they mounted their fiberglas shell on a base of wood or metal, Saarinen groused about "modern chairs with shell shapes and cages of little sticks below," the latter often "a sort of metal plumbing."[47] He dreamed of a sculptural chair combining shell and base in an integral whole. The resulting pedestal chair was introduced by Knoll Associates in 1957 and was known as the "tulip chair" for its blossomlike seat rising from a narrow tapered stem resting on a circular

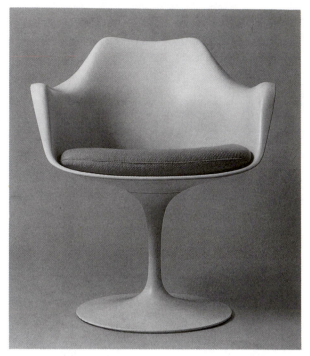

7–8. Pedestal or tulip chair with fiberglas-polyester shell and aluminum base, designed by Eero Saarinen, manufactured by Knoll Associates, Inc., 1957

pad on the floor (Figure 7–8). Saarinen tempered the chair's all-white purity with the organic grace of its form—and with a thin bright cushion in the seat. His goal with the project was "to clear up the slum of legs" of most furniture and "make the chair all one thing again." But he confessed that his chair was actually "half-plastic, half-metal" because a thin stem of reinforced plastic could not support a sitter's weight. The base was an aluminum casting disguised with a plastic coating so that it merged seamlessly into the fiberglas shell of the seat. Although Saarinen had achieved a formal triumph, he "look[ed] forward to the day when the plastic industry has advanced to the point where the chair will be one material, as designed."[48]

The pedestal chair revealed Saarinen as a designer whose commitment to a plastic aesthetic outran the material itself. His own architectural work, especially two dynamic sculptural airport terminals whose style defied the era's dominant Miesian rationalism, could easily have found expression in plastic had synthetic materials been structurally appropriate. In reflecting on choices faced by contemporary architects, Saarinen often used the word plastic. Given his experience with furniture design, it seems likely he intended a dual reference—to the traditional plastic art of sculpture and to the new synthetic materials whose example inspired him to stretch reinforced concrete to its limit. In 1957 he observed that "technology has made plastic form easily possible for us." The question of "when and where to use these structural plastic shapes" fascinated him. There were so many possibilities, all of them "equally logical" from a functional point of view—"some ugly, some exciting, some earthbound, some soaring"—that the "choices really become sculptor's choices." Although Saarinen warned that "we must beware of going too far," few designers and architects who moved to take advantage of plastic heeded his warning.[49] Whether for mundane form-fitting bus seats and wraparound bowling center equipment, or for more visionary structures, the ease of manipulating plastic inspired a jet-age extravagance similar to that of Saarinen's air terminals. That this artificial aesthetic often produced organic or biomorphic forms made it all the more acceptable to a public that remained ambivalent about the unnatural. A design magazine in 1958 approvingly quoted Dreyfuss's prediction that reinforced plastic would become "an enormously important building material" because "nature isn't rectangular." Echoing the contemporary disdain for anything "square," the magazine dismissed "bricks, boards, and beams" as "rectangular subdivisions" of "rectangular shells." By contrast, plastic yielded "softer forms . . . more in harmony with nature." This trend appeared to best advantage, it was suggested, in the Monsanto House of the Future, an experimental prototype constructed at Disneyland, appropriately located midway between Fantasyland and Tomorrowland, a building whose structure owed quite a bit to the lowly fiberglas chair and whose style paralleled that of Saarinen's TWA terminal.[50]

HARDWARE

Like many other people who made careers of plastic during the 1940s and 1950s, Albert G. H. Dietz became involved almost by accident and then discovered he was leading a trend. Although he modestly credited others with the design of the Monsanto House of the Future, it was Dietz who orchestrated the project, guided its engineering, and served as a focus for those engaged in creating an architecture of glass-reinforced polyester. Trained as a carpenter by his father, Dietz served as an instructor of structural engineering at MIT while completing graduate work during the late 1930s. After receiving a regular faculty appointment to focus on house construction, he became interested in plastic by way of synthetic resins used in plywood. During the war he helped develop nylon body armor. By then a major shift in his career was under way, and peacetime found him attracting corporate funding to MIT to pursue structural investigation of plastic. In the late 1940s, George H. Clark, a Formica executive serving as SPI president, solicited $500,000 from several companies for Dietz to run a Plastics Research Laboratory for five years as part of MIT's Department of Building Engineering and Construction. The venture proved so successful that funding continued later, and Dietz joined the industry's bureaucracy in 1950 when SPI asked him to serve as chair of its Committee on Plastics Education. As Clark asserted at an SPI board meeting, Dietz had proven that "MIT was definitely plastics conscious."[51]

The House of the Future evolved from a research project initiated by Monsanto in May 1954. Already a corporate sponsor of Dietz's laboratory, the Plastics Division of Monsanto approached Pietro Belluschi, dean of the Department of Architecture, with a proposition that MIT architects design an innovative house using Monsanto plastics as primary materials. Dietz and the Department of Building Engineering and Construction became involved immediately, and Richard W. Hamilton, a research associate in architecture, was appointed head of the project. They began with a systematic survey of current applications, published in June 1955 as *Plastics in Housing*, a seventy-page booklet whose introduction revealed that "from the start" they had planned "to design and build a house" to "demonstrate the architectural potentialities of plastic." In fact, they confessed, whetting appetites without disclosing any details, "our staff is currently engaged in the design of such a house."[52]

At that point plastic in construction was limited to superficial elements—electrical insulation and fixtures, floor coverings, Formica counter tops, Styrofoam insulation board, vapor barriers of polyethylene film, acrylic skylights, and translucent fiberglas sheets sheltering carports and patios. No truly structural uses existed. But Monsanto engineers were not alone in working toward the decades-old dream of an all-plastic house. Within a few months of Monsanto's overture to MIT, other firms publicized designs for two quite different plastic houses. One was a ranch-style beach house with exterior wall panels of fiberglas designed by architect Robert Fitch Smith and built in Florida by the Russell Reinforced Plastics Corporation. The other, though more innovative, was only a small model. Working for General Electric, industrial designer Eliot Noyes had envisioned a fiberglas tent or umbrella curving down at four corners to enclose a pavilionlike space whose four elongated side arches would be filled with vertical panels of plastic and glass (Figure 7–9).[53]

As the MIT team considered design strategies, they realized a difficult task lay

7–9. MODEL OF PROJECTED FIBERGLAS-POLYESTER WONDER HOUSE OF 1964, DESIGNED IN 1954 BY ELIOT NOYES FOR GENERAL ELECTRIC COMPANY

ahead of them. Dietz and Hamilton emphasized they "were dealing with a material the shape of which could be anything from flat to completely amorphous." Because they faced "a freedom that was all too complete," it made sense to consider the two solutions already proposed. Without naming names, they found both lacking. They rejected flat wall panels like those in Smith's beach house for accomplishing nothing that could not be done with wood siding. The "aesthetics" of such a structure, according to the final report, "were reminiscent of more standard materials"; in other words, "plastics again were a 'substitute.'" On the other hand, Noyes's tent or umbrella certainly looked innovative, and its curvature enabled a plastic roof-shell to serve a structural function by supporting itself. The design satisfied their insistence that the "ultimate form" be "peculiar to the plastics fabrication process." The Noyes model contributed to a growing certainty that this ultimate form would be found by "exploring compound curvature and shell structure," by devising "thin hull-like components" with a "minimum amount of material." But the MIT researchers decided that an umbrella or tent design yielded an enclosure "too large" for a typical family's "space budget." Such a house would be too expensive to heat, too open to the elements and to burglars. Shaped somewhat like a sail, the shell might even blow away. Eventually Dietz and his associates came up with a third approach, a house whose "total enclosure" was defined by "a continuous surface" running from floor to wall to ceiling in smooth molded curves. As they envisioned it, the plastic shell extended beyond a mere covering like an umbrella and became a self-contained, all-encompassing structure.[54]

Even this third solution was not a new concept. Late in 1954, after MIT's preliminary phase was well under way, architecture critic Douglas Haskell articulated the vision behind Dietz's work and that of other architects and engineers around the world who were seeking plastic's uniqueness as a structural material. Haskell asserted plastic buildings would be "all 'skin.'" With a seamless material as strong as steel at one-seventh the weight, architects could design structures "as thin as egg shells, as ribbed as leaves, as corrugated as sea shells." The same material would simultaneously provide both structure and surface, thus eliminating the assembly costs inherent in traditional construction systems. In "tomorrow's plastic order" of architecture, design and manufacture of structures would become "largely a laboratory job." He predicted that sculptural forms of architects like Saarinen and Pier Luigi Nervi who worked in reinforced concrete would soon be translated into "the new plastic mode." Although Haskell based his enthusiastic comments in *Architectural Forum* on irradiation, which briefly promised to transform polyethylene into a structural material, Dietz and his colleagues were already creating such a structure. As always, however, their chosen material of reinforced polyester supported the irony that its innovative engineering and space-age appearance depended on a slow, laborious process no different in concept from making papier-mâché.[55]

After deciding on an all-encompassing self-supporting shell as the best solution, the team began the actual design process, largely carried out by Marvin Goody, an assistant professor of architecture, and by a graduate student named Ernest

Kirwan—though Hamilton remained overall head of the project and Dietz continued with the engineering input that made the project possible in the first place. Conscious of making a prototype for mass production, an assumption that prompted them to ignore many traditional associations of house and home, they focused on "needs of the family of the future."[56] By that they meant the postwar nuclear family with its geographic mobility, its increased leisure time, its retreat into domestic activities. Traditionally a house had changed over generations by a process of evolution. Additions and renovations entailed expense and inconvenience and imposed subtle limitations on those who lived with them decades later. Mass production of housing promised to change all that by giving the house a life span no greater than that of the family station wagon. Despite the relative permanence of fiberglas, the plastic house envisioned at MIT would respond to changing needs. Identical modular rooms could be rearranged on site. A family could add rooms as needed and remove them when children left home. In its earliest, most solid incarnation, as the Monsanto House of the Future (Figure 7–10), plastic architecture thus promised flexibility and impermanence.

Final plans for the Monsanto House specified a square foundation of reinforced concrete sixteen feet on a side and about six feet off the ground.[57] The foundation enclosed a storage room and mechanical equipment; its upper surface became the floor of a kitchen and two small bathrooms. Cantilevered from each of the central core's four sides—jutting into space, in other words—was an identical modular room, also sixteen feet on a side, defined by a C-shaped structure of fiberglas-reinforced polyester. The outer structural surface of each C-shaped module consisted of four L-shaped fiberglas "bents," each eight feet wide. Two upper bents, placed side by side, projected out sixteen feet to form the roof of a module and then curved down about four feet at the outer wall, where they met two corresponding

7–10. Monsanto House of
the Future, Disneyland,
California, 1957–1968

1. CELLULOID AS IVORY, AS TORTOISESHELL, AND AS A FRANKLY ARTIFICIAL MATERIAL.

2. CELLULOID SOAP DISH IMITATING MARBLE.

3. OPERA GLASSES AND HAIR RECEIVER IN PEARLIZED CELLULOID.

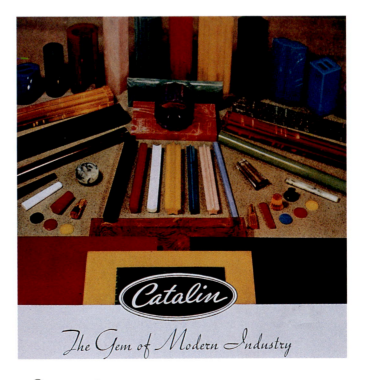

4. CAST RODS OF CATALIN PHENOLIC RESIN AND FABRICATED OBJECTS, 1936.

5. Salt and pepper shakers, biscuit cutter, toy car, and Westclox travel alarm molded of urea formaldehyde, 1930s.

6. Cellulose acetate resin flakes and color samples, 1939.

7. FADA STREAMLINER RADIO, 1941.

8. SEARS SILVERTONE RADIO, 1936.

9. MARBLEIZED, STREAMLINED SEARS SILVERTONE
RADIO, LATE 1930s.

10. TYPICAL PHENOLIC MOLDINGS OF THE 1930s.

11. Clocks molded of polystyrene, mid 1950s.

12. Acrylic purse, 1950s.

13. POLYSTYRENE GUM DROP TREE MANUFACTURED BY UNION PRODUCTS, INC., FROM THE EARLY 1950s TO THE PRESENT.

14. BOONTONWARE MELAMINE DISHES MANUFACTURED BY BOONTON MOLDING COMPANY, 1950s.

Quality furniture leads a charmed..and charming..life, surfaced with

Catalin LAMINATING RESINS

15. ALL-PLASTIC DINING ROOM, 1963.

floor bents curving upward. Where roof and floor L's met at the horizontal center line of the outer wall, they were joined seamlessly with bolts and epoxy cement. Four smaller fiberglas panels formed the roof of the utility core and tied the eight roof bents together above. The two nonstructural sides of each room or wing could be filled in with any combination of glass or siding. The designers intended the finishing-out of these open sides as an antidote to the boredom of identical mass-produced houses. Personal selection of materials would express an owner's individuality and afford "the male of the family" an outlet for "do-it-yourself creativity."[58] The futuristic prototype discretely avoided eccentricity, however, with floor-to-ceiling glass and white curtains. And while its exterior fiberglas surface required no protective finish, joint lines were smoothed with polyester gunk and the entire surface spray-painted a uniform eggshell white to emphasize the pristine modernity of plastic—at the insistence of Walt Disney's "imagineers."

Goody, Dietz, and Hamilton intended this basic fiberglas house of 1,280 square feet for mass production. Although the prototype cost Monsanto close to a million dollars, the company estimated it could manufacture the house in volume for about $20,000 apiece. The designers hoped to eliminate the expense of on-site collection and assembly of traditional building materials by providing a few major factory-built elements ready for assembly with minimal site preparation. The L-shaped bents of the Monsanto House nested for efficient shipping and could be lifted into place on site by crane. Actual assembly took only three weeks, not bad for a trial run, but it did not include shipping time from the Winner Manufacturing Company in Trenton, New Jersey. And it had taken three months to fabricate the parts in Trenton. Everyone hoped they could substantially reduce manufacturing time after moving into mass-production mode. But the process supervised by Winner's engineer C. G. Cullen was tedious. The company had considerable experience molding polyester and fiberglas, having worked with Saarinen to manufacture womb-chair shells for Knoll Associates. Even so, the processes used to mold the Monsanto House differed little except in scale from those used by the Eameses in their own studio—or by any California sports car enthusiast. At times it seemed everyone connected to the project had succumbed to the myth of an injection molding machine squirting out a finished house every fifteen seconds.[59]

The first step in making an L-shaped bent was to mold its exterior shell. A wooden form had to be constructed with the exact shape the completed shell was to take. Then a mold was created over the form by building up layers of polyester resin and fiberglas, reinforced at the core with an inch-and-a-half-thick honeycomb of kraft paper impregnated with phenolic resin. When removed from the form and inverted, the concave mold provided a surface for hand lay-up of the actual shell using polyester resin and ten plies of woven fiberglas mat, a process that took three workers about eight hours. In a variant of the bag method of molding, the laid-up material of the shell was pulled by vacuum against the inside curve of the mold and cured for about two and a half hours at a temperature of 80 to 90°C. When the shell emerged from the mold, it was nearly one-third of an inch thick. Then the shell was ready for insulating and stiffening with three and a half inches of rigid polyurethane foam sprayed on the inside. After hardening of the foam, a worker planed its surface to prepare it for bonding with a thin inner skin of glass-reinforced polyester applied by hand pressure and cured at room temperature.

A fully completed bent contained yet another plastic assembly. Each floor bent required an eight-by-sixteen-foot floor panel, about four and a half inches thick, composed of a layer of phenolic-impregnated honeycomb sandwiched between thin layers of fiberglas. Two similar floor panels lined the top of the concrete foundation. Each roof bent likewise required a ceiling section about three inches thick, made by a process similar to that of the exterior shells. In addition to the house's sixteen exterior shells, sixteen inner skins, ten floor panels, and eight ceiling sections, more than forty other miscellaneous panels and plates of reinforced plastic went into its structural system, four of them for the roof of the central section. Given the inexactness of fiberglas molding, all parts had to be trimmed with a hand-operated power saw. Because the saw's weight made it hard to maintain a straight cut along the sixteen-foot length of a bent, that cut was made outside the desired line and then ground down to the line by hand with a disc sander. All surfaces to be epoxied during final assembly had to be roughened by hand with a power sander. Structural parts of the house consumed twenty thousand pounds of polyester resin, twenty-five thousand pounds of woven fiberglas, and four thousand pounds of polyurethane foam. Glowing newspaper accounts made it sound as if two or three workers lowered sixteen identical L-shaped bents into place with a crane, bolted them together, and that was that. The reality, it should now be clear, was messy, complicated, and time consuming (Figure 7–11).

None of that ambiguous complexity was apparent to visitors trooping through the world's first all-plastic house when it opened at Disneyland in June 1957 with Sleeping Beauty's castle rising in the distance. The surroundings posed a different prob-

7–11. Assembly of L-shaped bents of the Monsanto House at Disneyland, 1957

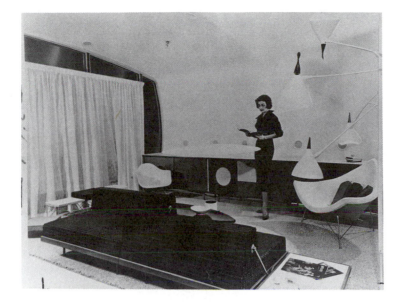

7–12. INTERIOR OF MONSANTO HOUSE OF THE FUTURE, 1957

lem by suggesting that the Monsanto House of the Future was just another Disney anomaly, as independent of everyday life as Mr. Toad's Wild Ride. On the other hand, Disney seemed to have his hand on the pulse of the future. His television show made Wernher von Braun a household presence and popularized the Aladdin's lamp of atomic energy. Disney's approval might be expected to bring the Monsanto prototype closer to reality in new suburbs spreading over pasture and prairie. After all, synthetic materials had come a long way since "The Plastics Inventor" satirized plastic utopianism at Donald Duck's expense. The paradox posed by the surroundings of the Monsanto House vanished as visitors entered to explore the interior (Figure 7–12), an activity that might be described, as someone once said of another such visionary experience, as "looking at Tomorrow and arriving mostly in the middle of Today."[60]

Some aspects of the interior marked the Monsanto House as unique. The far outer wall of each room curved gently from top to bottom. Small portholes for the air conditioning system created a feeling of space travel. But not much else distinguished it from other examples of ultra-contemporary styling familiar from magazines if not from actual middle-class suburban houses. Almost everything was plastic or covered with plastic. Synthetic carpeting gave way to vinyl flooring in kitchen and bathrooms. Colorful laminate covered minimal kitchen cabinets and counters and wall-hung storage units in bedrooms. Couches were formed from simple slabs of upholstered foam on concealed frames, and beds were minimal foam slabs. Lighting came from small recessed circular fixtures or from pole lamps with flared shades of translucent plastic, and in the kitchen from behind luminescent plastic panels. The family-room table was furnished by Herman Miller—as were its chairs, upholstered-foam upgrades of Eames fiberglas shells. An Eames fiberglas armchair graced the master bedroom, as did George Nelson's "coconut chair," a triangular wedge of fiberglas, foam, and cloth upholstery resting on metal legs. A Saarinen pedestal stool sat under a wall-mounted vanity. The Monsanto House offered an extreme version of an interior landscape already familiar from doctors' waiting rooms and new suburban schools.

Only in kitchen and bathrooms did the interior become visionary—if materialized visions can be so defined. The kitchen boasted an ultrasonic dishwasher, an oven rising from a cabinet at the push of a button, storage shelves and refrigerator dropping from cabinets at the push of a button, and a brightly colored industrial aesthetic. Even more unusual, each bathroom was assembled from two fiberglas shells, the lower of which included an integrally molded shower or tub. Push-button telephones, closed-circuit video monitoring, and a push-button selection of air-freshening scents completed the amenities of futuristic life. Planned by interior designer Victor G. Canzani, the finishes and furnishings of the Monsanto House harmonized with its radical structure but did not depend on it. A plastics engineer like Dietz might have viewed the interior as an afterthought—striking, even impressive, but also arbitrary. For many visitors, however, the interior offered a comforting familiarity that domesticated the unprecedented exterior form. The interior mediated between present patterns of suburban life and an uncompromisingly artificial future. This ultra-contemporary but also style-dependent interior ironically doomed the Monsanto House of the Future. In ten years it was out of fashion, reminiscent of a utopian moment whose mood had become a backdrop of middle-class life. Disney imagineers might have renewed the interior. But the exterior, its curving fiberglas shell, was also out of date, reminiscent of a popular plastic-as-plastic style of the mid-fifties epitomized in the fiberglas chairs of Eames and Saarinen.

Although Dietz and Hamilton had planned a flexible modular system reflecting the impermanence of families, the Monsanto House in fact was stubbornly permanent. Its solid contours belonged more to the conservative thermoset ideology of the 1930s than to the shifting thermoplastic transformations of the 1960s. The design team had not intended it that way. The Monsanto House was supposed to be "demountable." Goody had drawn alternate floor plans showing how its size could be increased by adding another foundation core and grouping five or six modules around the two of them. Techniques used to assemble the prototype house reinforced the goal of impermanence. The bents were connected on site with steel bolts so they could be taken apart and erected elsewhere. *Popular Science Monthly* breezily noted it would be "easy to remove a wing when you no longer need it" and dump it at a "secondhand-wing lot" whose proprietor would claim it was "formerly owned by an elderly schoolteacher who occupied it only on Sundays."[61] As it turned out, on-site assemblers had used epoxy adhesives for improved insulation. When the futurism of the Monsanto house wore off in 1968, eleven years and twenty million visitors later, it took the wreckers two weeks to bring it down. After wrecking ball, blowtorches, chain saws, and jackhammers all failed, the crew attached cables and literally "tore the modules into pieces small enough so they could be carted away."[62] In a sense fiberglas-reinforced polyester failed as a construction material because it was too perfect. It was too tough and resistant for an inflationary era that valued the ephemeral.

Although the Monsanto House did not live up to expectations of flexibility, it inspired others who hoped to use plastic in architecture and building. The Soviet Union, with its intense interest in prefabrication, erected a plastic house at Leningrad, a single heavily ribbed module of polyester and fiberglas about twenty feet square, with floor-to-ceiling windows on two sides. Perched ten feet above ground on a tiled concrete foundation, it avoided the soft consumer-appliance curves of the Monsanto House and instead looked industrial, hard-edged, like a huge radiator or giant section of ductwork.[63] Elsewhere around the world, architects, engineers, and plastic companies constructed dozens of innovative dwellings from reinforced polyester. The self-supporting shell proved most popular, but some designers developed versions of the panel and umbrella systems rejected by Dietz at MIT. A survey documented sixty-seven rigid plastic buildings completed between 1956 and 1969, with the annual number slowly accelerating. Most were experimental prototypes but a few entered limited production. The Futuro House, for example, was a fiberglas "flying saucer" on metal struts introduced in Finland in 1968 and later featured by the countercultural *Whole Earth Catalog* as a high-tech alternative to homemade geodesic domes.[64]

Despite fascination with the Monsanto House, the plastic industry experienced difficulty breaking into construction. The "problem of marketing" was "extremely complex," as Cruse reported in 1964 at the first meeting of a Plastics in Building Construction Council. Unlike the automobile industry, where only four or five companies had to be sold on plastic's virtues, in construction there were "literally hundreds of decision makers"—the engineers, the designers and architects . . . , the contractors, the sub-contractors, the material distributors, the home and building owners"—all of whom had to be "educated." Worst of all were local building codes controlled by conservatives bent on protecting traditional materials and the building trades that worked with them. In 1955 SPI had set up a Code Advisory Committee whose members attended meetings of the nation's three major code-writing associations. For years they endured the hostility of opponents who thought nothing of "hold[ing] a small sliver of . . . plastic . . . over a candle" to convince code officials to ban all plastics from construction sites. As even Dietz admitted, the conservatives had a point. There were no industrywide standards for plastic building components because none had existed long enough to prove their durability—their resistance to sun and weather—over a building's typical life span.[65]

Given these problems, most manufacturers of plastic building components tried to smuggle them in, if not in outright imitation of other materials then as one-for-one substitutes. This intention motivated a Formica World's Fair House cited by

Cruse as the most significant example of plastic in architecture at the New York World's Fair of 1964. In style a typical ranch house, it featured Formica on almost every interior surface, acrylic skylights, and an imitation-wood exterior "so expertly reproduced . . . one must touch the siding to be convinced . . . it is laminated plastic." The interior was decorated in a cloying modern manner (using "exciting new wipe-clean materials") with nothing more innovative than a flowery knock-off of Saarinen's pedestal chair. The company avoided any mention of "the Future" and emphasized current availability of all products. Formica cooperated with local contractors to scatter a hundred and fifty replicas of the World's Fair House across the country, many exhibiting Cape Cod or Williamsburg exteriors. By putting fairgoers in touch with regional contractors, the company reinforced use of plastic by ordinary contractors and ordinary clients. In a deliberate slap at the Monsanto House, then seven years old, Formica praised its own model home as "neither extreme nor bizarre" and bragged of avoiding "stark over simplified modular forms."[66]

Eventually Dietz recognized the inevitable. Writing in *Progressive Architecture* in 1970, thirteen years after he and Goody completed the Monsanto House, he lamented that plastic composed only 2 percent of the total poundage of materials used in building. While he encouraged experiments with reinforced plastic, he admitted that nonstructural uses would continue to dominate—pipe, trim, insulation, hardware, lighting, moisture barriers, and coatings.[67] Such uses enabled independent contractors to retain control over their projects. They remained free to pick and choose from among many products, to accept or reject. And workers could assemble or apply such products without changing their habits. These freedoms flew in the face of the major argument for mass-produced housing. As a British observer put it, eliminating skilled labor had provoked the dream of a "lightweight, maintenance-free, highly finished artifact" that could be "moulded off-site, swiftly transported, [and] slotted effortlessly . . . into place." In his view the dozens of experiments with glass-reinforced polyester, beginning with the Monsanto House, had all ironically failed because promoters could not eliminate skilled labor in the plant, at the point of fabrication. Despite a facade of rhetoric about "quality-control and mass-production," the working of fiberglas remained "an intensely craft-based industry."[68]

Beyond that, the Monsanto House failed because its precise machine-made appearance implied greater regimentation than American homeowners could accept. It was no accident that Soviet engineers rushed to experiment with what they had learned at Disneyland. In doing so, they abandoned the American idea of modular rooms added or subtracted at the whim of an individual owner; instead they designed a box the size of a Soviet apartment, intended for stacking in rows to form prefab apartment blocks. In fact the American prototype afforded precious little room for individual customizing beyond its side panels—mostly a condescending afterthought. Its very permanence, at one time the sine qua non of plastic, defied the promise of malleable plasticity. As an engineer put it in 1966, architects had to stay "fluid in ideas" to keep up with an "ever-changing world." Plastic enabled them to "meet the needs of an expansive future." Even an editor at *Modern Plastics* maintained the industry needed a "split personality" to "balance the hard-nosed realistic approach . . . of codes, unions, and education" with a "freewheeling, high-flying, imaginative approach." That unlikely combination "would create an architectural revolution around plastics."[69] For the moment, however, few people were thinking about codes. Instead they were dreaming of ever more insubstantial uses of plastic for inflatable vinyl domes, polyurethane foam yurts, disposable furniture, and personal dwelling pods. Lightweight, composed of insubstantial foams and films, of flexible vinyl and polyethylene rather than rigid polyester, the new plastic-as-plastic design of the 1960s and 1970s promised do-it-yourself environments fully malleable in the hands of individuals—and so ephemeral as to be disposable.

SOFTWARE

As a contributor to a new vision of plastic, Buckminster Fuller was something of an eccentric. Born into an old New England family in 1895 and inspired by the transcendental optimism of his great-aunt Margaret Fuller, he had earned a reputation during the Depression as a technological prophet and gained notoriety with his streamlined three-wheeled Dymaxion car. Like so many other inventors and designers he became obsessed after the war with factory-built housing. By the 1950s he was focusing his efforts on the geodesic dome, a self-supporting structure whose light metal struts formed triangular or hexagonal cells framing the panels of a lightweight skin. The small domes of the 1950s—experiments conducted with MIT students or for the military—led to ever larger projects in the 1960s. At St. Louis in 1961 Fuller erected a partial dome nearly two hundred feet across and seventy feet high to protect the tropical plants of the Missouri Botanical Garden. His geodesic triumph came in 1967, when a nearly full sphere two hundred feet in diameter enclosed the U.S. pavilion at Montreal's Expo 67. Sheets of acrylic plastic set in neoprene gaskets formed the transparent skins of these two domes. Such structures created an impression of airy insubstantiality. By comparison the Monsanto house and other buildings of fiberglas-reinforced polyester seemed heavily earthbound even though their sculptural curves evoked the concept of flight.[70]

While Fuller was predicting domes would someday enclose entire cities, other engineers and architects were experimenting with plastic films and foams to create even less substantial shelters. Their activities fit perfectly with Fuller's process of "ephemeralization," of "doing vastly more with vastly and invisibly less." Although he attributed this shift to an "invisible tidal wave of atomics, electronics, and computerization," plastics and alloys bore the ever lighter but no less material burden of ephemeralization. Fuller himself maintained that plastic was neither "artificial" nor "synthetic." Instead it was "inherently *natural*" because it was based on discovery of "complex structural behaviors permitted by Nature." Plastic indicated not an improvement on nature but a more efficient human adaptation to nature's "rules of structuring."[71]

Fuller's ideas appealed to a generation that was fearful of uncontrolled technological expansion rendering the earth unfit for human habitation by consuming resources and polluting the planet. The aging utopian crisscrossed the globe in the 1960s giving marathon lectures to young fans who idolized him as "Bucky." As hints of environmental consciousness surfaced before the first Earth Day of 1970, many activists opposed technology as a demonic force inherently destructive of nature. Others, including Stewart Brand of the decentralist *Whole Earth Catalog*, embraced Fuller's higher-tech vision of ephemeralization. Now that NASA had revealed the "whole earth" in photographic images suggesting that "we are all one," as a popular phrase had it, the space program shared in Fuller's mystique. Necessarily the most ephemeralized of technological fields, space exploration relied on high-performance synthetics, on Mylar, Teflon, and nylon, on heat-resisting composites and form-fitting foams. As *Modern Plastics* put it, success in the space program offered "the best testimonial yet devised" for establishing plastic's reputation.[72]

For some counterculturalists of the 1960s, innovative uses of plastic became associated with Bucky Fuller and geodesic domes, with holistic thought and decentralist lifestyles, with new frontiers of outer space and psychedelic inner space. Plastic's damp-cloth domesticity yielded to a boldly artificial "wet look" of shiny, bright-colored vinyl for boots, coats, skirts, and vests.[73] This "go-go" style revealed plastic's pop-culture affinity with the flawless reflective surfaces of NASA's space suits and gear. The plastic-as-plastic vinyl look was more than mere fashion, however. A whole new complex of images surrounding plastic suggested a resurgence of something like the old plastic utopianism. With high strength-to-weight ratios and relative cheapness, plastic films and foams promised to provide basic shelter with an unprecedented degree of efficiency and control but without sacrificing individuality. Because permanence was no longer a goal, these new plastics seemed perfect for enclosing the ever shifting artificial environments of the habitat Fuller referred to as "Spaceship Earth."[74]

Despite this new plastic utopianism, more ephemeral than thirty years before, ambivalence remained central to the American experience of plastic during the 1960s. At the beginning of the decade Frederick J. McGarry, assistant director of Dietz's MIT laboratory, criticized two major responses to plastic as equally harmful and misleading. Either people celebrated plastic with "evangelical fervor," expecting it to "solve the housing crisis, the school crisis, the transportation crisis, the urban blight." Or they rejected plastic goods because "they burn and break, they creep and fall apart outdoors, their colors are garish, and you can't tell one plastic from another—until it's too late."[75] McGarry overstated his case with such obvious extremes. Plastic was too much a part of everyday life for such unequivocal responses. Anyone who accepted the utopian position also experienced the negative underside.

Followers of Bucky Fuller dreaming of inflatable plastic yurts laughed like anyone else when hearing Dustin Hoffman being told that "Plastics" offered the key to life. That the same person could respond to both *The Graduate* and *The Whole Earth Catalog* indicated plastic was finally integrated into the cultural landscape. Both re-actions lay outside the structures of words and images used by the plastic industry to present itself to consumers. In each case the response to plastic functioned within a larger debate on the definition of the culture and its direction of development. At the same time, however, plastic was playing an independent role in shaping that culture—both directly, as a material presence, and indirectly, as a focus of other intentions. The Monsanto House of the Future was the last major promotion in which defining plastic remained in the industry's hands. Publicists soon lost control of plastic's image to enthusiasts and critics who took what was offered but molded it to whatever they feared or desired.

Nothing made from postwar plastics promised so much freedom and mobility, so much control over personal environments, as inflatable dwellings and furniture that could be deflated, moved, and reinflated at a moment's notice. Such artifacts re-mained novelties among the things of the 1960s but connected with a cultural trend that went beyond original intentions. Thomas Herzog, author of a technical treatise on "pneumatic structures," briefly abandoned his neutral prose to explain that in-flatables offered an opportunity to abandon modern architecture's "hard, cold, machine-produced surfaces." In their place rose a "sensuous plastic world" of "soft, flexible, movable, roundly spanned, 'organic' shapes . . . of great sensuous beauty." Dozens of photographs supported Herzog's claim.[76] By turns awe-inspiring or silly, warmly biomorphic or crudely artificial, all these pneumatic structures shared two properties: they relied on air pressure to inflate their plastic membranes, and they resembled no other buildings known to the human race. Herzog's images revealed huge globes, groups of interconnected domes (Figure 7–13), multiplying colo-nies of spheroids, arched structures ribbed and corrugated like vast air mattresses, visceral blobs and intestinal tubes, fields of massed breasts with nipples erect

7–13. Inflatable kindergarten designed by Gernot Minke at Ulm, Germany, 1972, one of the less provocative structures pre-sented by Thomas Herzog

217

7–14. FANTASTIC INFLATABLE PROPOSED BY GERNOT MINKE AND STUDENTS, 1971

(Figure 7–14), giant condom-sheathed phalli bunched like asparagus, smooth-surfaced cocoons, suggestions of alien pods and polyps from science fiction and soft space capsules for hedonistic weightlessness. Given these associations, it was not surprising that inflatables owed their existence to the aerospace industry. During the late 1940s the Office of Naval Research had sponsored development of the Skyhook weather balloons. Made of polyethylene film and inflated with helium to a hundred-foot diameter, they carried measuring instruments twenty miles from the earth's surface into the stratosphere. In 1951 a navy spokesman tried to puncture an early space-age myth by claiming people had mistaken sightings of Skyhook balloons for "so-called flying saucers."[77]

Most early uses of inflatable plastic were earthbound. At the end of the war the air force sought lightweight demountable shelters for radar installations. These radomes had to enclose large volumes in a free-standing skin whose material would not interfere with transmissions. They also had to withstand wind and cold in isolated arctic environments, had to be shaped to shed snow and ice, and had to shield equipment from unequal heating in bright sunlight. Engineers suggested geodesic domes and rigid structures of glass-reinforced polyester. Walter Bird of Cornell University's Aeronautical Laboratory at Buffalo had another idea. In 1948, after two years of work, he successfully inflated an experimental pneumatic structure at the Buffalo airport, where severe winter conditions approximated the arctic. Made of heavy nylon coated with synthetic rubber, Bird's first dome was a single-walled structure much like a child's balloon. Air blowers maintained the dome's shape by creating an interior pressure great enough to support the skin's weight and to withstand wind and snow. Double doors worked as an airlock to prevent loss of pressure. By the early 1950s hundreds of pneumatic radomes were scattered across the tundra as part of North America's early warning defense system. In 1956 Bird founded Birdair Structures to fabricate inflatable structures for nonmilitary customers. Ten years later his company was the leader of an industry generating sales of ten million dollars.[78]

Pneumatic structures soon became more complex than Bird's first balloonlike domes. In 1958 a competitor announced a "multipurpose, portable nylon dome-shelter" that required neither airlock nor blowers because it was "self-supporting" rather than "air-supported." This Geodome consisted of independently inflated panels or cells joined together to form a rigid shelter whose doors and windows could be opened without fear of losing air pressure.[79] Other companies eliminated airlocks and fans by inflating a sealed space between two membranes of a double-skinned wall; however, such systems risked catastrophic collapse in case of puncture. Eventually, in search of ever lighter and even transparent materials, some engineers specified vinyl film, much thinner than coated nylon and so weak it had to be supported outside by crisscrossing cables or by an unobtrusive nylon mesh.

In 1957 one of Bird's associates predicted an "unlimited" market for inflatable domes for "circus tents, fair buildings, exposition halls, portable hangars, farm produce storage depots, [and] sports stadiums." Other uses included temporary shelters for construction sites in winter and seasonal covers for swimming pools and tennis courts. A few entrepreneurs considered the inflatable dome a solution to the problem of housing an increasingly mobile population. For example, the Irving Air Chute Company exhibited a two-room Air House at the International Home Exposition in New York City in June 1957, the same month the Monsanto House opened at Disneyland. Intended as a beach or vacation home, the Air House consisted of two domes of vinyl-coated nylon connected by a short tunnel. A sausage-shaped tube filled with sand anchored each dome around the perimeter. The Air House enclosed about 1,585 square feet, with the larger dome measuring thirty-eight feet across and nineteen feet high, the smaller twenty-four feet across and twelve feet high. A revolving door airlock prevented loss of interior pressure. Although the company boasted that Frank Lloyd Wright, then ninety years old, had designed the floor plan of the Air House, its Herman Miller furniture and rubber plant suggested a dentist's waiting room. Even so, mainstream culture was not ready for anything quite so insubstantial.[80]

A few years later, during the late 1960s and early 1970s, inflatables took their place with geodesic domes as countercultural icons. One even popped up in suburban Maryland as a branch campus of Antioch College. Reflecting the Ohio liberal arts college's radical mission, most of some two dozen new branches served inner-city neighborhoods across the country. A three-quarter-acre dome inflated in 1973 at Columbia, Maryland, was an exception. The town was a planned satellite community attracting middle-class professionals and government bureaucrats fleeing Washington and Baltimore. As an exemplar of America's mushrooming suburban expansion, Columbia was an appropriate site for an inflated college with neither tradition nor definite agenda, given such shape as it had by impermanent plastic film. The *New York Times* reported that such domes were "gaining popularity as an inexpensive, deliberately impermanent and highly flexible means of providing shelter for education," in this case for about seventy-five students and their instructors.

The occasion for the *Times* article was a gathering of four hundred architects, academics, entrepreneurs, students, and "self-styled survival specialists" at the

219

National Conference on Air Structures in Education held at the Antioch inflatable in May 1973. The keynote speaker was British architect Cedric Price, who in 1962 had opened an era of manifestos for an alternative architecture with his design for a Fun Palace, a limitlessly extendable framework in which an endlessly pleasurable phantasmagoria of "expendable" and "changeable" environments and activities might be experienced. Invoking the psychologically liberating potential of an inflated environment (soft, hemispherical, visually unlimiting, opening outward from within), Price told the shaggy, psychedelically charged gathering of neo-plastic utopians that Antioch's bubble enabled people "to distort time and space for advantage." The dome's architect, Antioch professor Rurik Ekstrom, chose to express a more angry radicalism. Inflatable domes offered an antidote to the "rape of the landscape" by megalomaniac architects whose "brick and mortar monuments" flattered the egos of corrupt corporate and government clients. Implicitly attacking suburban Columbia, Maryland, itself, Ekstrom praised the impermanence of the pneumatic campus. "If they decide a few years later that this is a lousy place," he declared, then "they can pick the thing up and go somewhere else and the site will be a meadow again."[81]

Price's involvement at Antioch revealed that utopian celebration of plastic inflatables had become an international phenomenon independent of the plastic industry's public relations efforts. Responding on their own to new materials, designers and architects shaped an aesthetic of artificiality inspired by the dawning space age and the pursuit of pleasure among the rising middle classes. A conjunction of these influences informed the trendy science fiction film *Barbarella*, accurately pegged by British design critic Reyner Banham as a "triumph of software" with "an ambience of curved, pliable, continuous, breathing, adaptable surfaces."[82] The movie made up for lack of plot and acting by its sets, a series of sensuously polymorphous environments shaped from foams, membranes, bubbles, and coatings of polyvinyl chloride, polyurethane, polyethylene, polymethyl methacrylate, and synthetic fur of polyamide (a kind of glorified shag carpeting).

Few of the undergraduates attracted to midnight showings by rumors of Jane Fonda's weightless striptease during the opening credits realized they were witnessing a plasticized parody of Botticelli's *Birth of Venus*. As the film opens, Barbarella floats in the fur-lined control center of her spaceship, itself floating in a clearly aqueous medium. She is wholly encased in a black vinyl spacesuit resembling a deep-sea diver's outfit. Slowly, still floating, Barbarella removes long black vinyl gloves, long black vinyl boots, and then a reflective bubble helmet whose black mylarlike surface has just slid down inside to reveal the heroine's head in an acrylic fishbowl, face and hair posed like those of Botticelli's *Venus*. Finally she floats nude, blonde hair blowing up and back as in the painting, while a baroque trumpet flourish interrupts a song whose inane lyrics reinforce the association: "Barbarel-la, Psychedel-la, there's-a kind-a cockleshell-a bout you." There was something pneumatic about Barbarella as well, especially when wearing a tight-fitting wet-look corselet with transparent plastic cup enclosing a perfectly stylized artificial breast. Fonda, the only American in a cast of Europeans, unwittingly offered herself in fulfillment of *Fortune*'s thirty-year-old "American dream of Venus" in surrealistic plastic, an embodiment of the "Plastic Fantastic Lover" then being celebrated on a popular record by the Jefferson Airplane. *Barbarella* was released in the United States in 1968, at about the same time as *The Graduate*. Although the European film was not nearly as popular, it exhibited a sophisticated, accessible conception of plastic as an artificial substance so responsive to human desire as to inspire fetishism.

As Banham recognized, the designers of *Barbarella*'s environments had borrowed from an international architecture movement devoted to what he called "megastructures."[83] He defined a megastructure as a vast permanent framework into which modules could be slotted to meet changing social and individual needs. The concept was promoted by a group of British architects in a slam-bang sci-fi-comic-book-inspired magazine called *Archigram*, which took over from Price the challenge of developing an aesthetic of leisure. Most famous of *Archigram*'s colorful futuristic designs was Peter Cook's Plug-In City of 1964 (Figure 7–15), a set of infinitely extendable multilevel frames hung with transport and utility tubes and equipped with gantries and cranes for quick response to changing needs. The manic obsessiveness and bright primary colors of his drawings connected with a pop Carnaby Street sensibility, as did collaged images from slick magazine advertisements portraying the swinging life of Plug-In City and other *Archigram* projects.

A spin-off journal called *Megascope* summarized the effect by observing how "the rounded corners, the hip, gay, synthetic colours, pop-culture props all combine to suggest an architecture of plastic, steel and aluminum, the juke box and the neon-lit street."[84] Steel and aluminum constituted the hardware or permanent framework of a megastructure, while plastic formed the software, the moveable pods, modules, domes, and bubbles. As other architects followed *Archigram*'s original lead, their

7–15. Detail of Peter Cook's Plug-In City, 1964, with moveable plastic pods, crane, and steel frame

vision culminating in the megastructural Expo 67 at Montreal, those associated with Cook shifted emphasis away from hardware, "the *structure* part of megastructure," to plastic software with its "promise of permissiveness." As Banham described it, they desired an "autonomous living unit, of maximum flexibility, adaptability, mobility and non-monumentality, that could exist independently without assistance from megastructure or any other permanent support systems."[85] He playfully suggested in 1965 that such a lightweight temporary dwelling would satisfy all the trends of American domestic architecture—an emphasis on "cleanliness, the lightweight shell, the mechanical services, the informality and indifference to monumental architectural values, the passion for the outdoors." He predicted, only slightly tongue-in-cheek, that the typical American family would soon be living in a vehicle equipped with a "transparent Mylar airdome." Three years later, at the Milan Triennale, it was another Englishman, architect David Greene, who exhibited a prototype for a radically individualist piece of software. Taking the anti-megastructure trend to an extreme, Greene offered the Suitaloon (Figure 7–16), a plastic film garment worn by an individual who could inflate it to form a small personal pneumatic dome.[86]

Banham was not alone in regarding plastic domes and inflatable dwellings as antidotes to the weight of tradition. Similar ideas appeared in a "soft manifesto" issued by members of a countercultural commune called Libre to explain their decision to live minimally in a dome. Published by an architecture journal in 1971, the manifesto proclaimed that Libre's members were "trying to get free of our egos" so as to become "free to deliberately choose & build our environment & lifestyle." The commune itself was "a growing-living thing in space & consciousness" whose essence was not warped by conforming to rigid traditional buildings. They described their forty-foot dome as "especially fit for changes, being an empty unimposing shell." In over three years its interior had never "settled in" or become fixed. "Each month it changes again," they boasted, and it might "never stop shifting as new needs arise,

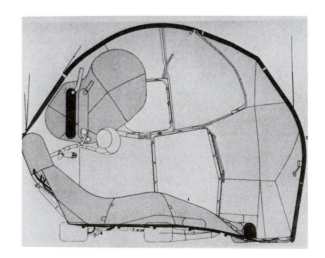

7–16. David Greene's Suitaloon,
a one-person plastic inflatable
proposed in 1968

new images are launched." For members of Libre, a dome offered a neutral interior landscape on which to project whatever identity seemed momentarily desirable. For other celebrants of domes and inflatables, the dimensionless sweep of a hemispheric interior proved a positive stimulus in itself. *Vogue* described an "inflatable, portable house" created in California by a design group called Chrysalis. Their twenty-five-foot vinyl Pneudome was a single-membrane structure supported by an air blower and ballasted by a perimeter tube filled with water. According to *Vogue*, the "soap bubble of a house" could be inflated by two people in less than an hour "on a hill, by the sea, in a garden," and then taken down and moved "without leaving a trace." Even more impressive was Pneudome's interior, "a shimmering magical environment." Dome dwellers were "Space Lib people" living "in wide open spaces even in the heart of big cities" and experiencing "a liberation of the human spirit as well as the body." As always, even when couched in countercultural rhetoric of self-actualization and social liberation, plastic promised unprecedented control over one's surroundings, in effect divorcing people from nature as it had always been experienced. One enthusiast insisted that "stadiums, shopping centers—even cities—may be encapsulated." Such possibilities shocked others who might have applauded had they overheard a negative reaction to Houston's air-conditioned, up-holstered, Lucite-enclosed Astrodome, which opened in 1965: "Who's kidding whom. This isn't baseball." The very concept of "Space Lib people" suggested a future identical to Barbarella's world.[87]

Oddly enough, one way of compensating for the disorienting transience of inflatables harked back to the stone age for its psychological effect. The polyurethane foam dome, a kind of petrified inflatable (Figure 7–17), invited comparisons to prehistoric cave dwellings or, in a more timely reference, to the hobbit holes of J.R.R.

7–17. Experimental polyurethane foam house sprayed up by Felix Drury and Yale architecture students

Tolkien. If an inflated vinyl or polyethylene dome projected an emotionless astral cool, then a solid polyurethane structure of rough textures and earth tones invited a sharing of sheltering warmth. A few architects had experimented in the mid-1960s with polyurethane foam as a structural material, but the idea did not gain much attention until about 1968 when it was taken up by Felix Drury of Yale University. A series of experimental houses ended with a showpiece completed in 1971 at Langdale, Alabama, sponsored by the textile firm West Point Pepperell and the magazine *American Home*. The basic technique was simple. Starting with a slab foundation, Drury inflated a dome of nylon fabric, sprayed a four-inch layer of polyurethane foam over it, let the foam harden into a shell combining structural support with insulation, and then deflated and removed the nylon form. Workers cut holes in the shell for doors, acrylic skylights, and polycarbonate windows, all sealed with neoprene gaskets. A coat of fire-retardant luminescent white paint finished the interior while latex house paint shielded the exterior from destructive ultraviolet radiation. The finished house had four lumpy white domes connected by rounded tunnels, with a single wooden wall running down the middle of each dome to carry wiring and orient the queasy by offering a vertical surface as a familiar contrast to curved walls. Similar domes occasionally popped up during the 1970s, often retaining the earthy reddish-brown of unpainted polyurethane foam and provoking dismay and delight among their suburban neighbors.[88]

Using language from fifty years before, *American Home* referred to polyurethane foam as "an almost miraculous substance." Unlike Bakelite earlier, however, it was a material that countercultural do-it-yourselfers could seize control of and shape. Easily sprayed on, it hardened to become "as light as meringue, as sturdy as concrete, as sculpturally free as the homeowner and the architect dare to be." If Barbarella and Plug-In City had a mass-culture parallel in the Jetsons, then the foam dome suggested the Flintstones—a reference not far from consciousness when *American Home* described Drury's creation as an "adventure of cave living, '70's style." But experiencing the interior of a foam dome was not so different from experiencing an inflatable. Both differed in the same way from traditional domestic interiors. As an enthusiast put it, "A lifetime's experience of squared rooms leaves one unprepared for the exhilarating space." To stare at the convergence of curved wall and flat floor was "like viewing a natural horizon." In addition to facilitating spatial liberation, polyurethane foam liberated designers to create forms of unprecedented plasticity. Beginning to tire of domes after doing ten in three years, Drury praised polyurethane foam because it was able to "do things no other material can do." Most important, it allowed architects "to freely work with curved surfaces." Drury perceived "drastic changes" under way "in man's use and sense of time and scale and place." With polyurethane an architect could participate in this transformation, could "experiment with conditions which might accommodate these changes." Despite foam's relative durability compared to vinyl or polyethylene, Drury repeated the litany of plastic's impermanence with a rhetorical question whose answer seemed transparent—"Why worry whether or not it's forever?"[89]

There was no single plastic-as-plastic aesthetic in architecture. Innovative uses of plastic took advantage of lightness, strength, and an ability to assume forms that functioned simultaneously as structure and sculpture. But plastic buildings differed greatly in terms of surface, texture, and degree of resiliency, not to mention technical methods of fabrication. A casual observer would have found sharp aesthetic differences if asked to compare a gossamer balloonlike pneumatic to a rough, heavy earthtone dome of rigid foam or a fiberglas shell resembling a consumer appliance in its industrially designed perfection. In each case materials and techniques diverged widely from those of traditional construction. But only the word plastic and an increasing impermanence united them in their differences. Rather than yielding a true aesthetic as in the 1930s, plastic fostered a succession of ephemeral statements, each quickly superseded and all witness to the responsive, even disruptive malleability of synthetic materials in the service of an inflationary culture. This volatility appeared most clearly in the work of furniture designers and manufacturers. After all, no one expected a chair or table to last as long as a house. Designers and consumers of furniture enjoyed greater experimentation.

Taking advantage of plastic's "total design freedom," designers in the 1960s offered a range of options—colorful chairs of rigid but resilient injection-molded polypropylene or ABS, beanbags of soft urethane foam covered with wet-look vinyl, modular couches of foam cut-outs, and inflatable chairs and couches of transparent vinyl.[90] The fiberglas chairs of Saarinen and Eames remained on the market, available from Knoll and Herman Miller. But the future of rigid furniture of the "frankly plastic" variety lay with one-shot injection-molded thermoplastics. As early as 1961 Aladdin Plastics of California was molding polypropylene bodies for a side chair and an armchair. Similar in shape to the classic Eames chairs, the Aladdin chairs flexed with a sitter's shifting weight. Initial production runs hit ten thousand a week, an improvement over labor-intensive fiberglas.[91] Such American ventures remained anonymous, however, limited to the market for inexpensive but durable institutional seating. The most spectacular examples of one-shot plastic furniture came from Italy, where Natta's innovations in the chemistry of rigid thermoplastics like polypropylene made an unbeatable combination with the Italian flair for design revealed in gleaming espresso machines, Vespa scooters, sculpted hairstyles, and Fellini's cinematic packaging of "la dolce vita"—all fascinating to cosmopolitan Americans during the Kennedy years and after.

Italian hegemony in high-style plastic design (as opposed to the economically significant mid-cult styling of American companies like Formica) rested with the furniture-molding firm Kartell. It was organized early in the 1960s by Guilio Castelli, an engineer whose father was associated with Natta at Montecatini. In 1964 Kartell introduced the world's first commercial all-plastic chair, a small child's model about twenty inches high entirely injection-molded of linear polyethylene. Like Saarinen with his pedestal chair, designers Marco Zanuso and Richard Sapper were unable to have the one-piece molding they wanted; the chair's polyethylene legs were molded separately. This project marked the beginning of an Italian plastic design style so artificial that the odor of the refinery seemed to linger about its products. Boldly colored, the chair evoked industrial architecture with its fat round legs and a slatted back like the side of a cooling tower—but all so exaggerated as to present a pop playfulness, and so warm and resilient as to counteract any suggestion of inhumanity. The chair's success encouraged Kartell to introduce technically more difficult pieces by Zanuso, Anna Castelli, Joe Colombo, and others.

As the decade progressed, Kartell and other Italian manufacturers introduced softer, more tactile furniture of polyurethane foam and vinyl—dedicated, as the family magazine *Look* enthused in 1968, to "celebrat[ing] the virtues of the material itself" by "treating plastic as plastic." In the following year the *New York Times Magazine* ran a two-page color spread of plastic furniture dramatically posed against the Manhattan skyline (Figure 7–18) with a caption explaining how "curving new forms . . . that flow naturally out of fluid plastic" can "help the urban resident in his pursuit of individuality." At about the same time *House Beautiful* offered evidence of what happened when "an exploding plastics technology" was put "in the hands of Italian designers." Five pages of color photographs revealed furniture with "seamless, all-of-a-piece upholstery, chairs and tables without legs, chunks of flexible foamed materials." Many pieces were intended "to stack, bunch, nest," and otherwise accommodate a mobile population; they all "express[ed] the ever-changing en-

7–18. ITALIAN-STYLE PLASTIC-AS-PLASTIC FURNITURE POSED AGAINST MANHATTAN'S SKYLINE BY THE *NEW YORK TIMES MAGAZINE*, 1969

7–19. Quasar inflatable furniture, imported from France, shown atypically with a young child as model, 1968

vironment of today." Italian furniture stimulated so many American knock-offs that *Time* could state in 1971, "Plastic furniture—no longer disguised as wood but blatantly and unashamedly plastic-looking—is now showing up in homes all over the United States." Suggesting "something straight out of *2001*," it often also seemed "uncomfortably like Motel Modern." By 1973 the formerly celebratory *House Beautiful* was complaining about "lots of poorly made little Parsons tables, lamps, and junky things." Americans had accepted plastic furniture, at least when it could be had for less money than traditional wooden pieces.[92]

Before long, inflatable furniture of transparent vinyl film had assumed the former trendiness of Kartell's rigid injection-molded pieces (Figure 7–19). Large and bulky, fabricated like air mattresses with multiple air pockets or cells, inflatable vinyl furniture seemed insubstantial, a ghostly parody of traditional overstuffed furniture. An inflated chair or couch could be deflated, folded up, moved to a different place. Inflatable furniture expressed the impermanence and hedonism of the pop movement celebrated by *Archigram* in the pods and "cushicles" of its later phase. There was something perfectly "60s" about an inflatable, its aura mingling incongruous elements of sexual liberation and unliberated lewdness. Publicity photographs of "blow-up" furniture often featured a nude female model posed as neutrally as possible. Her presence suggested a multitude of things: the visual nudity or simplicity of the form itself, the near-animate feel of firm pneumatic vinyl yielding to pressure, the inflatable as a sensuous support for lovemaking. Victor Papanek, an acerbic critic

of waste in design, asked his readers to "imagine the dismay of having some romantic interlude punctured by a sudden pillow blow-out." For Banham, on the other
hand, a huge inflated balloon encountered at a nighttime art "happening" in Los
Angeles was a "bulging, yielding, demanding plastic that wouldn't go away when
you pushed it," rather like "a plump, drunk, amiable, unstable girl at a party."[93]
Banham was articulating the fetishism of pneumatic plastic—a fascination that
brought moviegoers a view from below of Barbarella settling face-down onto her
transparent sleeping membrane, reached a low point with life-size inflatable "party
dolls" advertised in "men's magazines," and rebounded into respectability with the
spectacular success of vinyl water beds. As surface rather than structure, the inflatable suggested a more fundamental substitution of the artificial for the natural than
any prior use of plastic. For the most part such fantasies of control and simulation
went unstated or remained in a twilight state between pornography and science
fiction.

A New York firm called Mass Art Inc. garnered the most publicity among American companies producing inflatable furniture. Founded in 1966 to sell examples of
op and pop art, within a year it was concentrating on inflatable chairs, pillows,
mattresses, and jewelry of transparent vinyl. The firm's products reached the furniture salons of Milan, and its partners whimsically declared their intention of offering
an "inflatable fur chair" and a "disposable home." In 1968, with Mass Art's pneumatic chairs in production for over a year, the marketing director admitted they had
"invested a lot of money on the assumption that inflatable furniture will be around
a long time." Of course, it wasn't. It was disposable. That was the point.[94]

Even *Modern Plastics*, the industry's conservative voice, celebrated the centennial
of celluloid in 1968 with a projection of "plastics in the 1980s" emphasizing disposability. Encouraging the sort of blue-sky dreaming that had damaged plastic's
reputation twenty-five years earlier, the magazine solicited predictions from industrial designers. A McLuhanite envisioned a color television tube made "from a flexible plastic film" that would fold up "much as a plastic raincoat does today." Other
designers predicted "convenience-oriented" furniture discarded with changing
styles, modular rooms chemically programmed to self-destruct after a decade, zip-
out hospital rooms for hygienic incineration, and a domestic vacuum-forming unit
for making disposable dishes. No longer would families fight over who had to wash
the dishes—"just eat . . . throw away . . . and make some more as needed."[95]

Disposables, inflatables, the vinyl "wet look," bright polypropylene moldings, fiberglas saucer houses, foam domes, ABS appliance housings, artificial environments—these things and more came together in an exhibition on "Plastic as Plastic"
that opened late in 1968 for a two-month run at New York's Museum of Contemporary Crafts. The showing of 281 artifacts and architectural photographs was ambitious for a small museum, attracted wide attention, and marked the high point of
plastic's thermoplastic utopian phase. Funding for the exhibition came from the
Hooker Chemical Corporation, whose pollution of Love Canal near Niagara Falls

was soon to become a symbol of all that was wrong with America's urge for technological control and transcendence. Only in retrospect, however, does "Plastic as Plastic" take on a certain irony. Otherwise one is swept along by the enthusiasm of
curator Sandra R. Zimmerman's prose as she celebrated plastic for the ease with
which it could "be altered chemically to give almost any property desired in an end
product." More than ever before, a "creator" could "vary his material to fit his ideas"
and so enable "man to meet his environmental needs more precisely."[96]

To design "Plastic as Plastic" the museum hired Douglas Deeds of Los Angeles,
who specified a cavelike free-form interior of sprayed polyurethane foam, realized
with assistance from three of Drury's Yale students (Figure 7-2). That warmly inviting, reassuringly rough and imperfect environment offered a perfect contrast to the
mostly smooth, precise, physically refined high-tech objects of the exhibit. Beyond
the predictable domes, shells, inflatables, and injection-molded Italian furniture,
present mostly only in photographs, the museum gathered an eclectic mix suggesting plastic's expansive, inflationary presence in the material world of the late twentieth century. The category of apparel encompassed acrylic rings and bracelets,
wraparound sunglasses of cellulose acetate, safety helmets of polyethylene and fiberglas, sanitary medical gloves of polyethylene, transparent vinyl boots, and something described as a "fantastic body covering" in vinyl. In addition to more dramatic
utopian structures, architectural applications included fiberglas street lights, plan-
ters, and bus stop benches, inflatable ceiling panels, and a photograph of a boutique
facade designed in fiberglas by Hans Hollein. The furniture section contained multiple examples of current trends, while an eclectic display of housewares and appliances enabled visitors to compare a minimalist Braun electric toothbrush with a
commercial Hamilton Beach electric knife, and serving pieces by Kartell with a cutlery tray by Rubbermaid. The same spirit governed the selection of toys, ranging
from Creative Playthings to Fisher-Price, from the stylized plastic-as-plastic forms
of Lego to the imitative red-brick forms of Playskool's soon-to-be-discontinued
American Bricks. Perhaps the widest range appeared in the exhibit's presentation of
so-called technological applications—including orange traffic cones, washing machine agitators, a polyethylene minnow bucket, polypropylene hinges, several varieties of ductwork, an artificial heart valve and aorta, and an example of "circuits
encased in polyester film." Finally, Zimmerman had also selected thirty-seven art
works, most of them sculptures or constructions, many of them cast or fabricated
from transparent acrylic, their very existence testifying to plastic's pervasiveness in
contemporary culture.

"Plastic as Plastic" attracted favorable reviews. Writing in the *New York Times
Magazine* before the opening, Barbara Plumb promoted the exhibition as "plastic's
coming-of-age." Her photo spread, "Genuine Plastic," set out to prove plastic was
"no longer synonymous with cheap and tacky" but had become a favorite of "discriminating consumers who delight in its transparency, light-reflecting qualities,
and elegant shapes." Provoked by the exhibit, *Esquire* offered "a glimpse of your

plastic future" in a photo spread of artifacts, some from the museum, others not, all demonstrating that "plastic is wonderful as long as it's being itself and not imitating anything." Even more expansive, *Better Homes and Gardens* explained that plastic enabled designers to "create objects that could never be built before," things that "uniquely fit today's way of life—innovative, flexible, and mobile." Even Hilton Kramer, the skeptical art critic of the *New York Times*, praised "Plastic as Plastic," though he had little positive to say about its formal art works. He referred to plastic as "the answer to an artist's dream"—an "entire family of materials that can be made to assume virtually any size, shape, form, or color the mind of man may conceive"— but he faulted the exhibit's artists for not seizing this "almost Faustian freedom." They were uptight, too concerned with form, too obsessed with self-expression. Practical designers, by contrast, had already succeeded in "defining a new world of feeling."[97]

What a German critic has referred to as the *plastikoptimismus* of the 1960s attained its fullest expression at the Museum of Contemporary Crafts. A shift in public sentiment was already under way. Several months before the exhibit's opening, people of all ages were laughing at *The Graduate*'s line, reacting nervously to Buck Henry's epithet for a materialistic culture. Doubts about plastic were surfacing even in sources friendly to the industry and its materials. For one thing, it was too easy to make things out of plastic. Deeds, designer of "Plastic as Plastic," bragged it would soon be possible to "go into a store and buy a can and spray foam your own interior"; a glance at the offerings of any dime store revealed that many products were already molded just as casually. For another thing, it was possible to emphasize mobility and impermanence to the point of disorientation. Such phenomena as Plug-In City, *Barbarella*, inflatable domes, and disposable products raised the specter of a society that in seeking lightness had lost all awareness of gravity—understood both as a sense of rootedness to the past, to nature, to the earth itself, and as a sense of seriousness, of awareness of matters of weight, of consequence. Not only were "some of the wonders of the previous decade . . . beginning to look distinctly tarnished" with their "fading colours, crazed finishes and structural cracks" serving as reminders that "plastics do not grow old gracefully." Beyond that, an emphasis on impermanence and insubstantiality did not square with the facts of life in a supposedly disposable world. In 1970, shortly after the first Earth Day, furniture designer Paul Mayen complained about the havoc wreaked on the environment by an inflationary mentality. As he saw it, economic success of technical processes capable of fabricating plastic furniture "in ever increasing numbers" depended on popular acceptance of the idea of "expendable furniture." But in a supreme irony, while a fast-paced lifestyle led people "to cry out for more and more of these throw-away objects," such things had become "more durable than ourselves, our governments, and our society." Echoing Buckminster Fuller, Mayen argued that society must "metabolize all its activities and products." The industry had to stop making things "with such a high pollution potential" and instead develop plastics with "high reconversion potential."[98]

Such doubts multiplied, expressed not so much by insiders hoping to curb the industry's excesses as by outsiders who regarded the industry itself as an excess whose fallout was polluting the earth. At the very moment *plastikoptimismus* reached its apogee, opponents were raising such issues as toxicity, flammability, solid waste, and consumption of dwindling fossil fuels to feed insatiable refineries and molding machines. Other critics focused on more subtle discontents. For them plastic offered a metaphor of superficiality and dishonesty in contemporary life, and a measure of the degree to which technological "progress" divorced inhabitants of "developed" countries from simple human pleasures. These issues rose to popular consciousness during the 1960s and 1970s and coexisted with continuing celebration of synthetic materials and artificial environments as signs of a typically American plasticity, an incessant remolding of self and society. Most of these concerns, pro and con, appeared in the work of a handful of artists particularly sensitive to cultural trends and tensions. Their representations of plastic and plasticity afford a useful introduction to the debate.

SEDUCED BY ARTIFICIALITY

Asked to name the artist most associated with plastic in the 1960s, many Americans would have named Andy Warhol—more for a general aura than for any use of plastic in his art. A brief item in the *New York Times* on April 12, 1966, reported that a "summons for operating without a cabaret license was served on Plastic Inevitable, a discothèque operated on St. Mark's Place by Andy Warhol, pop art entrepreneur." Bored with painting and filmmaking, Warhol had invented the multimedia happening in a rented hall east of Greenwich Village. The Exploding Plastic Inevitable was the preeminent avant-garde event of the season. As strobe lights flashed, the Velvet Underground played feedback-laden proto-punk, leather-clad dancers improvised on stage, and an improbably manic Warhol directed simultaneous projection of several of his films onto shifting areas of the auditorium, sometimes with handheld projectors, sometimes distorted by gel on the lens. Simultaneously, uptown at the Leo Castelli Gallery on East 77th Street, people could experience a more calming Warhol environment, an installation called *Silver Clouds*. In an otherwise empty white room floated an array of pillow-shaped balloons, each about five feet long. Fabricated from metallicized polyester film and inflated with helium, these objects moved gently, randomly, reflecting vague patterns formed by each other, the floor, the walls, the lighting, and passing visitors. In either case people could make of the experience whatever they wanted by projecting their own emotions and desires.[99]

These two environments, so unlike one another, comprised all of Warhol's explicitly "plastic" output except for a set of one hundred tiny polystyrene boxes on which he silk-screened portraits of ten of Castelli's artists, including himself.[100] Warhol's association with plastic derived from the early Pop Art that made him famous. But it was Warhol himself who evoked a sense of plastic artificiality. "If you want to know all about Andy Warhol, just look at the surface," he told an interviewer;

"there's nothing behind it."[101] His image was continually shifting, becoming more artificial with each turn. As a young man Warhol altered his name and underwent plastic surgery on his nose. With fame he began wearing outrageously phony white or silver wigs that accentuated his unnatural pallor. In the mid-1960s he sent an imposter on the college lecture circuit. About fifteen years later this affinity for artificiality reached an extreme as he collaborated with a former Disney imagineer constructing an audioanimatronic Warhol robot of plastic and alloys—intended as the star of "Andy Warhol's Overexposed: A No-Man Show."[102] After Warhol's death in 1987, acquaintances recalled a passivity that enabled him to adjust like a chameleon to his surroundings. A biographer described him as "a void toward which others gravitated with their anxieties, their ambitions, and, occasionally, their useful ideas." He had a knack for remaining "detached while leaving himself open to an extraordinary range of influences, subtle or devastating, sophisticated or coarse."[103] Just as plastic, according to Barthes, was a formless material open to an infinite proliferation of effects, so too was Warhol capable of embodying his culture's shifting shapes.

By the late 1960s a number of artists were using plastic, mostly cast or fabricated acrylic. Critic Lucy Lippard reacted cautiously to an exhibition of works in plastic by more than a dozen artists held at the John Daniels Gallery in 1965. Commenting on a field that included sculpture by Donald Judd, David Weinrib, and Robert Smithson, she concluded that "plastic tends to take off on its own."[104] Four years later a major exhibition with works by some fifty sculptors demonstrated familiarity with fiberglas, inflatables, expanded foam, vacuum forming, and objects embedded in transparent cast resin. Organized by the Milwaukee Art Center, "A Plastic Presence" came with a naive catalogue essay blithely announcing that many plastics were "largely independent of specific raw materials" because they were "made entirely from gases . . . as accessible as the air." Critics avoided comment on that but generally found that artists had not yet mastered plastic. Robert Pincus-Witten complained in *Artforum* that they were using plastic to develop ideas already "worked out first in more traditional methods." However spectacular the results, plastic was leading artists into "imitative rather than innovative modes." Another critic argued the same point in *Arts Magazine*, noting that even Louise Nevelson's subtle private language of compartmentalization did "not change in passing from wood" to plastic. Artists had not learned to use plastic as plastic, for its own sake, nor had they come to grips with its cultural meanings.[105]

Bruce Beasley was typical of those who missed the point because he insisted on creating monumental sculpture in acrylic. Born in 1939, he worked in Oakland making large sweeping sculptures in bronze and aluminum. Looking for a medium that would reveal the knotted tension of arrested conflict, he discovered cast acrylic and began working with it in anticipation of entering a competition for a sculpture to be located between two state government buildings at Sacramento. Beginning in 1967, Beasley went through a series of experiments recalling Baekeland in their dependence on heat and pressure in an autoclave to cure liquid castings that often

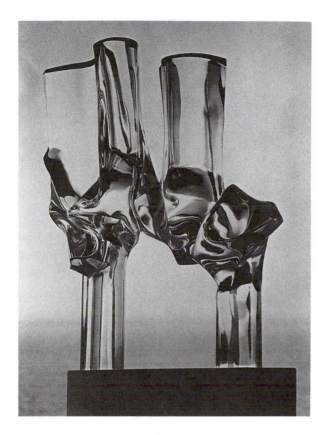

7–20. Bruce Beasley, *Stamper's Lighthouse, 1967*

emerged as "frothy plastic messes." After six months and $4,000 in loans, he obtained a small crystal-clear sculpture that made him a finalist and brought state funding to make a forty-two-inch hundred-pound model whose proposed realization was to extend fifteen feet in length and weigh thirteen thousand pounds. In 1968 he won the competition.[106]

In the meantime Beasley was showing some of his more successful experiments. *Artforum* singled out an untitled "organic free-form sculpture" as the most effective work in an exhibition of "Plastics West Coast" at San Francisco's Hansen Gallery late in 1967. He also exhibited at a show devoted to "Plastics and New Art" at the University of Pennsylvania early in 1969 and had his work *Stamper's Lighthouse* (Figure 7–20) included in "A Plastic Presence" at Milwaukee. The latter piece, twenty-nine inches high, depicted two crystalline columns deforming and joining with a fluid but clotted violence, before separating and continuing upward—the effect suggesting a simultaneously abstracted and sensuously photorealist vision in three dimensions of liquid chrome crushed and welded in some unimaginable crash.[107]

Such studies did reveal qualities of light that only transparent cast acrylic could convey. But increasing the size to a monumental scale added little to such explorations as *Stamper's Lighthouse*. After Beasley won the competition and began working on the large sculpture, his emphasis shifted from revealing the material's potential

233

to celebrating the heroic technological process involved in its creation. He quickly discovered his $50,000 award would not even pay for the resin the work would consume, nor had his experiments given him the expertise to handle such a large casting. Taking his model to Wilmington, Beasley convinced Du Pont's Plastics Department to provide cut-rate resin and technical assistance. From the company's point of view, as a product manager recalled, Beasley's project would yield "the largest acrylic casting ever made, and probably the largest transparent object in the world." They preferred that *Apolymon*, as it was called, "not be just a monument in acrylic, but a monument in our own 'Lucite' acrylic."[108] After arriving home, Beasley built the mold, bought a secondhand autoclave, and in the summer of 1969 cast the piece, a process that took four days of pouring and nearly two weeks of curing. He then spent eight months sanding and polishing the massive sculpture, which was unveiled in Sacramento in March 1970. Installed on a six-foot pedestal, the tortured yet luminous form of *Apolymon* spread horizontally across two irregular columns and suggested a monstrous bird in flight, or the heads of two Aztec beasts facing away from each other, or an expressionist crucifixion. Above all it drew attention to itself as a monumental sculpture whose very material defied the canon of monumentality. An *Art News* reporter captured its anomalous status when describing a confused woman who commented wonderingly to a friend, "It looks like plastic."[109] Beasley's drive toward the monumental led him to use plastic as if it were a traditional sculptural material.

Other sculptors working at the same time did use plastic in ways that revealed conjunctions of material and culture. Among them were Duane Hanson and Claes Oldenburg, who both addressed the complex of issues surrounding imitation, simulation, artificiality. Beyond that they had little in common. Oldenburg was a gregarious intellectual who moved in avant-garde circles and explored a wide range of media and themes. Hanson was an outsider who perfected a single vision and technique without really intellectualizing it.[110] After years of working in obscurity, Hanson's break came in 1966 with a small eighteen-inch fiberglas figure of a woman who died during an illegal abortion, a piece that caused a scandal in Florida where he was teaching. He followed that with a life-size but stylized fiberglas corpse in a plain wooden coffin, *Welfare-2598?*, a work criticizing bureaucracy's unfeeling response to human need. At that point Hanson adopted a technique that transformed his art— the taking of impressions for molds from bodies of living models rather than from clay originals. Each successive piece became more realistic as he vividly represented a variety of social concerns in fiberglas—a murder victim, a suicide, a rape victim, a crash victim, and eventually tableaux of American war dead, derelicts sleeping in the street, and men in a race riot. These pieces thrust the observer into the immediate presence of an unacceptably shocking life-size "reality," something known previously only through photographs.

After establishing himself with such works, Hanson shifted his focus during the 1970s. No longer striving to communicate a social message, he concentrated on making his figures illusionistically perfect. He became adept at placing models in positions of natural repose, took greater pains with the molds, lavished attention on the painting of the finished pieces, abandoned wigs in favor of a tedious procedure for placing individual strands of hair, and eventually abandoned fiberglas for cast vinyl, a material with the translucence of human skin. The simulation of life was so uncanny that gallery visitors would ask Hanson's uniformed guard for directions or beg forgiveness of the stout young woman they had just elbowed. Although Hanson's statements about his work indicated respect for his mostly working-class figures, he also charged them with satiric intent, as in the following: a blimplike sunbather spilling out of her beach dress, with a *National Enquirer* and box of Cracker Jacks beside her plastic-webbed beach chair; middle-aged tourists (Figure 7–21) in gaudy mismatched clothes, he with Hawaiian shirt and cameras, she in plastic net scarf, cheap pink sunglasses, carrying a vinyl bag imitating patchwork fabric; a bulging shapeless shopper with curlers, net scarf, vinyl purse, and supermarket cart filled to the brim with packaged foods. His plastic people, remarkable for their imitation of real life, for their simulation of reality, served to suggest a culture equally phony. Their incredible verisimilitude suggested that real Americans shared their vacuous concerns, their pathetic attempts at substance. The more real Hanson made them

7–21. DUANE HANSON, *TOURISTS 1*, 1970

look, the more insubstantial seemed the reality to which they aspired. To view a gallery of his plastic people offered a disconcerting shock of recognition.[111]

Claes Oldenburg, on the other hand, provoked a playful, exhilarating sense of recognition. If Hanson rendered the material abundance of the inflationary culture as an empty affair of shopping, supermarket tabloids, packaging, and obesity, Oldenburg gloried in the expanding material world and infused his sculptural representations of everyday objects with a sensuality that seemed extravagantly, messily human. While Hanson fabricated in plastic, Oldenburg sculpted with plasticity. While Hanson simulated people, then dressed and surrounded them with real things, Oldenburg created an alternate world of artificially exaggerated things. In that world, his most perceptive critic, Barbara Rose, has stated, "Objects are softer, harder, lumpier, smoother, bigger, brighter, than any real objects." For the opening paragraph of a book about Oldenburg published in 1970, the critic composed a riddle of sorts. "What is both hard and soft?" she asked. "What changes, melts, liquefies, yet is solid? What is both formed and unconformable? structured and loose? present and potential?" She was thinking of Oldenburg's art, but the answer might just as well have been plastic in all its multiplicity of forms. As Oldenburg once confessed, "I was seduced by artificiality."[112]

Educated as a painter, Oldenburg found himself caught up around 1960 in the happenings and environments of such New York artists as Red Grooms, Allan Kaprow, and Jim Dine, who were creating a temporary art of the moment from the debris of everyday life. Oldenburg himself created two environments, *The Street* and *The Store*, the latter filled with lumpy, gaudily painted representations of typical goods—clothes, shoes, delicatessen foods—made from muslin and plaster shaped over chicken wire. The next major development in his art, giant soft sculptures of brightly painted stuffed canvas, stemmed directly from his need to fill a large uptown gallery space that would have overpowered the small artifacts of *The Store*. Exhibited in the fall of 1962, these new objects—including a giant cloth hamburger, a giant ice cream cone, and a giant piece of cake—shocked an art world accustomed to abstract expressionism and helped move Pop Art out of the underground. Within a few months Oldenburg was experimenting with flexible colored vinyl sheeting as a substitute for canvas in the first of his so-called "soft machines."

There was something unthreatening, even warm and inviting, about the vinyl sculptures Oldenburg created during the mid-1960s. His soft machines included variations on a telephone, a typewriter, an electric mixer, a bathtub, a drainpipe, an old-fashioned electric fan, a toaster, and a 1934 Airflow Chrysler with some of its parts—radiator, engine, tires, and so on. Fabricated from sheet vinyl with panels visibly sewn together, the soft machines were huge, often several feet across. Clearly industrial with smooth, chemically bright reflective surfaces, the soft machines suggested all the same a yielding of rigid male technologies. Big, rumpled, shambling, too soft to support their own industrial functions, Oldenburg's vinyl representations portended a world in which, according to Rose, "the machine—no longer menac-

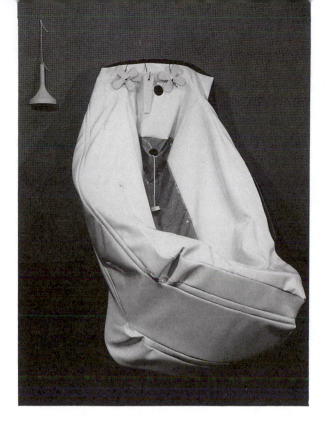

ingly strange, but reassuringly familiar—becomes more like man: unique, change-able, and above all vulnerable."[113] The soft machines were ripe with body imagery, as in the long vinyl labial folds of a *Soft Bathtub* (Figure 7–22) or in the collapsed vinyl phallic form of a *Soft Airflow* draped limply over a metal stand. Oldenburg explained such transformations by declaring that "anything soft, of course, becomes flesh-like."[114] One thinks of Banham floundering against an inflatable or, more par-ticularly, of his distinction between no-nonsense technological hardware and the polymorphous humanizing software that was replacing it. With his soft machines Oldenburg celebrated a perceived animation of the inanimate in a material environ-ment that seemed increasingly responsive to manipulation. For Oldenburg, nothing had greater reality than the artificial, and plastic best expressed the seductive artifi-ciality of things.

To consider the work of Duane Hanson and Claes Oldenburg is to touch on con-tradictory but related aspects of plastic's place in American culture during the final third of the twentieth century. For nearly a hundred years, since celluloid's commer-cialization, plastic had imitated other materials. That motive reached its final distil-lation in Hanson's fiberglas and vinyl figures. One of his grotesques brought to life could have played the role of the graceless booster who accosted Dustin Hoffman. Plastic was the material embodiment of everything phony in American life, its emo-tional shams and makeshift gratifications, its false fronts and Tudor suburbs, its white lies and patriotic flummery, its willingness to misrepresent, to distort, to let "cash value" determine the truth of things. But a shift in focus by a few degrees completely changed the picture. To misrepresent an unsatisfactory present no longer seemed an outrageous lie but the sign of a desire to reach for the future. The development of American culture rested on a faith in the unfolding of potential, a belief that success comes to those who help themselves, that the ideal can be made real, that, as William James put it, "faith in a fact can help create the fact."[115] For at least forty years, since Bakelite's commercialization, the plasticity of new materials had mirrored the plasticity of the New World. And if the land itself was played out, used up, or filled in, then new materials could extend its plasticity and renew a faith in the ability of Americans to remake or reshape their environments. Oldenburg's extravagant sculptures with their unique combination of chemical artificiality and idiosyncratic largeness of spirit suggested the potential truth of such an assertion. "My soft sculpture," he once declared, "is a remolding of the world."[116]

But that generous evaluation ignored another crucial element, the speed with which plastic was melted, deformed, or transformed, the apparent impermanence of materials whose physical existence paradoxically outlasted even the people who created and molded them. Oldenburg caught his soft machines in a slow process of dissolving, an illusion fostered as their creases and folds changed when he installed them in new positions. Aside from the drying and cracking of vinyl surfaces with age, each soft machine was frozen midway in an imagined trajectory from steel rigidity to complete liquidity. Although Oldenburg once said there were "two ways of experiencing matter: the solid state and the dissolving state," he failed in his own work to follow the logic of plastic to its final end.[117] One of his contemporaries succeeded more than Oldenburg in expressing the impermanence of plastic. That artist was Les Levine, an Irish-Canadian-American of shifting interests. Once so "familiarly known as Plastic Man" that *Life* magazine identified him as such in 1969, Levine soon dissolved into relative obscurity as if he were a creature of the ephemerality whose most certain medium was plastic.[118]

Born in Dublin in 1936 and educated at the Central School of Arts and Crafts in London, Levine emigrated to Canada in 1958 and organized a series of one-man shows in Toronto, progressing from spray paintings of chairs to the "wrapping" of wooden chairs in tight fiberglas skins. He mounted his first one-man show in New York in 1966, two years after relocating there. Beyond a white vinyl leisure suit worn as a trademark, Levine became known for two related concepts, "disposable art" and "environmental places," both depending on plastic for their realization. Synthetic materials attracted him for their cheapness, their ubiquity, their impersonality, and because their use bypassed the craft processes of traditional art. When specifying plastic, he could move from concept to mass production "without having to touch it by hand." Levine hoped to short-circuit the primacy of the unique high-art object by substituting for it a mass-produced "non-object" available to anyone for a few dollars, "disposable art" functioning as "a physical manifestation of a process which in itself is not complete." The idea seemed so revolutionary that a friendly critic referred to it with an unfamiliar term as a manifestation of "post-Modernism."[119]

Among Levine's first disposable art was a so-called "wall sculpture" exhibited in the "Plastic as Plastic" show at the Museum of Contemporary Crafts. This piece consisted of an arrangement of wall tiles, each fabricated from a thin foot-square sheet of rigid polystyrene foam. Each sheet was vacuum-formed over a common object—a wrench, a transistor radio, an inverted pie pan—from which it assumed a simple relief pattern.[120] In 1968 he developed the idea further with *Disposable Wall* at New York's Fischbach Gallery. The wall was composed of stacked modular cubes of transparent polystyrene, each with a kind of shadow box molded into its interior at a skewed angle, the whole ensemble creating a shimmering effect.[121] Visitors to the Fischbach could purchase sheets or cubes for two or three dollars apiece, little more than the price of a common dime-store item. Levine's concept of disposable art reached a limit of sorts in 1969 with an installation called *Windows*. The artist ordered commercial manufacture of several thousand identical die-cut stickers of reflective metallicized Mylar polyester, each a trapezoid about nineteen inches long. Peeling the backing from thirty of the mirrored stickers, he placed them around a room at the gallery—on floor, walls, ceiling. Others stuck up outside drew attention to the exhibit, a collection of "windows which see into the room rather than out through it." For Levine, selling individual stickers was more important than the installation. "Once the purchaser of a *window* puts it in the desired position," he explained, "it is used up and, although it can be removed and disposed of, it can never be reused." With all the disposables "the issue was not what the objects looked like, but that [they] could be thrown into the garbage after one got fed up with them." Levine hoped to encourage "active experience rather than a concern with ownership."[122]

Levine's other specialty, his "environmental places," proved more ephemeral than the disposables because experience was all one could take away from them. His most ambitious environment, *Slipcover*, was installed at the Art Gallery of Ontario in 1966 and then traveled to the Walker Art Center in Minneapolis and the Architectural League of New York. *Slipcover* (Figure 7–23) was a large room lined throughout with mirrored Mylar film. Two huge bladders of reflective plastic film continuously

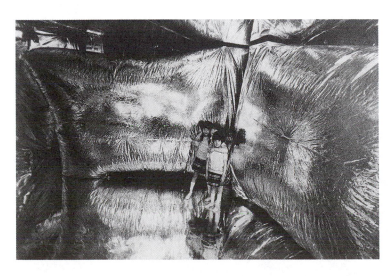

7–23. Les Levine, *Slipcover*, installation at the Architectural League of New York, 1967

inflated and deflated, changing the size of the space. As visitors walked through, six slide projectors cast on walls and people distorted images of the room during other exhibits of the previous year, while microphones picked up ambient sound and fed it back distorted. Interviewed for the catalogue before the environment's first opening, Levine professed a love for "clean things"; he was "fascinated by precision" and cited James Bond movies as an example. But there was more to *Slipcover* than the high-tech clarity of plastic. Its organizer hoped to induce a profound disorientation, a "fluidity," a sense of people "melting in the room because of what is being projected off the walls" onto their moving bodies.[123]

Levine's other major environmental work suggested precision more than disorientation but also aimed at a reduction of the art object to nothingness. Known as *Star Machine* in its basic form, which was installed at the Fischbach Gallery in 1966, it consisted of two large transparent acrylic panels, each about seven feet high and ten feet long, each with a huge vacuum-formed half bubble. The two panels were placed vertically parallel to each other, bubbles facing outward, so that people could walk through and experience the world from inside a plastic cocoon. A much expanded version was installed in the sculpture garden of the Museum of Modern Art in the spring of 1967. Funded by American Cyanamid, fabricator of the panels, *Star Garden* was made up of four groups of four panels each. Each group formed the outline of a rectangle, bubbles facing outward and curving toward those of the next group. The whole environment covered sixteen hundred square feet of the sculpture garden's upper terrace. People could walk in, through, and around all sixteen panels. Reflections of kibitzing visitors, trees, shrubs, sculpture, clouds, and sky multiplied in every direction. According to Levine the thing itself was of little importance. He wanted people to "sense space . . . the way you feel cloth or any textural thing." He explained that he intended *Star Garden* "literally as a place to be in, rather than an object," and observed that "it packages people in endless space and makes them look more beautiful, shiny and new." For an Associated Press reporter whose story about "plastic bubbles" was picked up by newspapers as far away as Ardmore, Oklahoma, Levine said simply that "it's supposed to make people happy, that's all."[124]

For all the talk of insubstantiality, *Star Garden* was so big and heavy it had to be shipped in moving vans to its next installation at the Walker Art Center. The AP story relied for its humor on Levine's worries about how to store or sell the piece when he ran out of display venues. Such concerns hardly afflicted another project, *Process of Elimination*, which combined Levine's two major forms, disposable art and the environmental place. The work "took place," for want of a better phrase, on a vacant lot across from Washington Square Park. It began on January 22, 1969, with Levine scattering three hundred "plastic disposable curves" over the ground—each a thin rigid sheet measuring thirty by forty-eight inches and curved to catch wind and reflect light. Every morning at 10:30 he removed ten sheets from the fenced lot and tossed them into a garbage can. The thirty days of their removal concluded precisely when excavations began for a new building, thus reducing the site to a condition less than zero. As Levine intellectualized *Process of Elimination*, it was "not about art as object but about process." The "work" occurred "on a mental level" and consisted of making people aware of the project by "moving an idea through the minds of a community." If the elimination took place without being photographed, written about, or otherwise recorded, it would "disappear back into the environment and cease to exist." The energy consumed in the process was "soft," Levine insisted, "much the same as radio or television and the electricity which flows through the air."[125]

Plastic Man's own process of elimination came to an end at just about the exact moment Levine dropped the last plastic sheet into the garbage can. He had already begun experimenting with video, a more ephemeral medium than plastic. Most of his subsequent projects fell under the headings of conceptual or information art. By 1976 he was referring to himself as a "media sculptor," an artist so obsessed with plasticity that even plastic was too intractable for his purposes. Although Levine continued to work and exhibit, he slipped from the prominence of his role as Plastic Man in the white vinyl suit. As a minor artist producing no objects to be possessed, he suffered from a metaphoric loss of existence. All the same, his involvement with plastic during the 1960s embodied cultural trends as surely as did that more famous plastic celebrity Andy Warhol. The latter had once dismissed Levine's work as "just nothing, it's just plain plastic."[126] But Levine accurately registered contradictions in society's use of plastic as the physical medium of an inflationary culture of disposable things and ephemeral phenomena. Although he did not stop to worry about the final "elimination" of his "disposable" plastic sheets after he had discarded them, he did have moments of doubt. For a show in 1968, at the height of his notoriety, he designed a flier printed on shiny, bright yellow, vinyl-like paper. Prominently displayed as on a cigarette pack was the following warning: "Caution: Some Plastics May Be Hazardous To Your Health." Whether intended as a joke, as a straightforward medical advisory, or as a broad metaphoric statement about ominous effects of plastic on psychological, environmental, even cultural well-being, that brief label reflected an ongoing reaction against plastic and all that it had come to mean.[127]

MATERIAL DOUBTS AND

PLASTIC FALLOUT

*E*arly in the twentieth century promoters of plastic realized they had to create images for popular understanding of new materials. Even Leo Baekeland tolerated the exaggerated rhetoric of John Mumford announcing Bakelite's material immortality. Within a few years the industry routinely celebrated plastic as a machine-age miracle material. When that vision dimmed as plastic became commonplace in postwar life, the industry adopted the prosaic theme of damp-cloth convenience. Soon, however, the Corvette automobile and Monsanto's House of the Future expressed visionary affinities to the promise of space exploration. When that plastic-as-plastic aesthetic expanded to geodesic domes, transparent inflatables, and vinyl-clad go-go dancers, industry leaders did not realize they had lost control of plastic's image. Celebrating celluloid's hundredth anniversary in 1968, SPI smugly relied on a Princess of Plastics who made public appearances around the country and granted interviews to reporters. A twenty-page centennial advertising supplement to the *New York Times* boasted that new plastics contributed "a fluidity, a grace, a technological beauty of line and purpose that is sure to become the hallmark of a new way of life and a new American culture."[1]

This self-satisfaction ignored plastic's presence in a bitter debate about the expansive American way of life to which it had contributed so much. While people used more plastic each year, its reputation steadily declined. By 1979, when *Modern Plastics* ran a series of articles defending the industry, the magazine had given up trying "to change the image of plastics" among the general public. Instead it merely presented "the facts" for readers to use as ammunition against family and friends who thought they ought to be "ashamed of working in such a terrible business."[2] It seemed so unfair. If critics had limited attacks to rational issues, to solid waste, toxicity, pollution, or depletion of petroleum reserves, the industry could have defended itself with facts and figures. But defenders were helpless against perceptions of plastic as a sinister presence embodying everything wrong with America.

LIKE FISH ON A VINYL FLOOR

Potential for a vague fear of plastic extended back to the 1920s when publicists celebrated the alchemical wizardry of industrial chemistry, promoting new materials and synthetic colors as products of arcane magic rather than rational chemical processes. By suggesting that organic wastes from the bowels of earth could be transformed into wondrous shapes and colors, chemical utopians encouraged ignorance and fostered a feeling that some things are best left unrevealed. While their rhetoric encouraged people to regard industrial chemistry as a democratic source of limitless material plenty, its subtext hinted at loss of control over forces of nature at the very moment of reaching to transcend them. This anxiety informed *Fortune*'s surreal 1940 images of Synthetica and the transparent plastic Venus. Ordinary citizens expressed doubts while touring Du Pont's chemical exhibitions of the 1930s. Some questioned the wisdom of trying to surpass nature. Others spread rumors about spinning nylon out of cadaverine from human corpses. Still others blamed nylon stockings for rashes, poisonings, and skin cancer. Despite familiarity with plastic during and after the war, such fears occasionally resurfaced, as in the case of the *Collier's* reporter whose 1947 visit to a chemical plant gave her "the creeps" because the "unearthly" material it produced made her feel "man has something quite unruly by the tail."[3]

By then her tone seemed more appropriate to a discussion of nuclear technology. Although plastic might be fueling a material cornucopia, organic chemistry no longer seemed the cutting edge of science. Whether people anticipated limitless atomic energy or worried about the bomb, nuclear physics commanded attention as the most important of the sciences. Even so, the goal of controlling or transcending nature remained the same—as did doubts about the wisdom of doing so. Fear of the bomb and of radiation not only gripped Americans for a generation but also cast doubt on the benevolence of science in general. Despite public debate about nuclear radiation, it remained especially frightening because it was invisible, revealing its presence only years later through irreversible effects on its victims. Some people objectified fear of this intangible threat by projecting it onto plastic, an unnatural material whose image, like that of atomic energy, reflected the ambivalence of a utopian hope afflicted with an undercurrent of dread.[4]

None of America's postnuclear intellectuals expressed more animosity toward plastic than Norman Mailer. "I hate to sound like a nut," he once said in a radio interview that reportedly blamed plastic for everything from pollution to drug addiction and blotchy skin. His opinions seemed gratuitously insulting to people who

243

had dedicated their careers to the industry's progress. But his criticism appeared prophetic as harmful effects of synthetics came to light in the 1970s. A journalist opened an article on toxicity of vinyl chloride monomer by invoking his notorious colleague. "Not long ago, cancer and plastic were associated with each other only in the writings of Norman Mailer," he wrote, "where they serve as symbols of decadence and self-destructiveness in the high-technology society." But the situation had changed. With medical reports revealing high incidence of liver cancer among vinyl workers, a symbolic association had become "real scientific fact." Mailer would have argued he was right all along.[5]

For more than two decades Mailer conducted a hostile campaign against plastic. The first attack came in 1963 when discussing nuclear apocalypse in a monthly *Esquire* column that explained the "existential" meaning of the Cuban missile crisis by referring not to proliferating nuclear weapons but to proliferating plastic. Speculating on the lack of rebellions and desperate orgies in the face of annihilation, Mailer concluded that "we gave our freedom away a long time ago." The era's lifeless conformity attained material embodiment in plastic. We had "divorced ourselves from the materials of the earth, the rock, the wood, the iron ore," he maintained; "we looked to new materials which were cooked in vats, long complex derivatives of urine which we called plastic." These new materials lacked the "odor of the living"; "their touch was alien to nature." In a suggestive word, plastic had "invaded" all aspects of social life (much as the nuclear threat had invaded the macro level of politics, or radiation the micro level of biology). The result was "a world in which all could live even if none could breathe." The following month's column extended Mailer's analysis by shifting its context from the threat of annihilation to that of an insidious death in life whose miasma radiated from the totalitarian "plague" of modern architecture. "Everywhere," he wrote, "we are assaulted by the faceless plastic surfaces of everything which has been built in America since the war"—all of it proliferating "like the metastases of cancer cells" and congealing in a "cancerous" environment where everything tends "to look a little more alike" and "a little less like anything very definite" (Plate 15).[6]

Diatribes on plastic as plague appeared throughout Mailer's critically acclaimed journalism on the social upheavals of the 1960s. Flying to San Francisco for the Republican convention that nominated Barry Goldwater in 1964, he sat next to an Australian journalist who wondered why "all the new stuff you build here, including the interior furnishings of this airplane, looks like a child's nursery." Pondering that question, Mailer considered the Astrojet as "another of the extermination chambers of the century—slowly the breath gives up some microcosmic portion of itself, green plastic and silver-gray plastic, the nostrils breathe no odor of materials which existed once as elements of nature, no wood, no stone, nor ore." If the plane's interior—and all of America's new schools, factories, churches, offices, homes—were indeed "like a child's plastic nursery," it was, he concluded in a passage whose breathlessness can only be suggested, "because we're sick, we're a sick nation, we're sick to the edge of vomit and so we build our lives with materials which smell like vomit, polyethylene and bakelite and fiberglas and styrene . . . we are one great big bloody nursery attached to a doctor's waiting room." In *The Armies of the Night*, which won a Pulitzer Prize in 1968 for its portrayal of an anti-Vietnam march, Mailer described the vast structure of the Pentagon as "isolated from anything in nature," rising "like an anomaly of the sea from the soft Virginia fields . . . , its pale yellow walls reminiscent of some plastic plug coming out of the hole made in flesh by an unmentionable operation." Even a discussion of language in 1972, of the era's penchant for jargon and neologism, led Mailer back to plastic. There he branded as totalitarian a widespread "need to inject non-words into the language, slivers of verbal plastic" that served not as language always had to connect people to life but to alienate them from it.[7]

To explain why plastic provoked Mailer is harder than to cite examples of his animosity. If traditional materials, those Mailer regarded as natural, retained a sense of earthly origin, then plastic was unique in retaining nothing of its raw materials' earthiness. Artificial surfaces, colors, textures, and odors bore witness to a dominant technocratic mentality of precision and efficiency in the service of instrumental goals. Plastic's odor more than anything else offended Mailer. The olfactory sense most directly connected the human race to its natural or animal origin. Plastic's hygienic aura, its synthetic mocking of odors of illness and death, suggested a willingness to divorce humanity from its past, to cross a line into a fundamentally non-human artificial reality. The materials Mailer abhorred and dozens of other innovations in late twentieth-century technology promised—or threatened—to liberate the human race from the millennia of its biological past. This potential transcendence chilled Mailer but also so stimulated the grudging admiration of the former Harvard engineering student that he dedicated a book to meditating on it.

His ambivalence as a critic of technocratic society emerged in *Of a Fire on the Moon*, an account of the first moon landing in 1969. Rather than immersing himself in the "right stuff" of the astronauts, as Tom Wolfe later did, Mailer focused on the technical systems whose complex orchestration and fail-safe duplications produced a result so predictable he could refer to it as "the event of his lifetime, and yet it had been a dull event." In the best tradition of the science reporter Mailer clearly translated reams of turgid PR material generated by the Apollo 11 mission. But as with the science writers of the 1920s, his account never strayed far from ideology; it often veered into private demonology. A long passage arguing the empty rationality of the American corporation—the space program's patron and profiteer—segued into a plastic-wrapped description of an Apollo capsule's interior. There all vectors of postwar development "came to focus in the bank of instruments eighteen inches over each astronaut's head as he lay in his plastic suit on a plastic couch—lay indeed in a Teflon coated Beta-cloth (laid on Kapton, laid on next to Mylar, next to Dacron, next to neoprene-coated nylon) space suit on his Armalon couch—plastic, that triumph of reason over nature!"[8]

That sentence's rush to peel back layer after synthetic layer of an astronaut's protective cocoon revealed Mailer's own synthetic syntax, but his ultimate critique of technocratic society came elsewhere in a passage whose climax directly invoked plastic. Looking from within the world of NASA and corporate America, he predicted a "society of reason" based on "the logic of the computer." Its operatives would rely on drugs for "accelerated cerebration." Its secure control would require "inchings toward nuclear installation, a monotony of architectures, a pollution of nature which would arouse technologies of decontamination odious as deodorants, and transplanted hearts monitored like spaceships." Its environment would be artificial, Mailer suggested, "obviously . . . plastic, air-conditioned, sealed in bubble-domes below the smog, a prelude to living in space stations." In such a society "people would die . . . like fish expiring on a vinyl floor." Although his Manichaean ideology required an opposing "society of the dropouts, the saintly, the mad, the militant and the young" subsisting outside those plastic domes, it was the dominant plastic society, divorced from natural instincts, embarked on a cancerous expansion, that he envisioned in the space program as "the Wasp . . . emerg[ing] from human history in order to take us to the stars." Unless it was just a pose, plastic inspired in Mailer a private pathological dread that became public because he so effectively expressed it for an oppositional culture, for people who already sensed an uneasiness they could not articulate.[9]

Mailer was not alone among postwar writers and intellectuals in sensing the world had changed fundamentally in 1945, nor in suggesting a conjunction of the atomic and the plastic. In 1950, long before Jack Kerouac gained notoriety for *On the Road*, his first novel transcribed a thinly disguised Allen Ginsberg ranting about an "atomic disease . . . a kind of universal cancer" spreading across the world. "Everybody is going to fall apart, disintegrate," he continued. "Everybody's radioactive and don't know it." Like Mailer with his vision of plastic spreading across America, Kerouac understood this "atomic disease" as a postwar phenomenon and caused his fictional persona to agree that "well, hell!—things aren't like they used to be before the war." Ginsberg's own rendering of this postnuclear landscape came in 1956 in the poem "Howl," one of the first intentional proclamations of a cultural opposition, in an image that combined plastic and the threat of nuclear destruction: "I saw the best minds of my generation . . . angelheaded hipsters . . . listening to the crack of doom on the hydrogen jukebox." Hydrogen bomb and baroque plastic music machine fused for Ginsberg to evoke a sick culture addicted to runaway technologies of death. Years later, in 1974, Gary Snyder, a friend of Ginsberg, echoed the latter's vision in *Turtle Island*, a Pulitzer Prize–winning collection of eco-activist poetry. A fourteen-line poem conjoined the atomic and the plastic as equally freighted images of false technological transcendence. "Death himself, / (Liquid Metal Fast Breeder Reactor) / stands grinning, beckoning" in the opening lines of the poem, which closes with "plastic spoons, / plywood veneer, PVC pipe, vinyl seat covers, / [that] don't exactly burn, don't quite rot, / flood over us . . . [to the] end of days."[10]

Not nearly so apocalyptic, psychologist Robert Jay Lifton in 1968 described a post-atomic "protean man" cut off by nuclear anxiety from the comfort of traditional social structures and malleable in the face of immediate pressures. For Lifton, no one epitomized protean man better than the Italian actor Marcello Mastroianni, the smooth ineffectual anti-hero of several of Federico Fellini's 1960s films on the decay of meaning among the postwar generation—a figure adrift among the shifting phenomena of "*la dolce vita*." Mastroianni himself accepted *Time*'s description of his "frightened, characteristically 20th century look, with a spine made of plastic napkin rings." Although Lifton remained ambivalent about protean man, lamenting and celebrating his willingness to survive by going with the flow (or "flux"), the image of "a spine made of plastic napkin rings" left little room for interpretation. Linked to the atomic age, to the threat of nuclear annihilation, to an insidious cultural weakening metaphorically akin to radiation sickness, plastic evoked a sense of death imitating life—and rather indifferently at that.[11]

AN ALMOST INVISIBLE PERIL

No one in the 1950s would have predicted the apparent level of anxiety tapped in 1984 by an article in the sensationalist *National Enquirer* about a "prisoner of the modern world." Opening with a headline declaring that "she's allergic to modern living," the article described a woman forced by "environmentally induced illness" to live in an isolated mountain cabin lined with aluminum foil. Allergic to phenol, to formaldehyde, and to the vast category of "hydrocarbons," she avoided plastics and synthetics. Leaving only on rare occasions, she used an oxygen tank to prevent being overcome by fumes from her car's plastic interior. By then ordinary consumers worried about being poisoned by plasticizers escaping from a baby's squeeze toy, or by Styrofoam decomposing in a cup of hot tea, or by urea formaldehyde insulation sprayed inside the walls of their very own homes. "Will I get cancer," people wanted to know, by drinking from a plastic cup, eating from a plastic container, or touching a plastic electric cord? In 1976 a public health official blamed this "great public anxiety and even panic" on irresponsible media attacks on the plastic industry. But he admitted "the possibility for toxic sequellae in consumers is great."[12]

For some people, a vague plastic anxiety settled around Teflon, one of the most unique new materials of the postwar era. In particular, they worried about the subsequent history of the otherwise almost miraculous no-stick coating that eventually disappeared from the surfaces of pots and pans in which they cooked their food. Du Pont chemists had also worried about Teflon's toxicity, so much so that they postponed introducing the cookware until 1961, long after it was technically and economically feasible. After a decade of predictions of nonstick cookware, Du Pont reluctantly authorized its manufacture in the United States. The company was threatened by imported French cookware that was damaging Teflon's reputation because its coating quickly scratched and sometimes peeled away from the surface.

247

Getting the slippery polymer to stick to metal in the first place was no mean achievement. The process required grit-blasting or acid-etching the surface before baking on two or three layers of Teflon at a temperature of about 425°C. Even then retailers had to teach consumers to abandon metal utensils for wood or plastic to avoid scratching Teflon's soft surface. Over the years Du Pont wrestled with changing specifications, quality marks, and certification programs that saw the initial Teflon replaced by Boeclad Teflon (using a fusing system developed at Boeing), Teflon II, and then SilverStone. With each new version consumers allowed a familiar promise of "no-stick cooking with no-scour cleanup" to override disgust at old Teflon pans scratched down to bare metal. When a marketing expert exclaimed that Teflon had "brought obsolescence to an industry where low turn-over has always been a basic problem," he ironically pointed out one of the two major problems with Teflon. The other was a nagging suspicion that the vanishing plastic was being consumed along with the bacon and eggs.[13]

Until forced into action by the low-quality French imports, Du Pont had been paralyzed, unwilling to commercialize Teflon owing to fears of health risks. Those fears were not groundless. Researchers who worked with Teflon at Oak Ridge during the war found that it decomposed at 350°C, releasing fumes that caused mild flu symptoms. As rumors spread, the stories became inflated, equating Teflon fumes with inexplicable, near instantaneous death. Since the mid-1950s, various industrial and military safety bulletins had circulated "a gruesome little anecdote," as *Newsweek* called it, about a machinist who had supposedly died, lungs swollen with fluid, five minutes after smoking a cigarette contaminated with Teflon.[14] Although no editor could pinpoint the source of the story, and each retracted it when Du Pont pointed out its apocryphal nature, it continued to make the rounds.

Finally, in 1962, fearing renewed attention to the story of the ill-fated machinist as Teflon moved from heavy industry into frying pans, Du Pont moved to establish its safety. The director of the company's Haskell Laboratory for Toxicology and Industrial Medicine, biochemist John A. Zapp, Jr., published a twenty-page pamphlet with a catchy Hollywood title exploring *The Anatomy of a Rumor*. Expressing dismay at "the baffling, almost frustrating, tenacity of rumors which continue to ascribe lethal characteristics to some industrial products in the face of repeated, authoritative denials," the pamphlet explained the rigorous testing that established Teflon as safe for cookware. In France, for example, guinea pigs had survived exposure to fumes from a pan whose entire Teflon coating was volatilized nearby. More to the point, Du Pont's own tests proved that fumes generated by an overheated Teflon pan were no more toxic than those produced by overheating an ounce of cooking oil.[15] The pamphlet effectively "zapped" the rumor as Teflon pans went to market. But consumers who had never heard of the apocryphal machinist occasionally paused to worry about their health even as they accepted the convenience of Teflon. At bottom their concern was the invisible poison of cancer, perhaps as fully concealed in the Teflon that vanished from their frying pans as in the strontium-90 whose radioactivity contaminated their children's milk. Early in 1962, at about the same time Du Pont was arguing the safety of Teflon, John F. Kennedy was drinking milk in the White House to demonstrate his faith in its safety.[16] Nagging doubts about Teflon, fueled not by industrial folklore but by those scratched and scarred pans, never became more than a minor irritant. Nor did anyone suggest parallels between Teflon and radioactivity. Only Mailer would have appreciated the irony that Teflon was first used at Oak Ridge. But those nagging doubts suggested that negative attitudes about plastic were associated with a more general fear of science that derived ultimately from nuclear anxiety.

Even industry publicists sometimes adopted an oddly alarmist rhetoric when generalizing about synthetic materials. An article on plastic in packaging opened with a passage on "creeping plasticism" nearly worthy of Mailer. "Even the most level-headed of men," it announced, "can begin to entertain the uncomfortable suspicion that plastics are taking over the world." Science fiction images of radioactive mutants and invasions of body-snatching pods glimmered within the title of another, more skeptical article with its promise that "Encasement Lies in Wait for All of Us." That enigmatic phrase, actually referring humorously to the embedding of decorative objects in transparent acrylic cubes, inadvertently recalled the dry cleaning bag tragedy of 1959, when about eighty infants and toddlers suffocated in polyethylene film that unsuspecting parents had used as mattress covers or left within easy reach of tiny hands. While Teflon projected an uncertain threat, it was "the plastic bag menace" that generated the most revealing antiplastic rhetoric. Long before it became fashionable to dislike plastic, when new materials still exemplified modern progress, the plastic bag crisis revealed how tenuous was plastic's reputation, how tangled it was in an undercurrent of doubt about the future, and how vigilant the plastic industry would have to become in defending its image.[17]

When dry cleaners first began returning laundry in bags of thin transparent polyethylene film, the new product seemed another example of plastic's ever more convenient presence. Du Pont's company magazine reported in 1956 that blankets "freshly cleaned and trimly wrapped in clear plastic" marked "one of the hottest promotions" ever run by New England's largest dry cleaning firm.[18] Suits and dresses in clear film bags soon lined closets from coast to coast. They kept clothes clean and revealed the contents at a glance, unlike the old paper ones. In 1958 extruders of polyethylene film sold a billion of the new bags with a gross income of $20 million.[19] But in that same year scattered newspapers reported that infants and toddlers were being suffocated by what Du Pont had once called "those handy plastic bags."[20] Public awareness increased in April 1959 as the American Medical Association issued a press release responding to a report that four children playing with plastic bags had died in Phoenix. A local doctor speculated that polyethylene film charged with static electricity could "literally grab" a child "through electrical attraction to his face."[21] By the middle of June, when articles in the national press reached epidemic proportion, the death toll exceeded fifty children, most under six months old. At least seven adult suicides had used the plastic garment bag as a lethal device. Within the next six weeks, the press reported that another thirty children had accidentally died and ten more adults had committed suicide.[22] The nation confronted a death toll swelling geometrically, a gruesome plague stemming not from natural disease but from artificial products of human technological ingenuity. The rhetoric of reaction to the tragedy exuded an irrationality suggesting people were responding to more than the obvious hazard. Among the industry, on the other hand, first responses suggested cynical concern for protecting a new market. Defenders of the industry blamed ignorant parents for allowing children access to a dangerous material—as if plastic bags were equal to sharp knives or loaded guns. A Du Pont spokesman "blamed parental carelessness in the deaths" and followed the conventional line that polyethylene film was "made and costed to be disposable." This buck-passing conveniently ignored the fact that the industry had promoted the bags as ideal for reuse at home. Du Pont's own announcement of the dry cleaning bags in 1956 had praised them for being reusable—"as housewives have discovered." As late as April 1958, shortly before the panic broke, *Modern Plastics* was still praising the bags because "they can be re-used by the customer." The industry chose to ignore its prior position, to maintain that reuse was misuse, and to blame parents for the tragedy.[23]

Reaction in the press blamed neither parents nor the industry but instead aimed a barely contained fury at the plastic bags themselves as entities of near demonic malevolence. According to one account, plastic bags were "murderous." Another, after mentioning that the menace had already "snuffed out" twenty children in other cities, announced that it had "struck today for the first time in San Francisco" when a local baby was "Killed By Plastic Bag." The *New York Journal* captured the dominant tone of popular reportage when it declared that "an almost invisible peril hangs loosely over the helpless heads of the nation's infants." That approach to the story became so common that Cruse of SPI felt compelled to insist there was no "mysterious built-in danger." Plastic bags did not "literally reach out to ensnare children." But his protest carried little weight because the plastic bag menace threatened a generation of parents who already feared loss of control over their children's survival. As an editor in Redwood City, California, pointed out, plastic bags killed far fewer children than cancer, which annually took about three thousand children under age ten. Already cancer deaths of some thirty children each year could be attributed to "fallout radiation" absorbed through the milk and vegetables that nourished them. Unlike the current epidemic of suffocations, however, these radiation deaths neither "excited nationwide reaction" nor sparked "alarms in press and legislatures"; instead they marked "just another chapter in a dreary story we do not want to listen to." In a few calm words this small-town editor, meaning to downplay the scope of the plastic bag threat, had unwittingly suggested the submerged link to nuclear anxiety that gave the suffocation tragedy such a poignant dread. A colleague at the *San Francisco News* more directly made the connection. His own editorial asked "How Many Must Die?" and ended with a direct reference to the antinuclear peace movement by exhorting citizens to "BAN THE BAGS!"[24]

To dramatize the threat, some newspapers posed models in cautionary photographs that occasionally suggested parental complicity despite the demonizing rhetoric of many reactions. The *Santa Ana Register* staged a scene with a mother bending down to smooth a plastic bag over her own daughter. Even more appalling was a photograph run by a Toronto newspaper that featured a three-year-old boy with a plastic bag clinging to his face. There was something fascinating about a human face in all its imperfect uniqueness wrapped in an imploding bubble of transparent plastic film. *Life* printed a photograph of such a face to warn people of the hazards of plastic bags.[25] This bizarre full-page image represented the agonized face of Dr. Leona Baumgartner, director of the New York City Board of Health, hermetically sealed in a "Thin Bag of Death" whose puckered folds converged into her mouth and nostrils. "After 20 seconds inside clinging plastic bag," *Life* informed its horrified but fascinated readers, "Dr. Baumgartner gasps for breath, is near suffocation." Presumably she survived the ordeal. This image, experienced directly on its own terms, left an impression of an unnatural technological plague striking at random and isolating even mature victims, rather than of a danger that careful parents might anticipate and prevent. More threatening in a real sense than Teflon poisoning but not so all-encompassing as Mailer's cancerous metaphor, the plastic bag panic of 1959 reinforced the uncertainty of Americans confronting synthetic materials. Even as the Monsanto House at Disneyland promoted a plastic-as-plastic future to thousands of visitors, the plastic bag scare stimulated an anxiety related to those first vague fears that industrial chemists were violating the natural order in their Faustian urge to create "something new under the sun." As the panic receded, so did the conscious anxiety about plastic, but it left an unconscious residue that became stronger with the rise of environmental concerns in the late 1960s and 1970s. That the panic did recede was the result of timely actions taken by the Society of the Plastics Industry, which emerged from the "plastic bag war" as a "nationally known organization" concerned with preserving plastic's good name among the general population.[26]

At the time of the plastic bag crisis SPI normally functioned only behind the scenes. Occasionally it became involved in questions of safety or reputation, as when it investigated flammable Christmas tree ornaments. But for the most part it mediated disputes between processors and material suppliers, or lobbied the railroads on freight rates and the federal government on trade regulations. In the spring of 1959 polyethylene film producers asked SPI to warn the public about the plastic bag hazard and to deflect blame from the industry. On May 7, after a meeting with film manufacturers and the National Safety Council, Cruse announced an educational campaign.[27] By the end of the month SPI had printed eight thousand placards for dry cleaners and two million copies of a warning pamphlet for distribution to hospitals, doctors' offices, and public health clinics. Eventually five million pamphlets and sixty thousand placards were distributed. SPI's effort reflected sincere concern for preventing loss of life, but Cruse also carefully solicited a statement from the director of the National Safety Council that the plastic industry was not "in any way to blame for these reported deaths . . . since all of the cases known have been due to hazardous misuse—through ignorance—of the material that is not of itself harmful." It seemed important above all not to lose the convenience of plastic film, "an advance in our modern way of life."[28]

These efforts soon proved inadequate. By the middle of June SPI was planning an advertising campaign to be orchestrated by Batten, Barton, Durstine & Osborn through daily newspapers and local radio stations. Initially funded at $500,000 with 80 percent of the voluntary pledges coming from polyethylene suppliers and the rest from film processors, the campaign and legal expenses eventually cost twice that much.[29] On June 17 Cruse announced SPI's intention "to blanket the country with a blockbusting campaign in the shortest possible time" so there would not be "one more death through ignorance."[30] The following day dozens of newspapers ran a full-page advertisement with "an important message to parents about PLASTIC BAGS." After reviewing the fact of infant suffocation, it exhorted parents to destroy bags after use and listed steps taken by the industry to spread awareness, including warning labels for dry cleaning bags. As the editor of Modern Plastics accurately maintained, SPI's ad campaign was "a sincere and forthright action, not a whitewash job."[31] By the time it had run its course, few American parents could have remained unaware of the dangers of plastic bags—dangers that now seemed manageable rather than yet another symptom of loss of control in a brave new postnuclear world. By the same token, there was now no need to ban the bags—though that is just what a significant number of editors and politicians were demanding.

In fact the whole matter escalated to a different level when local, state, and federal legislators began deliberating a welter of conflicting bills seeking to ban the bags outright, or to require warning labels, or to mandate thicker nonclinging film perforated with holes. With governmental bodies across the country discussing regulation of polyethylene film, the very survival of that branch of the plastic industry became problematic. To fight the legal battle, SPI called on Jerome H. Heckman, a partner in the Washington law firm of Dow, Lohnes & Albertson. As an expert on radio law, Heckman had previously assisted the society in a struggle to preserve certain radio frequencies for use in heating or edge-gluing plywood. By the time the plastic bag crisis finally wound down late in 1960, Heckman had responded to some sixty regulatory bills introduced in Congress, state legislatures, and county and city councils. He had traveled more than forty thousand miles to present the "unvarnished facts" at "every known public hearing," whatever the level, to ensure that the industry could count on uniform standards instead of a forbidding array of conflicting restrictions and outright bans. As it turned out, most legislative bodies dropped their proposals or limited them to printed warnings when presented with evidence of SPI's massive public relations campaign.[32] In November SPI announced a new standard for film thickness developed by a committee chaired by physicist Jules Pinsky of the Plax Corporation. While some dry cleaners had used film as thin as three ten-thousandths of an inch, the new standard called for film no thinner than a thousandth of an inch. According to Pinsky, the danger of suffocation disappeared "almost to the vanishing point."[33] With approval of the new standard by the Department of Commerce and a marked reduction in the death toll—whether owing to newspaper reports, the SPI campaign, or safer film—the dry cleaning bag menace finally receded from public consciousness.

The episode had lasting significance. It taught SPI that its responsibility extended out to encompass plastic's public image. The society also gained lobbying experience that proved invaluable when ecological concerns threatened the industry in the late 1960s and early 1970s. It was no accident that Heckman then continued to serve as SPI's chief legal warrior. In the meantime the plastic bag episode had little effect beyond the lives of those people who had lost children. It became common knowledge to withhold plastic bags from infants and toddlers. Otherwise there remained only a residue of distrust of plastic, nothing to act on when making consumption choices but a vague memory all the same, recalled years later when triggered by more insistent ecological doubts. American journalists would not have reacted so hysterically in 1959 had not fear of radioactive fallout already sensitized them to the issue of technological self-contamination. The malevolent imagery in their attack on plastic bags served as a link between early twentieth-century doubts and the outright hostility of later ecological attacks. For the moment, however, antitechnological hostility to plastic remained submerged. Even so, some people found more traditional reasons for questioning the virtue of synthetic materials. These doubts extended back into the nineteenth century, when celluloid's success as a commercial material had revolved around cheap imitations and substitutions. Plastic in its least innovative manifestations often seemed fake and shoddy. Such grand exceptions as the Corvette and the Monsanto House merely proved the rule. As the things of the mid-century consumer economy proliferated, so too did the negative image that had plagued plastic from the beginning.

ONLY TERMITES CAN TELL THE DIFFERENCE

For a material supposed to be a utopian wonder-stuff, plastic tried the patience of the people entrusted with maintaining its image. For forty years, from the end of the war into the Reagan era, lamentations by the industry's promoters regarding its shaky reputation changed little. It was hard to get anyone to say a good word about plastic even though everyone relied on it all the time. As late as 1985 the editor of *Modern Plastics* complained that after all those decades of success, few companies would own up to making plastic. "This industry does not deal in plastics," he observed sarcastically; instead it "molds miracle space-age polymer . . . extrudes tough impact material . . . foams durable synthetic and sprays up wonder composite." Sixty years after his own magazine's first issue coined a new word to describe a new class of materials, they continued to have "more aliases than a passer of bad checks." His complaint echoed both a predecessor who asked in 1963, "Why Not Call It Plastics?" and an even earlier editor who in 1951 suggested with greater optimism, "Let's Use the Word 'Plastics' with Pride!" While the editor in 1951 was critical of a manufacturer who referred to his material as a "synthetic," the euphemisms of 1963 included such gems as "quality materials" and "miracle resinous compositions." Names kept shifting to make up for slippages in reputation. If the industry could not stand behind its own product, how could the general public be expected to accept plastic?[34]

Most people who thought about the image problem during the 1950s would have agreed with a Harvard Business School report that traced it to "all the cheap, poorly conceived, badly engineered, sensationally promoted junk that first hit the market after World War II."[35] It seemed that plastic's reputation would improve as a younger generation that had experienced only first-rate synthetics came to maturity. Such analyses ignored the fact that plastic suffered not only for negative reasons but also because it lacked a positive identity. Most synthetic products quickly moved beyond the hall of infamy that enshrined tacky vinyl raincoats and broken polystyrene toys. Even so, new materials never established a positive identity more solid than the utopian projections that had first accompanied them to market. Even the plastic-as-plastic aesthetic of the 1960s impinged only occasionally on the awareness of ordinary Americans. Not many people owned a Saarinen pedestal chair, nor did many want an Eames fiberglas shell. Beyond such icons of "good design," plastic enjoyed hardly any positive identity. Even after plastic had reached an acceptable level of functional quality for consumer products, it was still hard to say exactly *what plastic was.*

This fundamental identity problem occurred to a journalist whose thoughts appeared in *Architectural Forum* about five years before Barthes's meditation was translated. "One cannot overestimate the versatility of this immaculate, gelid set of substances, the plastics," Walter McQuade wrote in 1967. They were "too adaptable, so glib that they have not yet acquired dignity as a material." No democrat in assessing American popular taste, McQuade asked his architect-readers to imagine "thousands of living rooms whose flowered furniture wears squalid, if spotless, plastic covers." But he reserved his greatest animosity for plastic's too easy imitation of other materials. Forty years earlier Slosson had celebrated plastic for its chameleonic ability to mimic other materials. For McQuade, however, that penchant for simulation prevented plastic from assuming a substantive identity that might inspire confidence. He observed that for many architectural applications the synthetic merely "replaces another material" by "encasing the other material, as in the plastic laminates of wood, which preserve the wood grain as a kind of desperate identification underneath." In this odd situation, which McQuade's normally smooth prose strained to describe, plastics wound up "smothering the identity of other building materials, without acquiring character of their own."[36] No matter how well it did its job, it was hard to like something that made a point of such cheap fakery. The delicious shock of being fooled by an intricate fraud had pleased consumers of celluloid at the turn of the century. But those effects had so precisely simulated tortoise-shell or ivory that some collectors decades later had trouble distinguishing them from their natural analogues. Imitative plastics of the postwar era, on the other hand, gave themselves away as if their makers could not be bothered with the effort necessary to carry off the effect. One did not have to be a modernist champion of "truth to materials" to grow tired of the charade and to prefer "the real thing" when it was available at an affordable price.[37]

The most gratuitous imitative use was artificial plants and flowers springing up in restaurants, hotel lobbies, offices, and waiting rooms. As a promoter observed in 1951, plastic flowers were convenient because they "require no care except for an occasional wiping, will last virtually forever, and are highly fire-resistant." At that time a Cleveland molding company was making two hundred varieties from aspidistra to philodendron by a painstaking process that rivaled the accuracy of the old celluloid techniques. Starting with real leaves and petals preserved in formaldehyde and frozen at extreme temperatures, the company made plaster casts and then production molds for use with translucent polyvinyl plastisol. Workers applied proper colors to the artificial leaves by injecting them with dye. The result was "authenticated by spectroscopes that compare the light reflections as well as the transparency of living and plastic foliage." Such scruples quickly vanished as injection-molded polyethylene became the material of choice. Uniformly colored flower, leaf, and stem parts snapped quickly into assemblages that from a distance approximated living plants. By the early 1960s the annual synthetic crop amounted to $120 million, about a sixth of what people paid for cut flowers, and production had moved to Hong Kong because a mold that cost $1,500 in the United States could be had for about a hundred dollars. "The blossoms aren't real," explained the heroine of the Broadway comedy *Any Wednesday* in 1964; "they're some kind of plastic." She had bought them at Bloomingdale's. "You can't tell, can you," she asked with a hint of guilt—"from here, anyway?" Anthropologist Margaret Mead got to the heart of the matter when she observed that people who used plastic flowers were "unwilling to be at the mercy of a fading flower."[38]

Much the same motivation, a desire to extend artificial control over natural

8–1. Plastic Christmas trees with polystyrene branches and polyethylene needles, 1959

things, lay at the root of the artificial Christmas tree (Figure 8–1). A variant on sale in 1957 was promoted as "the first all-plastics Christmas tree." Boasting a brown polystyrene trunk and green polyethylene branches that "cannot wear out, dry out, or rot," it would never "flare up from a spark or the heat of a Christmas tree light." And unlike the traditional evergreen hoisted up to the occasional accompaniment of unseasonal swearing, the plastic imitation with its numbered parts was "easy-to-assemble" and could be "dismantled in a few minutes" with no shedding needles. One could even purchase a small model and "add to it in succeeding years."[39] Despite claims that artificial trees looked "as real as nature's own greenery," people complained they were "just too symmetrical," or they lacked "the smell of a real tree."[40] Briefly during the 1960s, the popularity of bright reflective trees made from metallicized vinyl or saran marked a turn toward artificiality.[41] But for the most part Americans rejected such pure symbolism. If convenience or a promise of long-term savings prompted them to buy plastic, they sought a Christmas tree as real as human ingenuity could make it—and then felt vaguely cheated by it.

Most plastic simulations imitated not life itself but the natural materials from which people had traditionally fashioned their domestic environments. Professional contractors as well as do-it-yourselfers relied on an array of imitative substitutes whose "authenticity" ran along a continuum from highly convincing Formica to merely referential vinyl Con-Tact paper. During the mid-1950s a homeowner could decorate the game room with Decro-Wall paneling, a system of rigid two-by-four-foot sheets of vacuum-formed vinyl masquerading as brick or stone. The depressed lines that "simulate[d] the mortar lines around each brick" formed ridges on the back to which adhesive could be applied, and the projecting "bricks" or "stones" formed pockets of air for insulation. No more convincing was an artificial stone whose one-by-four-foot lengths of fiberglas paneling weighed one-thirtieth as much as a comparable surface of stone and could be nailed up indoors or out. Each panel represented seven variously shaped building stones "molded in dies cast from carefully selected specimens of fine quarried stone." Many do-it-yourself schemes, such as an "imitation wrought iron" molded from impact-resistant polystyrene or high-density polyethylene, quickly disappeared from the market. Other imitative substitutes soon became commonplaces of the American domestic landscape. So-called "cultured marble," for example, increased in sales from $3 million in 1964 to $77 million in 1970. These artificial marbles contained fillers (sand, kaolin, lime, stone, or shell aggregates) cast in a polyester resin matrix and opened up a new market for "luxurious" bathroom vanity tops. Just as successful, and more emblematic of a shift toward simulated materials, was the extruded vinyl siding that came on the market in 1964 and within a dozen years was applied to the exteriors of 750,000 houses. Despite fading of dark colors in the early years and a tendency to buckle in hot weather, vinyl siding solved the perennial problem of scraping and repainting wooden clapboards. But the simulation left much to be desired. While cultured marble might look "like marble" to people with no real marble in their homes, vinyl siding's stylized "wood-grain" embossing fooled no one. It was a substitute for people who did not want to hire a painter and do it right.[42]

It could be argued that the success of cultured marble and vinyl siding revealed little of people's attitudes about plastic imitations. After all, many people lived in apartments or rental houses and had no choice in the matter. Even homeowners relied mostly on interior furnishings to establish domestic identity, literally to make a home. For that reason the phenomenal sales increase in imitation wood furniture during the 1960s takes on special significance. A market study released in 1964 by the American Society of Industrial Designers suggested that the "too bright colors often identified with synthetic materials" were "retreating" in popularity and that "natural materials such as wood, leather, cork, marble, stone" were becoming "increasingly important in our environment." In that same report an in-house designer at Philco observed that portable radios and televisions with wood-grained plastic cabinets (probably pigment-swirled polystyrene) had met with "great public acceptance."[43] As the decade progressed, stereo components came with wood-grained cabinets of printed vinyl film fused over metal or chipboard casings. By the end of the decade, even expensive free-standing televisions had cabinets molded from plastic—no longer a hollow-sounding polystyrene but a rigid high-density polyurethane foam whose structural, tactile, and sonic qualities supposedly mimicked those of expensive hardwoods. As an engineer at General Electric explained, with furniture of polyurethane foam "a skilled carver need use his talents only once"; the result could be repeated indefinitely. In the words of another promoter, "intricate carvings, moldings, rosettas, routed overlays, and frets" could now be "mass-produced by injection molding techniques."[44]

This freedom to ornament at no expense violated thirty years of rhetoric pro-

257

8–2. Chest with drawer fronts of injection-molded polystyrene imitating wood, 1967

moting an artificial plastic-as-plastic aesthetic and fostered a revival of heavy-looking Mediterranean-style furniture exuding rustic charm. Even as design magazines promoted fiberglas shells and high-gloss injection-molded polypropylene, middle-class Americans filled their houses with plastic simulations of traditional furniture (Figure 8–2). While only a single molder exhibited at the National Association of Furniture Manufacturers in 1966, four years later there were eighty-five molders, material suppliers, and machinery suppliers.[45] Only occasionally did anyone directly object to simulated wood, as when *Progressive Architecture* snobbishly described how "Plasticamerica is offered pseudo-carved, pseudo-turned, pseudo-'Spanish Mediterranean' furniture of authentic cast polystyrene direct from the ovens of the High Point, North Carolina, furniture complex."[46] Although such sarcasm was rare, and boosters predicted that soon the only "non-plastics" furniture would be antiques, few people in the industry publicized the fact that they were imitating wood.[47] As a reporter for the *Wall Street Journal* admitted, "The consumer . . . isn't always let in on the switch"—an opinion confirmed by a Chicago retail executive who admitted "the less said about it the better." Unlike retailers, on the other hand, material suppliers had to admit to what they were doing because *their* customers were the furniture makers. "It looks so good, you'll swear it's wood!" proclaimed one company's advertising. "Only the chemist knows for sure," said another slogan, playing on the famous Clairol hair-coloring ad campaign, and in a variant, "Only termites can tell the difference."[48]

Some of the "traditional" uses to which molders put rigid polyurethane foam embodied an element of exaggeration or baroque innovation. In Newburyport, Massachusetts, for example, the Arborlite Company manufactured polyurethane replicas of the wooden eagles whose carved likenesses had graced the sterns of the town's famous clipper ships. Ranging in length from two to six feet, the hand-painted polyurethane eagles were "accurately detailed reproductions of century-old ship carvings," according to Arborlite's president. They were foamed in synthetic rubber molds made from fiberglas positives that in turn came from hand-carved wooden originals commissioned by the company.[49] More imaginative than any polyurethane television cabinet, Arborlite's eagles exhibited a new technology applied to create the semblance of a traditional form of expression whose enjoyment on a mass scale would not have been possible before the age of plastic. But to some members of the generation that grew up with synthetic materials, a plastic eagle above a suburban garage door—with its patriotic appropriation of a past imperfectly understood— epitomized everything false about their culture. Most plastics used in construction remained hidden as adhesives, as water barriers, as insulation. Many other plastics, those in Bic pens or Scotch tape, phonograph records or model airplanes, seemed inevitable and therefore also invisible. But it was a different story with the all-too-obvious simulations of the domestic landscape. Melamine wood-grain tabletops, polyurethane TV sets, rustic sideboards with wormholes molded in, cultured marble vanities, vinyl siding, polyethylene violets, walnut-grain vinyl film fused onto the stereo or recessed into the refrigerator handle—it all suggested a dishonesty so shameless it could be manifested on the very surface of things. Long before it became trendy to question plastic's ecological soundness, a generation of disaffected youth recognized "poor old Formica polyethylene 1960s America" as a sham.[50] To understand both the resonance of *The Graduate* and the instantaneous emergence of "plastic" as a synonym for fake or phony, one must bear in mind the degree to which plastic never escaped the stigma of the second-rate imitation. Slowly, unconsciously at first, plastic as substance and as idea took on negative meanings for people whose lives coincided with the material's expansion into the domestic environment. Perhaps, to paraphrase Marshall McLuhan, they feared they might become what they beheld.[51]

SLIVERS OF VERBAL PLASTIC

Long before *The Graduate*, a few observers of the American scene sensed a connection between the presence of synthetics and the new postwar mode of life. Plastic doubts were not limited to Norman Mailer, though his vitriolic crusade seemed at best eccentric. Around 1960 two writers who established their reputations on soundings of a hollowness at the core of suburban life almost casually referred to plastic as a carrier of meaning. For John Cheever, writing in a short story that revolved around a backyard fallout shelter, "the burden of modern life, even if it smelled of plastics—as it seemed to—bore down cruelly on the supports of God, and Family, and the Nation." Less insistent, his younger colleague John Updike made several references to plastic in *Rabbit, Run*, a famous novel of postwar anomie. After an opening scene in which former high school basketball star Rabbit Angstrom has intimations of physical decline while shooting baskets in the neighborhood, he runs up the hill to the duplex that traps him in a blighted marriage. Under the front step "a lost toy molders . . . a plastic clown." Later, whenever Rabbit enters the house, he is depressed by "the kid's toys here and there broken and stuck and jammed." His son's "broken toys on the floor derange his head," reminding him of his marriage. For all that, Updike also permits Rabbit to find comfort in plastic, in "that sweet tangy plastic new-car smell" that "cools his fear" as he leaves his wife for another woman, and, after their reconciliation, as he watches his son at a playground, in nostalgia set off by "the forgotten smell of that narrow plastic ribbon you braid bracelets and whistle-chains out of" at summer camp. Updike clearly intended plastic to carry symbolic meaning but remained as uncertain about it as about the possibility of finding moments of integrity and transcendence in contemporary life.[52]

Any ambivalence regarding plastic had vanished a few years later as the emerging youth culture identified it with everything they despised in American life. Entirely typical was a New Left manifesto of 1968 in which a group of white college students rejected as "meaningless" the "white honkie culture that has been handed to us on a plastic platter." New York City activists complained of mental health services at Bellevue Hospital where "they take your brain out and fill the hole with plastic." As psychedelic drugs spread from initiates to casual trippers, a broadside warned that LSD was "truly a product of this culture's technology and the trip can be as plastic as the system" that produced it if not taken with spiritual intent.[53] The word was so common a term of derision that when journalist Leonard Wolf interviewed San Francisco hippies in 1967, he asked a leader of the community whether he considered his family "up tight and plastic." Another tourist from the outside world, sociologist Lewis Yablonsky, devoted a book to exploring the hippies' rejection of the "plastic society" with its "massive hypocrisy and dishonesty" and their admirable search for "a more primitive society, . . . one that is honest and closer to the natural state of man's reality." Eventually disillusioned by a stint of participant observation, Yablonsky concluded that "much of the hippie philosophy and way of life is even less satisfying, more hypocritical, and more plastic than straight society."[54]

Not only disaffected young people and their chroniclers found in plastic a metaphor of a society that had abandoned the American dream. As intellectuals wrestled with the question of why so many of history's most pampered generation spurned their inheritance, the presence of plastic bulked large in their thoughts. One of the most popular explanations was offered by Yale law professor Charles Reich's book, *The Greening of America.* In his opinion American life was too artificial, too regimented, too predictable—in a word, too plastic. While most work was "pointless and empty," the leisure it earned was frittered away in a "grossly commercial" culture. "Our life activities," he wrote, "have become plastic, vicarious, and false to our genuine needs, activities fabricated by others and forced upon us." In a process of "impoverishment by substitution," the "corporate state" had weaned Americans from natural pleasures and made them dependent on artificial things. Owing to the profit motive and the urge for control, "the genuine is replaced by the simulated." For Reich it was "all epitomized by Astro Turf," the embodiment of an artificial culture "profoundly hostile to life." Fortunately the young protesters were pointing the way to "a culture that knows how and when to use technology, a culture that is not plastic or artificial but guides and uses technology in pursuit of values that are derived from human sources." Even their drab clothes, "browns, greens, blue jeans," demonstrated rejection of the "plastic, artificial look of the affluent society" and expressed an "affinity with nature." The book's final sentence predicted the "greening" of an America no longer condemned to be "encased" in "plastic." It would have curdled the blood of Cruse, Kline, or any other honest toiler in the plastic vineyard—all the more so because *The Greening of America* became a best-seller.[55]

Similar ideas gained more evocative expression in Theodore Roszak's *Where the Wasteland Ends*, an analysis of the culture of high technology focusing on "the expanding artificiality of our environment." Like Mailer before him, Roszak found special significance in the figure of an astronaut "sealed up and surviving securely in a plastic womb that leaves nothing to chance or natural process." Ultimately everyman and everywoman would share the situation of the astronaut as "all places become the same gleaming, antiseptic, electronic, man-made place, endlessly reproduced," and all people yielded to the addictive security of the "wholly controlled, wholly artificial environment." Although Roszak did not use *plastic* as a dirty word, a long description of society's impending decline ended with a ringing denunciation of plastic. If the human race avoided nuclear war, then it would slowly slide into a nightmare of "organizational confusion and bureaucratic malaise, constant environmental emergency, off-schedule policy, a chaos of crossed circuits, clogged pipelines, breakdowns in communication, overburdened social services." As for the latest technology, "the data banks will become a jungle of misinformation, the computers will suffer from chronic electropsychosis." Eventually "the scene will be indefinably sad and shoddy despite the veneer of orthodox optimism . . . rather like a world's fair in its final days, when things start to sag and disintegrate behind the futuristic facades, when the rubble begins to accumulate in the corners, the chromium to grow tarnished, the neon lights to burn out, all the switches and buttons to stop working." And then, finally, in the last manifestation of decay, "everything will take on that vile tackiness which only plastic can assume, the look of things decaying that were never supposed to grow old, or stop gleaming, never to cease being gay and sleek and perfect."[56]

For Tom Wolfe, a more meticulous observer of his society than either Roszak or Reich, much of America had already fallen to such a state—or had never risen out of it. As Wolfe portrayed the world in his "new journalism," plastic exuded a second-rate aura. A long catalogue of "the Rat lands of America" with their "drive-ins, mobile-home parks, Dairy Queens, superettes," and so on finally reached such details as the "$8,000 bungalows with plastic accordion-folding partitions and the baby asleep . . . in a foldaway crib of plastic net." Such phenomena fascinated Wolfe. He began a profile of Marshall McLuhan by describing a "trick snap-on necktie with hidden plastic cheaters" that the otherwise "distinguished-looking" media theorist wore at their first meeting ("I couldn't keep my eye off it"). Another profile featured Carol Doda, a topless dancer who "became the greatest resource of the San Francisco tourist industry" when she "blew up her breasts with emulsified silicone, the main ingredient in Silly Putty," and thereby, along with false eyelashes and bleached hair, became "the put-together girl." On another occasion he projected himself into the group mind of Ken Kesey's Merry Pranksters as they attended a massive Beatles concert: "It might as well have been four imported vinyl dolls for all it was going to matter." Most of the time Wolfe's references to plastic remained embedded in observed reality. But sometimes he could be as metaphorical as Mailer. In a piece on customized cars Wolfe lamented that, with Detroit sponsoring custom auto shows, "the manufacturers may be well on the way to routinizing the charisma, as Max Weber used to say, which is to say, bringing the whole field into a nice, safe, vinyl-glamorous marketable ball of polyethylene." Here, presuming on the reader's prior recognition of plastic as déclassé, Wolfe touched on the same emotionally empty security Roszak had evoked. Even so, Wolfe hardly considered plastic emblematic of everything wrong with society, which for him was far too fascinating to permit judgment to intrude. But he did lend his power of expression to the notion that plastic was second-rate, unauthentic, vulgar, and somehow obscenely humorous. His offhand remarks may have done more than any earnest polemic to destroy plastic's image for young Americans—though hardly as much as Buck Henry's infamous line.[57]

A glance at almost any work of fiction of the late 1960s or 1970s, whether popular or literary, reveals that plastic—as material and as concept—had become a pejorative used with confidence that the reading public would not object. Writers of detective novels and thrillers, exploring America's seamy underbelly, proved most casual in tossing off references to plastic. As early as 1941 James M. Cain had made his disparaging reference to a radio "in the bakelite style." In 1962 Ross Macdonald's investigator Lew Archer described a living room "furnished with the kind of cheap plastic pieces that you're still paying installments on when they disintegrate." Twelve years later Robert B. Parker's sleuth Spencer waxed apoplectic as he drove north from Boston into a ten-mile "plastic canyon" of "sub-sandwich shops, discount houses, gas stations, supermarkets, neocolonial furniture shops (vinyl siding and chintz curtains), . . . an automobile dealership attractively done in glass and corrugated plastic, an enormous steak house with life-sized plastic cows grazing out front. . . ." At about the same time, across the continent in a fading Montana, James Crumley's dissolute Milo Milodragovitch was subsisting on "plastic sandwiches" and becoming entangled with a client "from a simpler, better time" when "screen doors smelled like rain or dust instead of plastic." Apolitical and reactionary thrillers routinely invoked plastic. In one of Dick Francis's racetrack mysteries, a professor left the track early because "fifty plastic students were waiting for him to pat their egos." And in one of Simon Brett's theater-based mysteries, the characters were "little plastic people being manipulated" by a murderous director. By 1986 even one of Robert Ludlum's doomsday epics could describe negotiations over Hong Kong's future as marked by "tension beneath the civility, the verbal placebos, and the plastic smiles."[58]

Writers of more serious fiction reinforced the idea there was something wrong with plastic. The trend was more evident among authors whose political ideology or cultural stance placed them at odds with the mainstream. In Robert Stone's *A Hall of Mirrors*, for example, an old man rips into greedy speculators who "have done plasticated the entire country over"—a more or less physical description—and then metaphorically describes a corrupt politico as a "knock-nose hooley" with "two hundred and six bones in his body and every one of them plastic." Even more angered by a dishonest antihuman culture was the African-American writer Ishmael Reed, whose jazzily inventive counter-history *Mumbo Jumbo* in 1972 posited an age-old white man's control conspiracy organized by a secret masonic Wallflower Order. There was "nothing real" at the Order's headquarters; it was "a gallimaufry of synthetic materials . . . polyurethane, Polystyrene, Lucite, Plexiglas, acrylate, Mylar, Teflon, phenolic, polycarbonate." If the Order had its way, "plastic" would "prevail over flesh and bones," and the joy of "all night . . . dancing and singing" would yield to a sterile round of "build, drill, progress." This sense of cold abstraction pervaded other literary uses of plastic. Stanley Elkin, for example, in 1976 referred to the effect of multiple sclerosis on a character whose disease had "deadened others as well as himself," giving them "the dead, neutral texture of plastic." Another writer, Gilbert Sorrentino, parodied literary treatments of plastic. His *Mulligan Stew*, an intellectually trendy metafiction of 1979, boasted a writer called Antony Lamont who composed a letter discussing his own novel *Rayon Violet*, its "overriding symbol" of "plastic" chosen because it perfectly represented his characters, a "sterile and lost group of people."[59]

Widespread use of "plastic" to imply cheapness, falsity, or violation of nature in both popular and literary fiction suggested that readers in the 1970s and 1980s gave silent assent to such descriptions and figures of speech. Everyone understood a reference to a "plastic person" or a "plastic smile." No one, except for an irritated industry executive or publicist, would have questioned its appropriateness. Derogatory references filled the mass media and everyday speech. Businesses that used traditional materials had a field day. Swissair, for example, bragged in a *New York Times* advertisement that its passengers enjoyed "gourmet meals too good to serve on plastic." And Hobart boasted in *Good Housekeeping* that its KitchenAid mixer— "all metal, no plastic"—was so durable it was "often handed down from mother to daughter." The word plastic came easily to the lips of a participant in a sociological study who told an interviewer that his new condominium had "yet to glean any type of personality"; it remained "sterile, plastic right now." The fifth-graders of Four Corners Elementary in Greenfield, Massachusetts, balked when the school cafeteria began using plastic forks and spoons and protested that "real people deserve real silverware." More universal in scope was a column in *Atlantic Monthly* by L. E. Sissman expressing anger at a "rising tolerance, even appetite, for plastic in all our artifacts and ways." Moving from the real to the metaphoric, Sissman complained that "in the elevator and the supermarket and the airport waiting room, we are lulled and stupefied by plastic music; on the Interstates, we ingest plastic food; when we read a magazine or newspaper or watch TV, a flood of plastic English assaults our eyes and ears." Even designer Raymond Loewy, who had promoted use of plastic from the 1930s into the 1960s, somehow managed in 1979 to berate American civilization for becoming a "plastic world" of "cheap, sleazy junk." By then the plastic industry had suffered a series of blows so devastating that the word's semantic degradation seemed harmless by comparison. Ecological concerns increased so steadily after the first Earth Day of 1970 that insiders feared the crisis might "really end the industry." Plastic's reputation was worsening even as the stuff itself flowed from refineries and molding plants at an ever increasing rate.[60]

OUT OF CONTROL

By definition the plastic industry was everything ecological activists wanted to expunge from American experience. Since the early twentieth century, promoters of industrial chemistry and synthetic materials had boasted of transcending age-old limits of traditional materials by extending scientific control over nature. During the 1920s predictions of an expanding stream of inexpensive artificial goods had suggested material abundance as a basis for a utopian democracy. By the final third of the century that transcendence threatened to drain natural resources and pollute the society that supported it by generating a flow of irrecoverable, inassimilable matter—garbage, society's excrement. Far from engendering the steady-state equilibrium once suggested by thermosetting Bakelite, industrial chemistry had ushered in a thermoplastic flux whose artificiality threatened to poison or submerge nature if it could not fully replace it. Or so it seemed.

The most eloquent ecological critic of the plastic industry, Barry Commoner, came to prominence in 1971 with *The Closing Circle: Nature, Man, and Technology*. He argued the human race had "broken out of the circle of life, converting its endless cycles into man-made, linear events." This one-way process of divergence from nature afforded evidence of a "nearly fatal illusion"—the assumption "that through our machines we have at last escaped from dependence on the natural environment." Since that statement embodied the fundamental tenet of plastic utopianism, it was not surprising he had nothing good to say about synthetic materials. Commoner argued that of all species, human beings were "uniquely capable of producing materials not found in nature," resulting in the world's first known "intrusion into an ecosystem of a substance wholly foreign to it." Nature had no mechanisms for breaking down the "literally indestructible" plastics and integrating them into its eternal cycles. It was "sobering," he wrote, "to contemplate the fate of the billions of pounds of plastic already produced," some of it incinerated and thus polluting the air, some of it dumped in the ocean and harming marine life, the rest of it straining landfills or strewn across the landscape. Beyond that were unknown possibilities of carcinogenic plasticizers and other chemicals released during plastic's use. Dismissing the idea of a technological fix, Commoner attributed the problems not to "some minor inadequacies in the new technologies" but to "their very success in accomplishing their designed aims." Discarded plastics "clutter the landscape," he concluded, "*because* they are unnatural, synthetic substances designed to resist degradation—precisely the properties that are the basis of their technological value." Survival of the human race was threatened by ignorant exploitation of an unprecedented ability "to tear the ecological fabric that has, for millions of years, sustained the planet's life."[61]

Five years later Commoner expanded his analysis by describing the economic mechanism that he argued was fueling the expansion of petrochemicals. Published shortly after the energy crisis of 1973, *The Poverty of Power* sought to explain why society was entangled in a fatal reliance on oil. Between 1946 and 1974 annual production of synthetic chemicals had risen a hundredfold from 150,000 tons to fifteen million tons. The plastic industry's share reflected an average annual growth rate between 1948 and 1970 of 15.9 percent (compared with 4.3 percent for steel). For Commoner this expansion provided a measure of earth's pollution and a sign of the bankruptcy of a way of life based on dwindling resources. In his opinion the overriding economic motive for this expansion was the petrochemical industry's reliance on the high-investment continuous flow processes of the refinery. Successful competition with labor-intensive industries like wood, wool, or leather required pushing volume to the limit and finding a use for every by-product. "By its own internal logic," he maintained, "each new petrochemical process generates a powerful tendency to proliferate further products and displace pre-existing ones." The industry tended "not so much to serve social needs as to invent them" just to keep oil flowing from wellhead to refinery to extrusion machine. Although Commoner did not call it a pyramid scheme, his description suggested a house of plastic cards soon to collapse in a mess of debris and toxic waste as oil reserves ran dry. The solution required voluntary return to labor-intensive traditional materials except in those instances—phonograph records or heart valves—where plastic marked a true innovation.[62]

Even before "whole earth" became an idea to fight for, industry watchdogs had occasionally warned that disposing of disposables might prove costly. As early as 1966 Joel Frados of *Modern Plastics* advised readers to think constructively about the issue before "well meaning but misinformed authorities step in with homemade remedies and regulations."[63] He was writing during an explosion in the volume of plastic for packaging and other disposables—the culmination of a trend that got under way in the 1950s. In 1952 Americans had first experienced single-serving jelly "paks" of vacuum-formed sheet vinyl. Later in the decade they bought shirts packaged in clear polyethylene bags and vegetables packed in flimsy polystyrene trays or wrapped in thin film; they ate banana splits from "boats" of thin, rigid, vacuum-formed polystyrene sheet and drank coffee from Styrofoam cups.[64] The following decade witnessed polyethylene bleach and detergent bottles, polystyrene containers for cottage cheese and yogurt, recloseable polyethylene lids for cans of coffee and shortening and cat food, polyethylene squeeze tubes for suntan lotion, polyethylene swivel closures for lighter fluid cans and cosmetics, polyethylene bread bags, Styrofoam meat trays, polyethylene six-pack connectors, vinyl blister packs, green polyethylene garbage bags, and Ex-Cell-O's polyethylene-coated paper milk cartons, which eliminated annoying flakes of wax in the milk but were soon almost superseded by lightweight bottles of blow-molded polyethylene.[65] By the time Frados issued his warning, the only major consumer disposable still to appear in the next two decades was the soft drink bottle of PET (polyethylene terephthalate) developed by Du Pont's Nathaniel C. Wyeth, a maverick from the family of painters who resented that chemistry brought less recognition than art.[66] But if the major disposables were already introduced, the volume of single-use plastics continued swelling at a faster rate than plastic in general. In 1960 packaging accounted for 300,000 tons of resin, nearly 10 percent of total plastic production; in 1966, packaging increased to 1.3 million tons, nearly 20 percent of the total. By 1969 packaging absorbed nearly one quarter of all resin produced, and the tide kept rising. Frados knew what he was talking about.[67]

American consumers initially balked at the idea of disposable plastics. Throwing things away violated the image of durable quality the industry had built up since the 1930s. The few plastic packages of the 1930s, Ovaltine's Annie shakers and various cigarette boxes and cosmetic jars, won public acceptance as "premiums" that retained their value. In 1957, two years before the dry cleaning bag scare, sales managers complained about the difficulty of marketing "disposable and expendable" goods owing to a "disinclination of consumers to accept the fact that such merchandise has been designed to be, and therefore should be, discardable and destroyable." It was too bad the industry had emphasized "durability and re-use value" because now people expected them. Reeducation was necessary. As an editor announced at the National SPI Conference of 1956, "Your developments should be aimed at low cost, big volume, practicability, and *expendability*"—with a goal of winding up "in the garbage wagon."[68]

Actually consumers took quickly to the short-term convenience of plastic packaging and throwaway products. Within a few years, however, "expendability" provoked a vocal minority who had begun to worry about environmental decay. Soon after becoming editor of *Modern Plastics* in 1968, Sidney Gross protested that the "problem of garbage" flowed not from packaging or plastics but from "our civilization, our exploding population, our life-style, our technology." It was unfortunate that plastic's 2 percent by weight of the nation's solid waste comprised "the most visible garbage, and the most lasting." All too often the public condemned plastic as a "villain" to be "exorcised from the economy."[69] It certainly seemed that way in 1970 when a member of the liberal city council of Madison, Wisconsin, proposed a ban on nonreturnable food and beverage containers and a one-dollar deposit on every returnable container. Initially considered a joke, the ordinance was taken seriously after the proposed deposit was reduced to fifteen cents. Within a year fifteen state legislatures were considering bills to ban or limit plastic bottles or containers.[70]

Especially threatening was a tax of two cents passed in New York City in the summer of 1971 on every plastic bottle or container. After six months of opposition to the ordinance, SPI succeeded in having it declared discriminatory and thus unconstitutional. The legal battle was the first in a long series fought by the society in defense of disposable plastics, led by its attorney, Jerome Heckman.[71] Ironically, earlier that year New York's sanitation department had campaigned to convince residents to use polyethylene trash bags instead of metal garbage cans—a policy chosen after tests inspired by a successful program in Toronto. Some disposables were clearly better than others.[72]

As the industry heeded expressions of environmental dismay, sanitation engineers and plastics executives debated landfills versus incinerators for disposing of disposables. Because plastic remained inert in landfills and did not release toxic fumes through decomposition, it seemed an ideal packaging material, superior to traditional paper products. But as the prospect of overflowing landfills loomed, attention turned to incineration, and with it a fear that chemicals released by burning plastic would pollute the air, corrode incinerators, and leach into ground water. By 1970 people were discussing such options as recycling disposables and developing resins that would decompose in sunlight or in the presence of soil bacteria. Neither idea gained adherents except among environmentalists and newspaper editors. A disintegrating plastic would violate everything the industry had worked toward. Recycling, on the other hand, seemed impractical because it required sorting out dozens of different resin formulations from the general flow of garbage.[73] Not until the late 1980s did the industry take recycling seriously. The inhabitants of the United States then generated about ten million tons of plastic waste each year. That amounted to 7 percent of the annual flow of garbage *by weight*, the measure SPI typically used to minimize its impact, but it comprised a more impressive 16 to 25 percent *by volume*. With landfills dwindling, the industry adopted a system of resin codes for marking disposable containers for easier sorting. Material suppliers set up pilot projects for blending recycled materials with virgin resins and encouraged entrepreneurs to experiment with molding scrap into boards, flower pots, and other low-tech objects. By then it had long been true, as an engineering journal observed in 1979, that "plastics recycling" had "shifted from its emotional history . . . to an era of serious research and development."[74]

Long before that "emotional history" had run its course, however, the industry experienced a series of nonstop one-two punches during the early 1970s. The initial uproar over solid waste came to a head in 1971 with dozens of regulatory bills introduced across the country. At that time the issue that most provoked vocal opposition was not the overriding problem of garbage but the aesthetic problem of litter. With landfills nowhere near bursting, it was easier to arouse people about bleach bottles washed up on beaches and Styrofoam cups tossed along the road. While paper litter quickly disintegrated, the plastic stuff remained as a visual reminder of an inflationary culture.

But not all litter was harmless. In March 1972 Edward J. Carpenter of Woods Hole Oceanographic Institute in Massachusetts opened a new issue by announcing he had found tiny bits of plastic in Long Island Sound at a density of one to twenty samples per cubic yard of water. Not only were toxic plasticizers released into the marine food chain, he suggested, but bits of plastic provided surfaces for bacterial growth and blocked digestive tracts of smaller fish. Other investigators reported the Sargasso Sea clogged with plastic—minute spheres of Styrofoam, worn discs of polyethylene, bits of bottles, torn sheets of film, strapping bands—most of it dumped from ships, battered by waves, and becalmed in an area formerly romanticized as a ship's graveyard.[75] Carpenter proved willing to compromise, to work behind the scenes without embarrassing the industry. In April he informed Ralph L. Harding, Jr., SPI's new executive vice president, that fresh spheres of polystyrene resin in Long Island Sound indicated dumping by a plastic processor. Carpenter hoped he would be able to report along with his initial findings "that we have had the cooperation of the plastics industry in locating the source" and that recent checks revealed "decreasing concentrations of plastic spheres in coastal waters." A letter from Harding to polystyrene producers flushed out the anonymous culprit, ended the spills, and Carpenter seemed "willing to drop the matter," though he and the U.S. Coast Guard continued to monitor the Sound.[76] This episode of wary cooperation hardly ended the problem of ocean pollution, however. As the tide of trash swelled at sea, with seven million tons dumped in 1980, most of it from ships, thousands of fish and birds died from ingesting particles of plastic, thousands of sea turtles and whales died from swallowing plastic bags or sheets of film, and miles of plastic driftnets discarded by high-tech fishing ships entangled seals, sea lions, and birds. At the end of 1987 the U.S. government finally signed a United Nations treaty prohibiting ocean dumping from ships, but the problem was far from solved.[77]

By 1972 many Americans were sensitized to environmental issues, enough that they might support taxes on disposable containers but not so much that they would give up the convenience of plastic. The industry proved so successful in defending itself against legislative attacks that some executives assumed the worst was over. Even Harding, whose job it was to coordinate the defense, admitted in August that they had entered a "lull" except for occasional "anti-plastics" attacks in the media. He cautioned against the attitude "that our troubles are behind us" and referred to the relative calm as "the eye of the storm."[78] Little did he suspect the size of the approaching storm. The environmental debate over plastic had focused on issues that did not concern anyone's immediate welfare, but that soon changed. By the end of the year the Federal Trade Commission was investigating reports that polyurethane foam and other synthetics were more flammable than industry testing standards indicated. Several airline accidents in which dozens of passengers were asphyxiated revealed that many plastics released toxic fumes when burned. In some cases chemicals used as flame retardants compounded the effect. The threat of toxic fumes from burning plastics extended beyond aviation into the everyday lives of people who encountered polyurethane foam in furniture, bedding, and auto interiors, and whose domestic surroundings contained dozens of other potentially dangerous synthetics.[79]

Despite an eleventh-hour meeting of representatives of the FTC, SPI, Underwriters Laboratories, the American Society for Testing Materials, and various safety and insurance groups, the FTC in May 1973 filed a class-action suit against twenty-six material suppliers, SPI, and ASTM for making false claims about nonflammable or self-extinguishing qualities of polyurethane and polystyrene foams. Harding of SPI and Gross of *Modern Plastics* cried foul, arguing there was no need for persecution when the industry was already responding to inadequacies in outdated ASTM tests and standards.[80] Within the next two years a study at the University of Utah's Flammability Research Center confirmed that laboratory rats who could survive exposure to concentrated wood smoke would succumb to fumes from burning polyurethane, and the National Academy of Sciences convened a committee headed by pioneer polymer chemist Herman F. Mark to investigate improvements to aviation synthetics.[81] But the damage was done, confirming a popular image of the modern airliner as a "potential gas chamber" and threatening an annual output of half a million tons. A comment from an insurance executive suggested that plastic's utopian slant had gotten it into trouble. "Hell," he declared, "the wood industry doesn't claim its products don't burn." But some plastics salesmen "seem to be trying to make you think their materials are almost inert." The revelation of hidden danger at the heart of a high-tech material dedicated to unearthly perfection proved especially damaging to its public image. "I expect a plastics witch hunt," declared an engineer.[82]

Before the industry had time to recover from that blow, it took another that threatened to be even more serious. While most people could expect to avoid being in a fire, no one could avoid daily exposure to a plastic as common as polyvinyl chloride. By the middle of 1973, with the industry reeling from accusations of fraud about flammability, it was common knowledge that the newly organized Occupational Safety and Health Administration (OSHA) was investigating reports of workers in vinyl processing plants suffering at an excessive rate from angiosarcoma, a rare liver cancer. Sixteen American vinyl workers had died of the disease since 1961. Another ten had died in Europe, where the correlation first came under investigation. Tests with laboratory animals confirmed early in 1974 that the carcinogen was vinyl chloride monomer, the gas from which polyvinyl chloride was polymerized. By all accounts, many chemical companies producing vinyl resin were lax about meeting even a minimal exposure standard of fifty parts of gas per million parts of air. It was not unusual for workers to descend unprotected into polymerization vats to scrub them out. Six thousand American workers were routinely exposed to vinyl chloride gas. The AFL-CIO's health director denounced vinyl producers for a "barbaric attitude that death and disease are part of the sacrifice that must be made for food, clothing and shelter." In October 1974, in the first major regulatory decision of its existence, OSHA mandated a new exposure standard of one part per million averaged over each eight-hour shift, with temporary exposure to five parts per million allowed for periods not to exceed fifteen minutes. Although resin suppliers, mostly the major chemical companies, claimed the new standards would destroy a $1.5 billion industry, they managed to comply. Editor Gross later privately accused the PVC industry of being "slovenly" in its processing methods and noted that the extra monomer saved by OSHA's standards more than paid for reengineering a stricter process.[83]

After OSHA's revelations, a fear of vinyl extended beyond the men and women engaged in making it—and with good reason. People living near processing plants worried about the cancer risk from uncontrolled discharges of gas over the years. Some experts speculated about exposure to vinyl monomer through the burning of discarded vinyl at garbage dumps or the baking of auto upholstery in the sun. The Food and Drug Administration became concerned about vinyl monomer migrating from food containers. In May 1973 the FDA had banned polyvinyl chloride for liquor bottles for that very reason, and some evidence suggested the monomer also migrated into vinegar, apple cider, vegetable oils, and mineral oils. In the summer of 1975, as the agency prepared a stringent "nondetectable" vinyl chloride limit of less than fifty parts per billion for any food or drug packaged in polyvinyl chloride, consumer advocate Ralph Nader's Health Research Group accused the FDA of dragging its feet and thereby promoting cancer. SPI's general counsel Heckman attacked this "cheap shot" because Nader could not have been unaware that the FDA was about to act. Gross feared, no doubt accurately, that Nader's announcement would provoke "fear, panic, and confusion in the public mind." But with even NASA preconditioning its high-tech synthetics in a vacuum to force "off-gassing" of chemicals that otherwise might pollute air breathed by astronauts, ordinary citizens could be forgiven a suspicion that litter was not plastic's most serious environmental impact. Fear of cancer, the era's most inexplicable medical killer, added an irrational element to a negative image transcending the old disdain for the cheap and shoddy. Just as plastic had stood in for radioactive fallout, it now became a highly visible stand-in for often nebulous products and by-products of the chemical industry—whose reputation was so low, partially owing to use of napalm and defoliants in Vietnam, that Du Pont in 1972 abandoned its "Better Things for Better Living" slogan.[84]

As if flammability and toxicity were not enough to engage the industry's defenders in 1973, they also confronted the Arab oil embargo and the resulting energy crisis. Initial worries revolved around supplies and prices. Some resin suppliers limited deliveries to long-standing customers. Others increased exports to Europe, where they could charge five times as much as domestic price controls allowed. "It has taken five years off my life," complained a resin buyer for the Mattel toy company. As the supply of petroleum feedstocks became more uncertain, alternate sources of raw materials became an issue. Kline, retired from the National Bureau of Standards but still technical editor of *Modern Plastics*, predicted (inaccurately, as it turned out) that coal would replace petroleum as the main source of plastic before the year 2000. In the meantime the industry counseled conservation of petroleum. As Gross put it, while thanking Congress for retaining a controversial 55-mph speed limit, "Every gallon saved is resin in the hopper."[85] Designers became adept at "resin-stretching" by molding products with thinner walls or by specifying new structural foams that "put air bubbles where material used to be." Such innovations saved oil twice—by using less resin and by decreasing weight-based transportation costs.[86] Only after confronting immediate problems of shortages did some people in the industry realize they were open to attack for wasting petroleum to make throwaway products. What started out as in-house economics threatened to become another nail in a plastic coffin.

A typical attack on the plastic industry for diverting scarce petroleum appeared in a *New York Times* op-ed piece in June 1979. Robert C. Lohnes described an "ocean" of oil "gulped by plants manufacturing millions of disposable plastic knives, forks, plates and scores of other products." The plastic industry sucked up "millions of barrels of oil" each month, much of it for "disposable products . . . littering our streets and parks." Maintaining that a return to paper and wood as primary materials would contribute to solving the energy crisis, Lohnes said he would "trade a plastic cup for a gallon of gasoline any day."[87] The argument was powerful to anyone already disposed to think of plastic as carcinogenic, toxic when burned, dangerous to air and water, an indestructible blight on the landscape, a shoddy substitute whose very mention evoked laughter. It all came together to form a powerful antiplastic mentality that had little impact on legislation or on actual consumption of plastic but that perpetuated a compelling negative image.

Gross later recalled the energy crisis as "a very happy issue for us" because its major antiplastic argument was "easily pricked."[88] Although the industry seemed to be wasting energy by using oil as the raw material for its products, it was actually saving energy. As a lighter material than metal or glass or even wood, plastic consumed less energy in processing, manufacture, and shipping. The amount of petroleum consumed both as material and as energy in the making and transporting of a plastic bottle, for example, was less than the petroleum consumed as energy in the making and transporting of a glass bottle. Defenders also liked to point out that in theory plastic could be recycled with little energy expenditure. When its recycle life was over, it could be burned as fuel in an energy-generating incinerator, thereby returning much of its petroleum content full circle to the energy-producing role from which it had been borrowed. A convincing argument for the energy-conserving aspect of synthetics stemmed from the auto industry's effort to use ever more plastic to make cars lighter and thus boost gas mileage. As a Texaco advertisement boasted, "We're putting more petroleum in the body of your car so you can put less in your tank."[89]

Support for the industry's defense came from a seemingly unlikely source, the ecologically minded Worldwatch Institute. In 1980 Christopher Flavin completed a fifty-five-page report, *The Future of Synthetic Materials*, advocating continued use of plastics because "many natural materials actually consume more oil and natural gas than synthetics do." In the United States, synthetics annually consumed 900 million barrels of oil, only about 3 percent of the total consumption of petroleum and natural gas. Two-thirds of that small amount went for feedstocks, one-third for energy used in processing. Making plastics was "one of the most valuable uses" of oil and natural gas—more beneficial than burning it up. Even so, Flavin came down hard on ecological deficiencies, especially toxic manufacturing wastes and the fact that a quarter of production almost instantly became garbage. He also observed that the postwar "synthetic materials revolution" was "intimately connected with the world-wide economic expansion" of the same period, both being "fueled by abundant and cheap supplies of oil and natural gas." Now that those supplies were curtailed, with every prospect (he wrongly predicted) of oil reaching a hundred dollars a barrel by 1990, Flavin believed that intelligent exploitation of synthetics offered a way of maintaining the material gains of that temporary expansion as the world adjusted to the reality of a steady-state or diminishing economy. "In an oil-short world," his report concluded, "synthetics are likely to be much more essential than they are today."[90]

When the *New York Times* introduced a summary of Worldwatch's favorable findings by observing that use of plastic was "often discouraged and even disparaged . . . as symbolic of a throwaway society," the tone clearly implied that such critics did not know what they were talking about.[91] By 1980 the plastic industry was close to full cultural rehabilitation, with details regarding solid waste and toxicity left to responsible technical experts. But in 1973, when the energy crisis first emerged, it confronted a state of siege. A beleaguered Gross complained that the industry was "facing more major crises than it has ever confronted before—all at once." Among them were "supplies, flammability, toxicity, environment, and OSHA," all demanding "to be dealt with almost simultaneously." Most committed to the fight to salvage plastic's reputation and to maintain a climate in which companies could profitably do business was SPI, which Gross characterized four years later as having transformed itself from a "loose grouping of personalities" into a "streamlined organization of professionals." Serving throughout the critical period of the 1970s as executive vice president of SPI, Harding focused the society's energies and twisted the arms of executives who often saw no reason to get involved.[92]

As the society geared up to fight New York City's proposed tax on plastic containers, it solicited contributions for a legal defense fund, but the effort proved difficult. By then SPI had already committed $87,000 to a study of the impact of incinerating plastic conducted by two engineering professors at New York University.[93] Harding reported in April 1971 that most "packaging members" had contributed to the Environment Program but complained about a difficulty convincing other companies of their "substantial stake in solid waste matters."[94] Only two months later Harding was organizing a meeting of representatives from twelve major corporations, mostly material suppliers like Du Pont, Mobil, and Union Carbide, to begin raising a million dollars beyond SPI dues for a voluntary Plastics Waste Management Fund.[95] After several name changes reflecting widening membership and purpose, the group emerged as the SPI Public Affairs Council (PAC). During a presentation in December, Harding described PAC's mission "to fight off restrictive legislation everywhere, to correct mis-statements, and most of all to get across a positive point of view about plastics." With only $300,000 pledged by fifty companies—less than 5 percent of SPI's corporate membership—and legal expenses of $360,000 incurred in New York, Harding insisted that everyone join the struggle against an amorphous foe threatening them all.[96] A note of despair characterized the effort. At a PAC meeting people talked in "plain English" about "mobilizing the industry" against a monolithic "attack on a 'plastic' society." The keynote address featured the president of the U.S. Brewers Association, who ominously described himself as a man from "an industry that was once legislated out of business." Don't think, he told his audience, that "it can't happen here."[97]

SPI's financial report for the fiscal year ending May 31, 1972, revealed Harding had wrangled pledges worth $942,000 and had collected on more than half of them. But as expenses multiplied and the industry's environmental entanglement became more complicated, he realized fund raising would remain a constant worry. In the spring of 1973, as new crises emerged on a daily basis, SPI's executive committee removed PAC from voluntary funding and instituted a mandatory dues surcharge. Minimum dues went from $250 to $350 (for a company with annual sales of $875,000 or less), and maximum dues went from $16,500 to $50,000 (for a company with sales of $125 million or more). Members received a number of new services for their increased contributions. The staff of PAC monitored the legislative process in state governments, in Washington, and in major cities to keep abreast of proposals with implications for the plastic industry. When necessary, PAC helped organize local lobbying efforts. The council also relied on Jerome Heckman's legal firm, Keller and Heckman, which was formed in 1962 partly in response to the latter's expanded business with SPI. Heckman's firm not only responded to legal challenges and coordinated local attorneys but also provided expert testimony by an in-house team of professional chemists. The council engaged in a low-profile campaign to reply to derogatory press notices, plant feature stories in local newspapers, and meet with local journalists, broadcasters, and environmentalists. Occasionally PAC funded outside researchers to prepare ostensibly neutral reports on such subjects as migration of vinyl chloride monomer and the relative environmental impact of various materials. Finally, the council invested in public opinion polls to determine who were the industry's "enemies" and "where they get their information." An expressed desire to "pinpoint problem areas, problem people and problem press" and to defuse them before they had a chance to "mold majority opinion" indicated the degree of paranoia afflicting the plastic industry—whether reasonably or not—amid the polarized society of the early 1970s.[98]

An address Harding delivered to a meeting of plastics distributors in August 1972

273

revealed a sense of isolation from much of society. He expressed genuine concern for environmental issues, citing the Club of Rome's study of overpopulation, *Limits to Growth*, and hinting that solid waste would remain a problem long after air and water pollution were resolved. But he focused primarily on the industry's image. As if speaking to people who might not believe him, Harding insisted that "there really are a lot of people in this country who think plastics are bad." According to PAC's opinion surveys, the general public was "not very excited about ecological matters." But among "a large number of the educated, the active, the community leaders," there was much "opposition to plastics." Many considered plastic "somehow bad—ersatz, phony, substitute." Harding took pains pointing out to possibly skeptical distributors that SPI had trouble "mobilizing" the industry precisely because "our people, particularly those with technical training, would look at some of the statements that people were making about plastics and they couldn't take them seriously." Trying to put the best light on a bad situation, he joked that the phrase "plastic society" differed little from earlier phrases like "glass jaw, paper tiger, wooden nickel, rubber check, tin ear"—indicating that "people traditionally pick on a material in this way." Sadly there was still "massive misunderstanding or lack of understanding about plastics." The public had to be educated, informed of plastic's contributions to "better living."[99] Harding's description of public ignorance differed little from complaints voiced back in the late 1940s, when teaching housewives to distinguish polystyrene from polymethyl methacrylate alone seemed to matter. But the synthetic landscape had changed. It had lost its naive brashness, its smooth promise of damp-cloth convenience, its veneer of postwar optimism. Harding's choice of phrase suggested nostalgia for an era of simpler misunderstandings. It also suggested that for all his awareness of current problems, it was as hard for him as for the rank and file to understand what had happened to their industry.

An exasperated sense of betrayal also marked Gross's editorials in *Modern Plastics* during the late 1960s and early 1970s. Writing soon after *The Graduate*'s release, he complained that plastic's reputation "remained about as low as it can get" despite a phenomenal growth rate. "Somebody must like the stuff" was his grumpy conclusion. "If they hate plastics so much," the magazine groused a year later in 1970, "how come they're buying more and more of it?"[100] The reasons for public disaffection—and thus for the betrayal felt by insiders—were complex. At the most general level the plastic industry suffered from animosity directed at business as part of "the establishment." More specifically it suffered from a "chemophobia" generated by environmental concerns and by involvement in the Vietnam War.[101] Finally, the old ambivalence about plastic remained strong. Even sympathizers occasionally expressed plastic doubts. During the critical year of 1973, for example, SPI's public relations firm arranged an "off-the-record . . . informal brain-storming session" with four academic historians attending the annual meeting of the Society for the History of Technology. Topics for discussion included emergence of new technologies in the past, trade-offs of earlier technologies, the energy issue, and the environment. One of the consultants was John J. Beer, author of a history of the German synthetic dye industry. In his notes for the meeting Beer outlined many familiar advantages of plastic—for example, that it "helped us transcend" natural materials. He also observed disadvantages "in the area of safety and waste" and noted that "synthetic dyes" had suffered from a "bad name" for "fully two generations." For the most part Beer's comments revealed a balanced neutrality. But then he shifted into personal experience. His language became more emotive as he described his irritation "whenever I must pick up a plastics 'McDonald' straw or cup cover from my lawn," or "my hostess tells me to throw away the plastic cocktail cup and tableware." He complained about broken control knobs in his car and about a "closetful of milk and juice containers" with leaky lids. He revealed, in other words, that his sense of dissatisfaction with plastic was no different from anyone else's—it was personal and there was not much to be done about it.[102]

If a sympathetic historian working in the field of organic chemistry had problems with plastic as it impinged on his daily life, then it is not surprising that much of the population had little good to say about it, and that the word had become a synonym for fake or superficial. By 1990, on the other hand, it seemed possible that a younger generation that had grown up with plastic of quality might not think so poorly of it. Even into the late 1980s, however, a new editor of *Modern Plastics* emphasized the image problem. In the "post-Bhopal era," Bob Martino wrote in 1985, "coping with the public's perception of us requires strategy, not catharsis." Instead of "contemptuously dismissing" the critics, he maintained in language reminiscent of Roland Barthes, "we must attempt to demystify plastics for them." Two years later, on the occasion of SPI's fiftieth anniversary, Martino observed how odd it was that people still used "plastic" as a singular term when the industry had expanded into so many different polymers and applications that it encompassed all of manufacturing. "People may say they're against trash bags or foam cups or plastic appliance housings or polyalloy auto fenders," he wrote, "but what they really oppose is the idea of plastics." Polymer technology—and the word plastic itself—offered "a powerful unifying factor," an "essential oneness both as a fact of chemistry and as a perception by the rest of the world—lawmakers, environmentalists, journalists, and citizens." The challenge to the industry was "to reconcile the astounding diversity of plastics with this oneness"—and in doing so, once again to try to mold a positive identity and present it to the general public.[103]

Martino's predecessor Gross would not have wasted his time with another foray into familiar but unproductive territory. Gross had decided that a poor image had little practical effect because consumers could choose only from among goods presented in the marketplace. If manufacturers used plastic—whether for versatility of design, durability, lower cost, greater profit, or whatever reason—then consumers had no choice but to go along. Even those who thought they despised plastic would buy it and use it, often without even recognizing it. "If we were as evil as our adversaries claim," Gross once wrote, then "we wouldn't be where we are today."[104] He

would have appreciated a science fiction tale, *Mutant 59: The Plastic Eaters*, published in 1972 at the height of plastic's period of ill repute. A clever disaster novel,
it described a strain of bacteria mutating after exposure to a new biodegradable
plastic and thereby gaining an ability to feed on *any* plastic. As solid plastic turned
to slime everywhere, toys ran amok, clothes melted, heart valves malfunctioned,
subway trains crashed, and an airliner dissolved in midair. "Good God, just think of
it," someone declared. "Take out plastic from a modern city and what do you get—
complete breakdown." Like it or not, as he put it, "we're totally dependent on it."[105]
Eventually the technical experts devised methods for isolating and neutralizing the
mutant bacteria, for reasserting the control that plastic had always promised. But
with the melting into slime of most plastic objects, a kind of ultimate reputation for
shoddiness was established. And with an irony Mailer would have appreciated, the
novel ended with a contaminated space probe landing on Mars, ready to dissolve
any future expansion of plastic beyond the bounds of spaceship earth. In the real
world, however, plastic's expansion continued over the next twenty years, as did the
inflationary culture of which it was substance and image. The throwaway society
kept on expanding as Americans learned to live with more and more of less and less.
They glimpsed the outlines of a new relationship to things, or a more tenuous conception of things, as the physical yielded to the digital, the material to the immaterial, the plastic presence to the process of plasticity. As that transformation began,
or as intellectuals posited such a transformation, plastic's meaning began to shift
almost beyond any correlation with material things. Once again, after several decades, plastic expressed a sense of limitless shape-shifting.

BEYOND PLASTIC: THE CULTURE

OF SYNTHESIS

*E*uropean intellectuals of the 1970s and 1980s tripped over each other to represent Americans as obsessed with plastic artificiality. Visits to several garish California wax museums convinced Italian semiotician Umberto Eco that Americans actively sought out experiences of "hyperreality"—defined as imitation or replication of things assumed to have existed in other times and places. Not to be outdone, French poststructuralist Jean Baudrillard praised Disneyland for cleverly obscuring the fact that all America consisted of "simulacra"—a procession of ungrounded self-referential simulations. Both Eco and Baudrillard implicated synthetic materials in this desire to substitute complex fabrications for ordinary reality. For Eco, America was a place "where Good, Art, Fairytale, and History, unable to become flesh, must at least become Plastic." For Baudrillard (less articulate or a victim of translation), "plastic" was a miraculous "simulacrum where you can see in a condensed form the ambition of a universal semiotic." Eventually he upped the ante. Celebrating America on the edge of the third millennium, he envisioned the new world as an astral plane of hyperspace, a vast shimmering hologram. American readers who experienced holograms mostly as tiny illusions of depth on plastic credit cards could not evaluate this assertion, but it dovetailed nicely with a "holographic paradigm" of the universe. As an enthusiast explained in 1987, the idea of a holographic universe suggested an immaterial projection supporting unlimited freedom to "alter the fabric of reality." Such wishful thinking recalled plastic utopianism and echoed the theme of the Horizons Pavilion at Disney World's Epcot Center: "If we can dream it, we can do it."[1]

European commentators were only repackaging what Americans had already dis-

covered for themselves. Back in 1962 historian Daniel J. Boorstin clearly foreshadowed simulacrum and hyperreality in *The Image: A Guide to Pseudo-Events in
America*, with its angry thesis "that what dominates American experience today is
not reality." Boorstin believed many people lived in a cocoon of fabricated images
sheltering them from real life with its real risks and real satisfactions wrested from a
difficult environment. He understood technology, according to a title-page epi-
graph, as "the knack of so arranging the world that we don't have to experience it."
Most people enjoyed the comfort of "pseudo-events" organized and reported by
television, movies, advertising, and slick magazines. In this gutless world the "celebrity" replaced the "hero" and was "known for his well-knownness" rather than for
any inspiring accomplishment. Surface calm remained unbroken by the shock of the
unexpected. Even travelers, a breed formerly known for enduring discomfort to
confront the unknown, had become tourists on package tours, people for whom
"the Grand Canyon itself became a disappointing reproduction of the Kodachrome
original."

Despite such hyperbole, Boorstin convincingly maintained that "the making of
the illusions which flood our experience has become the business of America." The
nation had entered an "age of contrivance" in which, according to Boorstin's rhetoric
of reversal, "the artificial has become so commonplace that the natural begins to
seem contrived." Although he described the images dominating American culture as
"planned, contrived, or distorted," he confessed that they seemed "more vivid, more
attractive, more impressive, and more persuasive than reality itself." In a revealing
phrase he explained that "more and more of our experience . . . becomes invention
rather than discovery." That formulation seemed to cast image making and pseudo-
events in a more positive light as part of the American tradition of building a civilization out of the boundless resources of an empty continent. But Boorstin criticized
his generation for forgetting that the environment had also shaped its settlers and
limited their accomplishments. The mid-century faith in an ability to fabricate experience marked an apostasy from the beliefs of the nation's founders. Acting as a
historian, Boorstin reminded his readers that "the Laws of Nature and of Nature's
God" had "governed an orderly universe" to which humanity had submitted. In
former times, at least, "for neither God nor man was the world wholly plastic."[2]

At the end of the twentieth century, however, American experience became ever
more artificial, ever more a product of synthesis—whether chemical or digital or
the hybrid variety of such plastic media as film, tape, and disc. The implications of
synthesis had not changed since the early 1900s. The basic issue remained the ease
of manipulating the stuff of experience. The old disdain for imitative plastics remained strong. But so did the urge to celebrate plasticity. The history of plastic from
billiard balls to inflatable domes revealed a widening desire to transcend nature by
rendering the resistant malleable. Partially as a result of their experiences with plastic, people assumed an increasing ability to mold experience, whether externally by
fabricating environments or internally by restructuring personalities. With "disney-

fied" landscapes spreading from theme parks to shopping malls and historical preservation zones across the continent, and ordinary people as willing as celebrities to
embrace fantasy role-playing fostered by a therapeutic culture of multiple lifestyle
options, it was not surprising that European observers found it all a bit unreal. As
middle-class consumers molded the environments around them, and reshaped their
own personalities, they assumed plasticity as an American birthright or an element
of nature. As they transformed their world and themselves in myriad ways, they
remained indebted to past experiences with plastic and to its continuing presence in
their lives—as material, as prototype, as metaphor. While champions of modernism
had struggled heroically to construct utopias of rational order, inheritors of post-
modernity found the world so responsive to incompatible desires that it fragmented
into individual facets generating meaning only through juxtaposition. Ambivalent
hopes and fears that had defined the history of plastic spread outward to engage a
larger culture of postmodernity and in doing so to clarify the cultural meaning of
plastic.

SYNTHETIC PLACES

According to legend Walt Disney conceived Disneyland as an antidote to the seedi-
ness of the carnivals he visited with his daughters. He sought total control over his
parks. When expanding into Florida with Disney World, he bought enough land to
exclude the cheap motels, wax museums, and sleazy attractions that had sprung up
around the California park. He employed unobtrusive groundskeepers to sweep up
litter almost before it touched the ground. At night they yielded to teams of painters
who endlessly repainted both parks to ensure they remained as timelessly perfect as
the day they opened. Rumors hinted that Disney's mania for control extended beyond the grave, that he had ordered his body cryogenically maintained for a secular
resurrection. At Disneyland, someone once noted, "no raw edges spoil the picture."
It appeared perfect, seamless, and whole like some complex one-piece injection
molding.[3]

Many commentators on Disneyland have noted its intimate relationship to the
movie industry in which Disney's animation first gained him fame. Architect Charles
Moore praised the park's lively pedestrian environments in 1965 and observed that
"everything is as immaculate as in the musical comedy villages [of] Hollywood." Six
years later alternative historian William Irwin Thompson suggested "it was Walt
Disney's insight that one could make a fortune, not by making movies with a studio
lot, but by charging admission." But Thompson was intrigued by more than the
opportunity at Disneyland to inhabit landscapes previously seen only at the movies.
He recalled a description of a Hollywood studio lot in Nathanael West's novel *The
Day of the Locust* where dozens of outdoor sets represented a disorienting array of
times and places, culminating with a "Sargasso of the imagination" into which workers dumped fragments of discarded sets by the truckload. For Thompson, Disney-

9–1. Piazza d'Italia, New Orleans, designed by Charles Moore (with Allen Eskew and Malcolm Heard, Jr., of Perez & Associates and Ron Filson), completed in 1979, as it appeared in 1984

land offered a similar "shattered landscape in which the individual moves through a world of discontinuities: Mississippi riverboats, medieval castles, and rocket ships equally fill the reality of a single moment."[4] Disney had intended a set designer's "purified simulation" of such scenes as Main Street and World's Fair, to name two objects of his nostalgic perfectionism. And critics described Disneyland as a "programmed paradise planned by people who improve upon history or nature whenever the need arises."[5] But Thompson, by looking beyond the frames of individual "sets," as early as 1971 experienced something not intended by the control-minded Disney: an essentially postmodern condition of fragmentation and incongruous juxtaposition.

That experience became common in the 1980s, as in Moore's Piazza d'Italia in New Orleans (Figure 9–1) with its flat, colorful, stylized pastiche of Italian architectural forms inserted like literal stage flats into the modernist wasteland of an unfinished urban renewal project; or in Terry Gilliam's film *Time Bandits*, with its wrenching shifts from one historical pastiche to the next. At a more popular level Tom Shales of the *Washington Post* in 1986 described people of "the ReDecade" as "sail[ing] through time zones . . . among not just the looks, the fashions, the fads, the personalities and events of other times, but also the sensibilities." From dozens of cable channels one could sample syndicated reruns of programs from television's entire history—and the entire range of history *represented* by television and film. For Shales the next step was to apply the principle of the music synthesizer to "make new stuff from fragments of old stuff"—to do with immaterial images what Disney had done with material fabrications at Disneyland. By 1992, twenty years after Thompson's insight into Disneyland's "shattered landscape," new ways of experiencing television made it possible for architecture critic Michael Sorkin to conceptualize Disneyland as a "channel-turning mingle of history and fantasy, reality and simulation." But defining Disneyland in terms of immaterial electronic media did not entail its material obsolescence. For Sorkin, Disneyland was "the place where the ephemeral reality of the cinema is concretized into the stuff of the city."[6]

By the 1980s it was a cliché to refer to the American landscape as ever more disneyfied, as in the practice of renovating derelict commercial or warehouse districts rather than bulldozing and rebuilding in the modernist manner. From early projects like St. Louis's Gaslight Square to later developments like New York's South Street Seaport, these renovations aroused nostalgia for the imagined past of Disney's Main Street U.S.A. Some critics objected to an ahistorical aesthetic of exposed brick and brightly painted trim; others to the use of renovated districts for middle-class entertainment and consumption—also the primary activities of Disneyland. But disneyfication soon spread beyond renovating the old to fabricating wholly artificial environments. Franchises like Spaghetti Warehouse enthralled diners with streetcars, English telephone boxes, and tin advertising signs arranged to the point of sensory overload in a form of historical pastiche praised by a trade journal for maintaining authenticity in the face of a spreading "plastic culture." As franchising engulfed the restaurant industry, disneyfication affected such chains as Long John Silver's with its mock wharf, and Wendy's, where a critic observed it had "degenerated into plastic pseudo-Victorian beads, horse-and-buggy wall murals, and tabletop reproductions of late-nineteenth-century newspaper advertisements" (in custom-designed melamine laminate).[7]

These developments witnessed a desire to inhabit a reassuring past, to celebrate a history so "expurgated and sanitized," according to cultural geographer David Lowenthal, that it remained mere entertainment, "makeshift, flimsy, and transient, obsolete from the start," and not very satisfying.[8] But synthesis of environments expanded beyond those imitating or extrapolating from the past. Making commodities out of services like medicine and banking required a new architecture for which the only prototype was the fast-food outlet, the building as package with bright, smooth, literally plastic surfaces, so perfect in appearance that the slightest wear would make it obsolete and so inexpensive it could be torn down and replaced when the aura of newness wore off. As the "have-a-nice-day" syndrome spread from franchise restaurants to other areas of life, people were invited to mistake surface for substance. Even the cynical began to accept at face value the universal cues of late-twentieth-century synthetic environments programmed to inspire confidence that even risky undertakings had become risk-free. One observer noted the public's complicity in the matter, its avid acceptance of "the routine substitution of fantasy for reality in the everyday environment, and the willing confusion of which is which."[9]

Nowhere outside of Disneyland did people experience this confusion more fully than at their local shopping mall. Constructed during the 1960s and 1970s, these private developments offered suburbanites a deceptive semblance of public space. Oddly self-contained during an expansive era, the mall promised finite limits within its enclosing walls. As a microcosm it offered some hope of comprehending a complex, messy, disorienting society. To derive its form, one took a highway with its scattered businesses, curved it into a circle, and inverted it to place the shops at the center, thereby relegating the automobile, the icon of American mobility, to a peripheral or invisible position outside. Like the enclosed environments of the New York World's Fair of 1939, the mall excluded awareness of the outside world. Its synthetic surfaces enclosed consumers within an artificial cocoon boasting perfect climate control, closed-circuit surveillance, the attractions of a (relatively sedated) crowd, and the distractions of an array of commercial facades whose bright thermoformed elements echoed Disney's pastiches of past and future.

For William Kowinski, who published *The Malling of America* in 1985, the mall was "the culmination of all the American dreams, both decent and demented." His book, a perceptive mix of travelogue, pop sociology, merchandising history, and new journalism, frequently implicated plastic both as material and metaphor. According to Kowinski, cheap petroleum had "fueled the new American way of life" by making everyone dependent on the automobile and by offering up "petroleum-based synthetics and plastics" as the basis for a "new mass-production consumer economy." Without plastic goods to fill plastic display fixtures, there would have been no malls. At one point, assuming the perspective of a cosmopolitan critic of mall culture, Kowinski described the shopping mall as a perfect target for anyone intending to savage suburban life. The malls springing up everywhere were "ugly, vulgar, pallid, and pretentious; as uniform as the subdivisions that supplied them with customers"; they were "founded on blithe tastelessness and full of plastics." In his travels Kowinski encountered many people who expressed similar feelings. One person described a nondenominational mall chapel as "totally inoffensive to any religion"—"it was plastic, indescribable." Another recalled a vertigo that Kowinski dubbed "*Plastiphobia*, or the fear of being enclosed in a cocoon of blandness." This informant, hanging out at a mall, suddenly realized he "was in a plastic place with plastic people buying plastic products with plastic charge cards," and he "had to escape." On the other hand plasticity had its attractions. The mall owed its success to a characteristic it shared with the plastic that formed its commercial facades. As Kowinski phrased it, "the mall is, in a word, malleable." It was a "Never-Never-Land" that could easily become "virtually anything" its consumers desired.[10]

As the reference to Never-Never-Land suggested, discussions of synthetic environments were bound to lead back to Disney. "It's beautiful," said a shopper about a new mall outside Hadley, Massachusetts; "it's just like Disney World." And from there the connection looped back to the issue of the real and the unreal, the natural and the artificial. Disney World was indeed a wonderful place, admitted a woman who had wintered in Orlando since before the Magic Kingdom emerged from the swamp. "But of course," she added with some distaste, "it's all man-made." As the epithet of "plastic" followed Disney through the years, a supreme irony emerged. Whatever the larger significance of Disney World's recombinations of pseudo-historical images, the place actually relied as much on chemical as on cultural synthesis for its success. Much of it was literally made of plastic, with fiberglas-reinforced polyester facing the exteriors and interiors of most structures. The apparently solid stone of Cinderella's eighteen-story castle was composed of panels of reinforced polyester, less than a quarter of an inch thick, fastened to structural steel. Soaring blue spires that vaguely suggested slate were fabricated from sheets of acrylic and vinyl designed to withstand 120-mile-an-hour winds. A year after opening, with most attractions completed, Disney World still absorbed two and a half tons of polyester resin every week. As *Modern Plastics* proudly observed, plastic was "ubiquitous" in the Magic Kingdom.[11]

Reinforced polyester marked an improvement over the original Disneyland's traditional building materials. Using plastic as a facing material nullified Florida's destructive climate. Maintaining immaculate perfection became easier for Disney's gangs of painters. More important, plastic enabled imagineers at Disney's Burbank headquarters to retain greater control over special effects. They could specify the precise shapes of the molds from which fiberglas facade panels were formed. A single panel could incorporate simulations of several materials, wood and stucco for example, that otherwise would have been assembled on site using expensive traditional construction techniques. A modular system of panels fabricated from a homogeneous material yielded simulacra—to use Baudrillard's term for copies with no originals—of unique traditional structures whose apparently natural imperfections and irregularities constituted much of their charm for visitors. The switch from wood to fiberglas enabled "the 'cloning' of a streetscape at its most extreme—and most obvious," as an architect observed. The material's plasticity yielded "infinite facility in creating the most complex of designs." Pseudo-Victorian gingerbread in reinforced polyester exceeded the talents of the most skilled jigsaw operator. Simulation at Disney World called attention to itself to evoke amazement just as the earliest celluloid imitations of ivory or tortoiseshell had evoked pleasure at the cleverness of the effect. The fundamental message of Disney World—as of plastic and the synthetic process in general—was the conviction noted by Eco "that technology can give us more reality than nature can" (Figure 9–2).[12]

9–2. THE ULTIMATE SYNTHETIC PLACE, ENVISIONED BY CARTOONIST DON WRIGHT IN 1989

9–3. ITALIAN COUNTRY HOUSE WITH STAINED, CRUMBLING STUCCO IN FIBERGLAS-POLYESTER, EPCOT CENTER, WALT DISNEY WORLD, FLORIDA, AS IT APPEARED IN 1987

Simulation entered a new dimension at Epcot Center, which opened as part of Disney World in 1982. Fanning out from a geodesic dome whose surrounding streamlined structures bore the heavy, burnished look of a George Lucas retrovision of the 1939 World's Fair, Epcot Center offered the attractions of a corporate-driven, high-tech utopian future and the pleasures of a risk-free photogenic visit to romantic countries around the globe. One small corner of Italy's section, facing the geodesic dome of Spaceship Earth across the World Showcase Lagoon, revealed the degree of simulation possible with reinforced polyester. Across a tiny Piazza San Marco from a Doge's palace in faux marble and a miniature campanile stood a two-story stucco country house with wrought-iron balconies, as much out of place in the marshes of the Adriatic as in those of central Florida (Figure 9–3). Its water-stained walls bore evidence of an indeterminate orange-brown wash. Trowel marks textured its stucco, as did several substantial cracks and scattered pockmarks. At one corner the stucco had fallen away, revealing rough handmade bricks and crumbling mortar beneath.

Not as dramatic as an audioanimatronic Lincoln, this precisely designed effect in reinforced polyester stood for thousands of similar Epcot Center details. It revealed an urge to fabricate a synthetic past so perfect that an observer could imagine spring rains carrying rust stains down the face of the stucco from the iron railing above. In such details Disney imagineers exhibited extraordinary confidence. They were so certain of control over past, present, and future, so convinced of reality's plasticity, that they dared simulate the very process of decay Walt Disney had banished from his Magic Kingdom. To simulate decay in a state of arrested motion or timeless perfection, as if preserved in Baekeland's synthetic amber, suggested humanity had

finally perfected its control over time and space, past and future, history and nature—though a doubter might be forgiven for imagining the horror of a fiberglas-
polyester simulation of crumbling stucco and mortar itself cracking, crazing, or dis-
coloring in a way unique to plastic.

Disney's enthusiasm for simulation revealed genuine nostalgia for a turn-of-the-
century boyhood and a need to reconstruct it outside the vicissitudes of time. But
most Disney visitors came of age surrounded by the inflationary surrealism of postwar consumer culture—with desires stimulated by television's flickering images
barely outpacing their material embodiment in plastic's proliferating forms. As the
artificial eclipsed the natural, synthetic environments became the norm everywhere—in malls, in stores, in franchise restaurants, in such hotels as Morris Lapi-
dus's Miami Beach extravaganzas or John Portman's for the Hyatt chain, even in the
home. Cultural geographer Edward W. Soja suggested that a "second wave" had
"carried hyperreality out of the localized enclosures and tightly bounded rationality
of the old theme parks and into the geographies and biographies of everyday life,
into the very fabric and fabrication of exopolis." As the trend progressed, Disneyland
and Disney World seemed outdated in their mania for precise imitation. During the
1980s a postmodern preference for frankly synthetic experiences replaced an older
progressive or modernist emphasis on nature, history, and utopia. Although often
assembled from fragments of those earlier visions, as in MTV rock videos or Terry
Gilliam films, the new synthetic mode of orientation inhabited a foregrounded present with no connection other than style to past and future, no sense of direction or
purpose, no sense of movement in a solid-state entropic situation, not even a modernist avant-garde ability to shock.[13]

The synthetic culture of the 1980s and 1990s relied on plastic both as provocation
to and as medium of an overstimulated but enervated sensibility. Ability to recom-
bine and broadcast a vast array of images, both visual and aural, depended on plastic
for photographic film, audio- and videotape, laser discs, computer discs, and other
reproductive media. Beyond that, plastic's proliferation in everyday life reinforced a
conviction that not much was really changing. The continually repeated act of
throwing away disposable plastic pens, razors, bottles, tubes, and other packaging
seemed to suggest a continuous movement from the present into the future. But each
of these objects, even something as mundane as a polyethylene shampoo bottle, was
miraculously renewed each time an identical replacement appeared. The repetitive
act of throwing away lost its meaning in the face of a more insistent stability through
instantaneous replacement. A forever vanishing world of objects was forever renewed with identical plastic clones—occasionally mutating in style enough to suggest novelty but not enough to produce an expectation of substantive change. Much
the same was true of the multiplying images purveyed by plastic media. With each
image instantly replaced by another, similar enough to avoid instability but different
enough to maintain interest, a superficially thermoplastic culture avoided the boredom of a thermoset utopia.

The desire for a shifting array of slightly varied experiences achieved popular

satisfaction in channel surfing but found more permanent expression in the archi-
tecture and design of the 1980s. The signature structures of the postmodern movement represented as great a range of cultural moments as did the debris in West's
"Sargasso of the imagination." Only a digitally cool eclecticism united such structures as Philip Johnson's AT&T building with its transposed Chippendale silhouette,
Helmut Jahn's unbuilt Southwest Tower with its nod to Buck Rogers futurism, and
Robert Stern's sybaritic Long Island poolhouse with its echoes of Cecil B. DeMille.
That their historical references derived from popular culture indicated the preemi-
nence of reproduction. High-style furniture design revealed a similar obsession with
retro modes, as in the case of the Italian firm Memphis, headed by Ettore Sottsass,
Jr. Inspired in part by the jet-age patterns and bright colors of plastic laminate, its
decorative pieces celebrated the wacky exuberance of American plastic design of the
1950s rather than the heavy corporate utopianism of the forgotten Monsanto House.
Soon after the first showing of Memphis furniture at Milan in 1981, its stridently
artificial colors and ornamental motifs began showing up in graphics, fabrics, and
knock-off domestic accessories. By 1984 the new wave in design had captured
middle-class markets with the Swatch watch, the ultimate consumer product of the
synthetic age. Manufactured on a robotic assembly line with fewer parts than most
watches (and those few mostly injection-molded), the Swatch was a true disposable.
Promoted as "the watch to wear when you're wearing more than one," the plastic
Swatch remained unchanging in a functional sense but appeared in a constantly
expanding array of restyled faces and bands. Most varieties were available only a few
months. The Swatch was an accurate timepiece, solid, waterproof, as dependable as
its battery. Although it offered such traditional plastic benefits as durability and low
assembly costs, it satisfied consumers because its shifting surfaces encouraged
people to change it, again and again, like changing channels.[14]

The ideal postmodern environment had to be as malleable as the Swatch watch
in the face of subtle shifts in outside pressure or personal preference. It also had to
indicate its lack of traditional cultural grounding. The imitative fakery that formerly
sought to escape unnoticed (as in vinyl siding), or to inspire admiration for its immaculate perfection (as at Disney World), now instead called attention to itself as
superficial affectation. Several months before Memphis's initial showing in 1981, the
style section of the New York Times ran an article about use of imitative materials by
interior designers. Now that postmodernism had "boosted the popularity of the
past," reported Suzanne Slesin, even former "minimalist" designers were beginning
to "look differently at ornamentation." She expressed amazement that the "fake
stone patterns once popular only for bathroom vanities are now favored for their
undisguised fake look." To document this trend Slesin cited a prominent art critic,
Robert Rosenblum, whose redecorated Manhattan loft was equipped with a built-in
cabinet surfaced with plastic laminate in a marble pattern. "This is the age of reproduction," Rosenblum noted, "and it's vulgar and witless to show real materials." In
his opinion plastic laminate was no longer "chintzy" but had become "classy." More

to the point, his wife declared, "I like anything that's fake."[15] An environment that indicated its own artificiality thereby admitted that it could be easily replaced. It was arbitrary, no better or worse than anything else. To choose one from an infinity of choices indicated personal whim.

Someone once remarked that the demise of the modern movement in design "coincided roughly with the death of all who can remember a life without machines."[16] One might also argue that the filtering in of postmodernism during the 1970s and 1980s coincided roughly with the coming to maturity of people who had grown up with plastic during the age of synthesis. Inhabitants of a fragmented late twentieth century possessed neither the coherence of progress toward utopia nor the avant-garde urge to transcend the present. Satiety seemed at hand. Proliferation of plastic environments, whether material or electronic, made it difficult to discriminate among them, to distinguish one from another. Even Edward Relph, a geographer who criticized "disneyfication" of contemporary "plastic" landscapes, sometimes had trouble clarifying his reactions. In 1976 Relph offered two fascinating but conflicting analogies connecting these environmental phenomena to the personalities who created them and inhabited them. He described such landscapes as "other-directed" places, a phrase borrowed from David Riesman's analysis of the postwar American personality, implying that plastic places too easily conformed to hollow desires of plastic people. But Relph also borrowed from Robert Jay Lifton's discussion of a creatively adaptive "protean man" by referring to landscapes as equally protean.[17] As people consciously took control of an endless process of remolding their own psychologies, it became clear that shifting environments embodied dreams and desires projected by equally shifting personalities. The concept of plasticity could define inner as well as outer experience.

PLASTIC PEOPLE

Middle-class Americans growing up during the 1950s and 1960s received more attention than any other generation in the nation's history. After suffering depression and war, their parents wanted to protect them from life's uncertainties and provide them with material advantages. When the postwar generation came of age, many had lost their innocence to assassins' bullets, the flames of Watts, and the slaughter of Vietnam. Youthful idealisms of psychedelic drugs and political revolution yielded to pessimistic assessment of limits of social change. Approaching middle age, some baby boomers tried to retain something they could reform by focusing on narrowing parameters of the self. They sought to achieve personal growth or realize human potential or liberate themselves from co-dependency. Hostile social critics described them as self-centered and narcissistic, or during the Reagan years as selfish and greedy. A clear-eyed sympathizer might have concluded they were trying to prove themselves worthy of all the attention. An early expression of this generational tension came in *The Graduate*, whose passive protagonist was showered with gifts merely because he completed a rite of passage experienced by thousands of other people.

This portrayal of the postwar generation is too harsh but suggests why young moviegoers laughed uneasily at *The Graduate*'s "plastics" scene in 1968. There was no way to deflect the epithet from Dustin Hoffman's character Ben and his parents onto their domestic environment. Nothing in the scene was made of plastic. Furniture, lamps, draperies—all indicated a high standard of upper-middle-class taste and nothing revealed origins in a chemical refinery. The epithet "plastics" could only refer to Ben's parents, their friends, and the pseudo-reality of suburban life. Despite vague attempts at rebellion, Ben was doomed to the same patterns. It was one thing to refer to Disneyland or the suburban landscape as plastic. It was quite another to hint, as Buck Henry's script did, at the existence of plastic people with unreal personalities. Because the famous scene contained no obvious plastics—no polyethylene pineapple pitchers or polystyrene wall clocks, no Tupperware or Eames chairs—it invited viewers to perceive their own parents, teachers, friends, even their very own selves, as somehow plastic.

The term "plastic person" became so common that it obviously satisfied a cultural need. By 1984 to refer to someone as plastic was such a cliché that only baroque elaborations avoiding the word itself remained convincing. In that year a review of Frederick Barthelme's *Second Marriage* invoked various plastics and their qualities—without ever mentioning the word—to generalize about the novel's representation of life in sunbelt suburbia. The reviewer, novelist Ron Loewinsohn, described *Second Marriage* as possessing the "arresting, slightly over-detailed superrealism of a group of Duane Hanson sculptures." The novel's characters were "thoroughly ordinary" people captured "in the midst of actions that are quintessentially unimportant." They inhabited "a thoroughly recognizable contemporary urban America, with its subdivisions, shopping malls, apartment complexes and drive-ins, a world of surfaces, all of which are either disposable or easy-to-clean—nothing will take or leave a lasting impression." As Loewinsohn presented Barthelme's America, not only the contemporary landscape was plastic. Frequent divorce and remarriage yielded unstable "ersatz" families whose people, Barthelme's characters, were "no thicker than Formica" though they "hunger[ed] obscurely for some continuity with the place and with each other." The review observed that the novel's "world of surfaces remains so sterile because it is so rarely penetrated by any kind of authentic experience." Although Loewinsohn could no longer refer directly to plastic people without seeming trite, his metaphors suggested that intellectuals still considered the concept valid.[18]

Not all references to plastic people implied they were as fake as vinyl siding. Charles Reich, hardly a social critic with a high opinion of synthetics, suggested in *The Greening of America* that people raised in plastic surroundings would enjoy a special plasticity. He asked readers to imagine "a boy or girl . . . brought up in a plastic home, with rugs and furniture that have no more individuality or character than the furnishings of an expensive motel." Unlike children living in houses with "genuine character" who became rooted in their surroundings, the "child of a plastic home" would be "equally 'at home' anywhere." Lacking attachments or loyalties, such a "rootless and truly liberated individual" would disdain its parents' "slavish and passive dependence on consumer goods" and might even become a threat to an authoritarian "corporate state." Reich's theory suggested a dialectic moving from a plastic society's rigidity to the plasticity of free individuals embracing reality without fear of change.[19]

Among the first to observe a new plasticity of personality were sociologist David Riesman and his associates. In 1950, as the first postwar children emerged from cotton diapers and plastic pants, they took the national pulse in *The Lonely Crowd: A Study of the Changing American Character*. In this famous monograph Riesman described three types of social character—all based on an assumption that success-ful societies require conformity. A "tradition-directed" person, unknown in America since the eighteenth century, was one "whose conformity is insured by [a] tendency to follow tradition." An "inner-directed" person, the mainstay of entrepreneurial capitalism, was one "whose conformity is insured by [a] tendency to acquire early in life an internalized set of goals"—a person whose "psychological gyroscope" kept him balanced regardless of others' opinions. But Riesman found few gyroscopes. Postwar Americans, adrift in the present, were "other-directed persons . . . whose conformity is insured by their tendency to be sensitized to the expectations and preferences of others." They moved through life guided by "radar," picking up cues from others as to "rapid adaptations of personality" needed for survival. Riesman feared Americans would "lose their social freedom and their individual autonomy in seeking to become like each other," as if, abandoning his metaphors, they were all extruded through the same plastics die.[20]

To the extent that Riesman's analysis was accurate, postwar Americans were learn-ing to be plastic in a different sense—to remold themselves as circumstances re-quired. Self-confessed "chameleon" Jake Horner, one of the 1950s' more honest fic-tional characters, phrased it well in John Barth's *The End of the Road* when he said, "We're always reconceiving just the sort of hero we are"—or, on another occasion, "A man is free not only to choose his own essence but to change it at will." Horner had a limited stock of roles to choose from, but other people benefited from scores of possibilities offered by the mass media. Jerzy Kosinski's novel *Being There* (1971) presented a simple tale of Chance the Gardener, a man who emerged full-grown from a hermetic existence into the so-called real world knowing only what he had seen on television. Befriended by an elderly industrialist and his young wife, Chance wondered how to behave and finally settled on "the TV program of a young busi-nessman who often dined with his boss and the boss's daughter." As a basic premise he believed that "by changing the channel he could change himself." At a passive extreme, the idea of personal plasticity yielded the character created by Woody Allen in the film *Zelig* (1983), a chameleonlike figure whose appearance, demeanor, per-sonality, and even skin color changed instantly to reflect those of the people he happened to be among.[21]

The career of Richard Nixon provided insights into the plastic American. Known earlier as a vigorous opponent of communism, Nixon as president renounced his past by opening diplomatic relations with China and embarking on détente with the Soviet Union. During his chameleon's progress through postwar America, political analysts often announced "a new Nixon." Garry Wills recalled that reporters frequently pondered the question, "*Is there, then, a new* new-Nixon?" For Wills, Nixon epitomized the self-made man, "the true American monster," or to put it more neutrally, the other-directed man. Instead of creating "something outside himself—a chair, a poem, a million dollars," the self-made man was forever perfecting himself. "He must ever be tinkering, improving, adjusting; starting over," Wills explained, "fearful his product will get out of date, or rot in the storehouse." For this reason Nixon was "the least 'authentic' man alive, the late mover, tester of responses, submissive to the discipline of consent." In reality there was only one Nixon, though there seemed to be "new ones all the time" because he continually tried to adjust himself to "what people want." Four years later, addressing the Watergate scandal, liberal editor Irving Howe observed that Nixon had always projected a sense of "deep unreality," a sense "that there was *no one there* but a puffy concoction . . . put together to appeal to everything inauthentic in American life." Nixon was "made of plastic," he suggested, "and under the heat of revelation, it melted away." Earlier, discussing the phenomenon of endless new Nixons, Wills had concluded that Nixon stood "for all that the kids find contrived, what they call 'plastic' " (Figure 9–4).[22]

In comparing Nixon to the younger generation, Wills's analysis took a curious turn. Nixon seemed plastic or phony because he recast his behavior in a new mold to appeal to popular expectations. His countercultural opponents were "the opposite; plunging, ready to take risks." Wills described how they "move up as close as they can to each experience, flow out to it, undergo it for its own sake," and assess

9–4. A LATER PORTRAYAL OF THE POLITICIAN AS PLASTIC PERSON. COPYRIGHT 1987 BY MARK ALAN STAMATY, REPRODUCED WITH PERMISSION

its significance in "their own intensely private evaluative process."[23] He might have suggested they were indeed more plastic than Nixon, more playfully open to the plasticity of experience. While men and women of the postwar generation tried to shape themselves to exterior constraints in a painful process requiring frequent re-adjustment, many of their sons and daughters abandoned themselves to experimentation. Taught by psychedelic drugs to experience the universe as a kaleidoscope of flowing sensations, they chose to "go with the flow" wherever it took them—into rural communes to live on the land (sometimes with fiberglas domes), into Marxist cells to plot revolution in the streets, into solipsisms of oriental mysticism and western occultism, and into deceptively gentle authoritarian systems of Transcendental Meditation and Scientology, where plasticity sometimes hardened into rigid structures.

By the mid-1970s, when this explosion was beginning to subside, to cure but not entirely to harden, middle-class Americans approached life with a playful spirit of liberation. "Finding oneself," formerly an adolescent phase, became a pursuit for all ages. Gail Sheehy's best-seller *Passages* instructed Americans in 1976 that life was a series of transformations. Like a crustacean, every adult periodically outgrew his or her "protective structure." When this happened, she maintained, "we are left exposed and vulnerable—but also yeasty and embryonic again, capable of stretching in ways we hadn't known before." People, in other words, were thermoplastic. Tom Wolfe took a dim view of people flocking to Esalen Institute or to Primal Scream therapists to find themselves, but he was right on the mark when defining "the alchemical dream of the Me Decade" as "changing one's personality—remaking, remodeling, elevating, and polishing one's very *self*."[24] Although this description recalled Wills's analysis of Nixon's personality, there was a fundamental distinction. Ideally the *new* plastic Americans changed for their own satisfaction—not to meet other people's expectations. This new plasticity yielded the concept of "lifestyle"— whose quick passage from two words through hyphenation to one word indicated wide acceptance of the idea one could transform one's personality at will. If adopting a new lifestyle merely meant acquiring new clothes, a new jargon, a new circle of acquaintances—or a new body, through aerobics, power-lifting, or plastic surgery—the new plastic American was as phony as the old. And yet the concept of lifestyle embodied a faith in genuine plasticity extending forward from Emerson and James into a future of indeterminate shape and limitless potential.

Intellectuals envisioned the new plasticity several years before the concept of life-style emerged into popular consciousness. Lifton's influential essay on "protean man" abandoned traditional psychological concepts such as "character" and "personality" because they wrongly suggested that "fixity and permanence" defined individual identity. Instead, emphasizing "change and flux," Lifton focused on the "self-process" of the late twentieth century. Identifying the "self" as a "person's symbol of his own organism," he defined the "self-process" as "the continuous psychic re-creation of that symbol." Although people had formerly agonized through no more than "one ideological shift per life," it was now common, or so Lifton claimed, "to encounter several such shifts, accomplished relatively painlessly, within a year or even a month." Lifton credited this acceleration not only to nuclear anxiety but also, in a prophetic comment, to a "flooding of imagery produced by the extraordinary flow of postmodern cultural influences over mass communication networks." Protean man, divorced from tradition and possessed of an unprecedented number of suggestive images, was free to undertake "an interminable series of experiments and explorations—some shallow, some profound—each of which may be readily abandoned in favor of still newer psychological quests."[25]

As early as 1936 Walter Benjamin had suggested that such an inflationary proliferation of images destroyed the "aura" formerly surrounding unique objects. The traditional assumption of "permanence" yielded to a perception of "transitoriness" as the basic condition of existence. Mass reproduction of images in the twentieth century and their storage and distribution by film, videotape, and CD-ROM created an expanding cultural storehouse beyond the scope of West's Sargasso of the imagination. As philosopher Jean-François Lyotard observed in 1979, this realm of experience, twice-removed from traditional notions of reality, assumed the role of "nature" for "postmodern man." It was among the millions of discontinuous fragments, shards of past experience (real or imagined)—often distorted and endlessly recombined in the electronic data banks of the "Encyclopedia of tomorrow," as Lyotard referred to it—that protean men and women rummaged as at a cultural flea market for images and concepts from which to fabricate or with which to embody their shifting identities.[26] This concept of *bricolage*, of gathering disparate scraps and recombining them into new synthetic arrangements, almost as Baekeland created an artificial material from organic wastes or Carothers constructed a superpolymer from simpler molecules, became a common postmodern motif. Whether discussing inner or outer experience, artificial landscape or protean man, critics converged on the centrality of this expanded potential for cultural synthesis. For some, the prospect was attractive. For others, it was frightening. Meditating on architecture and on personality, Charles Jencks and Christopher Lasch came to diametrically opposed conclusions in the late 1970s.

Jencks's influential survey of postmodern architecture defended the "radical eclecticism" of buildings by Moore, Stern, Johnson, and others, revealing not a coherent architectural movement but "something you'd expect *after* a movement has broken down." Released from modernism's rigid dogmas, architects could respond to the desires of a new "Everyman" equipped with a "well-stocked, indeed over-stocked, 'image-bank' . . . continually restuffed by travel and magazines." This "*musée imaginaire*," reflecting a "plurality of other cultures," enabled cosmopolitan urbanites to "make choices and discriminations" from a "wide corpus" rather than being "stuck with what they'd inherited." Jencks praised postmodernists for "learn[ing] to use this inevitable heterogeneity of languages" to make an architecture of eclectic fragments. Why, he asked, "if one can afford to live in different ages and cultures, restrict oneself to the present, the locale?" A playful eclecticism, well represented in the color photographs of successive editions of his book, revealed the verve with which architects were "reproduc[ing] fragmented experiences of different cultures." For Jencks, seemingly unaware of the conundrums within his phrase, postmodern architecture revealed "the natural evolution of a culture with choice."[27]

This celebratory portrayal of contemporary architecture paralleled exactly Lasch's pessimistic portrayal of contemporary personality in *The Culture of Narcissism*, an impassioned jeremiad and unlikely best-seller of 1978. As a historian whose left-leaning sentiments tended toward a traditional conservatism, he criticized the typical American for being a "new Narcissus" engaged in "unremitting search of flaws, signs of fatigue, decay." Americans viewed life as a "work of art" and were obsessed by a self-centered goal of "shaping" their own personalities. But the materials they had to work with hardly inspired confidence. For the "performing self," as Lasch described the other-directed person ever alert for reactions from others, "the only reality is the identity he can construct out of materials furnished by advertising and mass culture, themes of popular film and fiction, and fragments from a vast range of cultural traditions, all of them equally contemporaneous to the contemporary mind."[28]

Here, viewed in a distorting funhouse mirror, was the exact situation Jencks celebrated, a reliance of postmoderns and their culture on artificial images, often flat and two-dimensional, sometimes hyperreal, fragmented and disconnected, images collected and distributed by anonymous agents with a myriad of commercial and ideological agendas. It remained uncertain whether a process of artificial synthesis of such fragments could sustain a liberating culture of plasticity as Jencks believed, or whether it could yield only a sham culture of plastic fakery as Lasch feared. Although plastic itself attracted less attention as people focused on electronic media, the issue was defined by a century of experience with plastic as the stuff of artificial synthesis, the preeminent medium of "materialized fantasy."[29] The most compelling presentation of this central issue of postmodernity came in a work of fiction, *Gravity's Rainbow*, published by Thomas Pynchon to nearly universal acclaim in 1973. One critic pronounced it an ambitious "encyclopedic narrative" similar to those of Dante and Melville in representing an entire culture in a time of intense crisis.[30] Especially because Pynchon employed plastic as a central metaphor, an examination of *Gravity's Rainbow* extends our understanding of the cultural meaning of synthesis in the twentieth century.

A CHEMIST WHOSE MOLECULES ARE WORDS

In the midst of the Bicentennial celebrations of 1976 Gore Vidal savaged works by Pynchon and several other writers as "American Plastic." Admitting technical ignorance, Vidal quoted briefly from Roland Barthes's meditation on plastic to attack much contemporary fiction as the lifeless, artificial product of English department

R&D labs. Using Barthes's words, Vidal described this fiction as presenting a "flocculent appearance, something opaque, creamy and curdled, something powerless ever to achieve the triumphant smoothness of Nature." A writer of realist fiction, Vidal referred to *Gravity's Rainbow* as "plastic" to indicate disdain for a work that did not imitate reality but instead extrapolated from it artificially. Pynchon intended his vast fiction to provoke recognition not of literal reality but of the human urge to fabricate, to plot, to shape and mold the artificial from the natural. More appropriate lines for describing *Gravity's Rainbow* were those in which Barthes expressed "perpetual amazement . . . at the sight of the proliferating forms of matter" engendered by plastic—a phenomenon offering "the euphoria of a prestigious free-wheeling through Nature." Vidal's chosen quotation dismissed much recent fiction as unredeemed fakery with an unnatural "hollow and flat" sound and "only the most chemical-looking" of colors. On the other hand, the ambiguity of the word *prestigious* suggested an exhilarating performance by a consummate magician—often merely tricky but also revealing a genuine alchemy beyond the reach of sleight-of-hand.[31]

As the author of an encyclopedic narrative, Pynchon simulates the language of innumerable twentieth-century points of view, including the plastic utopianism of such works as Slosson's *Creative Chemistry* and Haynes's *This Chemical Age*.[32] While describing Imipolex G, which turns out "to be nothing more—or less—sinister than a new plastic, an aromatic heterocyclic polymer" developed at IG Farben in 1939, he offers a mini-history of organic chemistry and plastic's development. The industry's "grand tradition and main stream" leads to Wallace Carothers of Du Pont, "known as The Great Synthesist," whose "classic study of large molecules spanned the decade of the twenties and brought us directly to nylon." With such polymers, Pynchon tells us in the language of the 1930s, came "an announcement of Plasticity's central canon: that chemists were no longer to be at the mercy of Nature." Instead "they could decide now what properties they wanted a molecule to have, and then go ahead and build it" (p. 249). Scores of journalists from the 1930s onward have recited this litany of plastic promoters, and Pynchon emphasizes it.

In this case, however, the outcome is hardly utopian. Imipolex G—"the material of the future," as a minor character refers to it in the jargon of public relations (p. 488)—is the first erectile plastic, a simulation of life by inert, dead matter, its molecular chains "grow[ing] cross-links" under electronic stimulation and shifting the material "from limp rubbery amorphous to amazing perfect tessellation, hardness, brilliant transparency, high resistance to temperature" (p. 699). Developed by Laszlo Jamf, a fictional IG Farben chemist, and similar in name to the company's actual Mipolam polyvinyl chloride, Imipolex G is central to *Gravity's Rainbow*.[33] With most characters involved in plots and counterplots concerning the German V-2 rocket during and immediately after the Second World War, Imipolex G becomes the focus of a more esoteric quest as the material used to fabricate a subassembly for a single mysterious rocket. The reader eventually learns that a crazed rocket battery commander used the subassembly as a shroud to enclose his passive love-object, a young male officer, as he was launched near the end of the war into an orgasmic death-wish flight over the English Channel—a flight whose parabolic arc symbolizes the pull of gravity, of earth itself, against the twentieth century's urge to transcend nature by means of technologies of control.

The unnatural bias of Imipolex G and plastic in general becomes more apparent as another character, a German film actress, recalls being taken to a plastic factory, a place of dread where images of excremental viscosity and ethereal otherworldliness existed in surreal juxtaposition. After noticing a "strong paint-thinner smell," she watched as "clear rods of some plastic came hissing out through an extruder at the bottom of the tower, into cooling channels, or into a chopper." An oppressive heat made her think of "something very deep, black and viscous, feeding this factory." After dragging her into a warehouse where "great curtains of styrene or vinyl, in all colors, opaque and transparent . . . flared like the northern lights," her faceless male escorts "stretched [her] out on an inflatable plastic mattress," dressed her in a crotch-less black costume of Imipolex G ("it felt alive on me"), and watched as one of them "strapped on a gigantic Imipolex penis over his own" and brought her to a delirious unconsciousness from which she emerged days later, outside, naked, in the midst of a vast "tarry kind of waste" devoid of life (pp. 487–488). It was as if Pynchon had discovered the *Collier's* article of 1947 in which a journalist declared there was "something unearthly" about plastic, something that gave her "the creeps" when she visited a plastic factory and made her feel as if "man has something quite unruly by the tail."[34] As Pynchon's Greta recalls of one moment of her not entirely unpleasant excursion into plastic's dead fetishism, "Someone said 'butadiene,' and I heard *beauty dying*" (p. 487).

Synthetic chemistry, with its power to organize matter in artificial structures, pro-vides Pynchon with a metaphor of the twentieth century's organization of human beings into deadening economic and bureaucratic structures. He delivers his mes-sage through the voice of the spirit of industrialist Walter [*sic*] Rathenau speaking to a director of IG Farben during a prewar spiritualist seance. The coming reign of the kingdom of death began, the spirit intones, with the synthesis of mauve dye, "the first new color on Earth," from out of the coal tar left behind after the manufac-ture of steel. Sounding much like John Mumford writing in 1924 on Bakelite's derivation, spirit-Rathenau characterizes coal as "Earth's excrement . . . ancient, prehistoric . . . growing older, blacker, deeper, in layers of perpetual night." Coal is "the very substance of death," a "preterite dung" from which "a thousand different molecules" will be synthesized in a movement "from death to death-transfigured." He warns, however, that "polymerization is not resurrection." And that IG Farben itself, prototype of industrial cartels, is as artificial a structure as its polymers. "The more dynamic it seems to you," the spirit tells his audience, "the more deep and dead, in reality, it grows." He leaves them with an injunction to investigate "the real nature of synthesis" and "the real nature of control" (pp. 166–167). But the men of

IG Farben are already locked into synthetic structures that Pynchon isolates as motivating forces of the Second World War.

These ideas are echoed near *Gravity's Rainbow*'s conclusion as a businessman turned occultist realizes "that Earth is a living critter." Beyond that, it has always metabolized its organic wastes by breaking them down and embracing them as part of its own substance. He marvels "that Gravity, taken so for granted, is really something eerie, Messianic, extrasensory in Earth's mindbody . . . having hugged to its holy center the wastes of dead species, gathered, packed, transmuted, realigned, and rewoven [their] molecules [into coal and petroleum]."[35] But there is nothing alive, nothing sacred, in synthetic chemistry's transcendence of these honest earthbound molecules as it transmutes them into dead substances endowed with nonbiodegradable immortality. The narrator, or Pynchon himself, despairs of the possibility of earth's ever redeeming all this "plastic trivia" with its "slick persistence." Wondering if it is possible to find among the proliferating litter "the meanest sharp sliver of truth in so much replication, so much waste," he offers an obscure incantation that resonates for anyone who has reflected on plastic's meaning: "plastic saxophone reed *sounds of unnatural timbre*, shampoo bottle *ego-image*, Cracker Jack prize *one-shot amusement*, home appliance casing *fairing for winds of cognition*, baby bottles *tranquilization*, meat packages *disguise of slaughter*, dry-cleaning bags *infant strangulation*, garden hoses *feeding endlessly the desert*" (p. 590).

For Pynchon, synthetic plastic signifies death masquerading as life. To the extent that we surround ourselves with it, or incorporate it into our bodies, or dream of a securely controlled artificial environment in some utopian future, we become more dead to life, less in touch with what Pynchon would call the natural. Oddly enough, however, for an author who has raised plastic to a fetish in the pornography of death, Pynchon's fiction exhibits astonishing plasticity of theme, image, and technique. A plastics engineer turned literary critic might conclude that Pynchon had become obsessed with thermosetting Bakelite, so like a perfect fascism with its molecular structure rigidly set for all eternity, but that he had overlooked his own affinity for the thermoplastics with their endless capacity for transformation. Pynchon writes as if under the influence of the very euphoria celebrated by Barthes, as if by his prose's exuberant plasticity, by remolding and transforming the raw materials of his culture, he might construct "a plot . . . too elaborate for the dark Angel to hold at once."[36]

Just as synthetic chemistry moved beyond nature to create an endless variety of unprecedented molecules, polymers, and materials, so does Pynchon use the stuff of ordinary experience to create a synthetic multiverse. From raw historical facts and fragments of popular culture, functioning for him as the chemist's proverbial lump of coal, glass of water, and whiff of air, Pynchon synthesizes realms of imaginative possibilities. If at times he mocks his own endeavor, as when referring to "the insanely, endlessly diddling play of a chemist whose molecules are words" (p. 391), his inventive prose continues all the same to flow forth in structures so flexible, so shifting in voice and phrasing, often so apparently innocent of ultimate purpose, that it is hard to take the measure of the chain of words. The raw materials of Pynchon's synthetic process, his novel's overabundant surface details, are usually accurate enough, whether it is a matter of popular song lyrics, plots of obscure films, scientific theories, wooden coffee spoons, the double-S shape of the underground factory in which V-2 rockets were manufactured, bomber noses of Perspex, or vials with red Bakelite caps "bearing the seal of Merck of Darmstadt" (p. 621). But despite hundreds of historical details great and small in Pynchon's chain, many of whose sources have been identified by scholars as if by literary spectroscope, he continually entices readers to move with him beyond the surface into artificial extrapolation and fabrication.

To cite some examples: What if those two future victims of assassination, John F. Kennedy and Malcolm X, once confronted each other over a shoeshine at Boston's Roseland Ballroom? What if military intelligence recruited those prominent British academics who conducted ESP experiments during the blitz? What if adventurous world traveler Richard Halliburton, supposedly lost at sea on a Chinese junk in 1939, appeared before a onetime fan six years later as a drug-addled stowaway in a military cargo plane? What if Richard Nixon once fantasized about being smothered by a wind-blown dry cleaning bag while driving in a convertible on the Santa Monica freeway? And finally, to end a list that might approach the length of *Gravity's Rainbow* itself, what if two plastics executives got together shortly after V-E day "to finalize plans for the Postwar Polyvinyl Chloride Raincoat?" ("Imagine the look on some poor bastard's face when the whole *sleeve* simply falls out of the shoulder—" "O-or how about mixing in something that will actually *dissolve* in the rain?" [p. 615].) These examples reveal Pynchon reshaping his culture's raw materials and thereby reinvigorating them. Such extrapolated factoids are literally artificial but often ring culturally true. Just as synthetic chemists "improved on" nature when they developed plastic, so does Pynchon "improve on" history by offering a subversive parahistory that reshapes the recent past to create alternative narratives more real than the official history of the twentieth century.

But however artificial Pynchon's parahistory, it is not plastic in the sense of a compression-molded Bakelite radio, solid and unchanging, with (to extrapolate now a bit from *Gravity's Rainbow*) its "streamlined corners" and "simple solid geometries of the official vision." Instead, Pynchon's parahistory possesses a plasticity that might have been "set up deliberately To Avoid Symmetry, Allow Complexity, Introduce Terror" in readers too absorbed by fantasies of control to understand the reasons for recalling and redeeming inconsequential things "from their times and planet . . . the wine bottle smashed in the basin, the bristlecone pines outracing Death for millennia, concrete roads abandoned years ago, hairdos of the late 1930s, [even, with complete catholicity,] indole molecules, especially *polymerized* indoles, as in Imipolex G—" (p. 297). Ultimately only these details matter, not the chains of significance into which they might be integrated. While tempting readers to integrate

clues, to construct inert patterns, to reach the center of a supposedly global conspiracy many of his characters are determined to crack, Pynchon makes it impossible to do so. Gradually we learn that each character's version of events is flawed,
marred by faulty data and skewed by personal desires and angles of perception.
There is no coherent plot, no closed construction whose parts fit intricately together
like the linked chains of atoms in a vast polymerized molecule. While the chemist
Jamf, a man of "pale plastic ubiquity," ends his career dreaming of indestructible
new synthetics based on the "mineral stubbornness" of the ionic bond, the wander-
ing army deserter Slothrop realizes there are "many other plots besides those polar-
ized upon himself," and that "this network of all plots may yet carry him to freedom"
(pp. 490, 577, 603). As the novel's own apparent plot disintegrates, its final two
hundred pages explode in a kaleidoscopic rainbow of outrageous fantasies, bur-
lesques, mock-histories, pseudo-mythologies—their randomness infuriating some
readers by inviting plasticity of interpretation.

More than a century earlier William James overcame despair in the face of a rela-
tivistic existence in which traditional religious doctrines no longer afforded stability
by asserting, as an individual, an independent "will to believe." The universe was, as
James conceived it, open, unfinished, in a never-ending process of development, an
incomprehensible process to which the human race could only incompletely contribute in a localized manner. Pynchon embodies James's philosophy in *Gravity's
Rainbow*, whose very form asserts nature's plasticity, its openness to transformation.
Pynchon shares with James a sensitivity to the liberating potential of disintegrating
structures—whether religious, scientific, historical, or literary. It is this disintegration, already begun but hardly perceived when James experienced his dark night of
the soul, that yielded the late-twentieth-century culture of plasticity.

When Lasch complained that contemporary culture's "only reality" was that "construct[ed] out of materials furnished by advertising and mass culture, themes of
popular film and fiction, and fragments torn from a vast range of cultural traditions,"
he did not mention that Pynchon had just constructed such a reality, one of an
infinite variety of possible realities, in *Gravity's Rainbow*, a work of fabulation and
fabrication positing the liberating potential of postmodern decentering. We are now,
apparently, abandoned to our own devices. Everything we create—landscapes,
buildings, political and social structures, literature, personality itself—is of our own
devising: plastic, artificial, and unsupported by transcendent certainties or natural
laws. Even science, formerly presumed to be a passive reflector of ultimate reality,
has been described as a continual shifting of artificially constructed paradigms. To
retreat to the past by endowing artificial structures with rigid permanence invites a
state of death-in-life through self-strangulation. But to accept infinite plasticity
opens up a state of uncertainty that seems as threatening as it does liberating. Pynchon's work poses but does not resolve the question of whether postmodern culture
is a culture of plastic—fake, phony, ultimately unsatisfying—or a culture of plasticity—exuberant, extravagant, and life-furthering in the Jamesian sense. Nor does

Pynchon ever really reconcile his disdain for plastic as an artificial substance with his dedication to creating a fiction of substantial artificial plasticity.[37]

IMMATERIALITY TAKES COMMAND

As the twentieth century wound to a close, the paradox of plastic revolved around two apparently opposing tendencies. The molding and extruding plants kept pouring out an accelerating volume of stuff. At the same time that stuff seemed ever more ephemeral. A famous phrase of Marx bore on the situation. "All that is solid melts into air," he wrote in 1848, by which he meant that continued expansion of the bourgeoisie depended on continual turnover in technologies, means of production, types of goods, even ideas and opinions.[38] At least since the 1920s synthetic materials had contributed to that process. Each year they became cheaper to produce, stronger by weight, more varied in color and characteristic compared to competing materials. Their proliferation not only stimulated expansion, even inflation, of the economy and the culture. It also gave expression to those processes and to a resulting perception that nothing is forever. In a sense, however, plastic also became a victim of that perception. When other technologies, particularly those of communication and information, came to the forefront as carriers of technology's aura, plastic melted into cultural invisibility—even while continuing to proliferate materially, actually serving as essential media of recording, storage, and display of information and images.

Paradoxical disappearance of plastic as its annual volume continued to increase attracted the attention of a variety of commentators in the late 1980s. Among the most provocative was an Italian designer, Ezio Manzini, whose poetical, technically accurate treatise, *The Material of Invention*, was published in 1986 under the auspices of the Italian petrochemical firm Montedison. Tracking the shift from material to immaterial, or from substance to image, in the everyday life of the late twentieth century, Manzini asserted that "the proportion between what we know as a deferred image and what we know by direct experience is shifting more and more in favor of the deferred image." As a result, he maintained, in an echo of McLuhan's theory of sensory balance, "matter no longer appears to the scale of our perceptions as a series of given materials, but rather as a continuum of possibilities." In other words, people no longer valued material objects or things in their own right as unique physical entities but for the experiences they could convey, for their service as the means to other ends. Materials became mere "stage machinery" when they attained "such high levels of performance that we perceive their effects almost without noting the source." What seemed at first glance a matter of perception alone soon also encompassed the underlying material reality. Ongoing miniaturization had relied on plastics at least from the 1940s as a source of strength without the weight and bulk of metals for which they substituted. But by the time Manzini was writing, miniaturization no longer referred only to smaller components but to a more radical trend, an "elimination of components" altogether. He defined the new technology as one of "less matter, less energy, more information." Civilization had reached a point where, in the words of another observer, "immateriality takes command."[39]

The prospect aroused prophets of all sorts, no different really from those who enthused over the alchemy of synthetic chemistry seventy-five years earlier. If the vision then was of a cornucopia of material plenty, the vision of the new plasticity was of a fully digitized culture capable of responding instantaneously to the most casual desires with "virtually real" satisfactions. Electrons and computers replaced molecules and molding machines, but the vision continued to reflect a faith in technology's capacity for transmuting the inherent imperfection of the natural into the dazzling perfection of the artificial. The rhetoric was just as breathless as before, and the willingness to ignore reality while projecting total control over it just as extreme. In 1985, for example, an Associated Press story quoted an animator working with a Cray X-MT, then the world's most powerful computer, as saying there would soon be "no fantasy that can't be put onto the screen with a degree of truthfulness that would be awesome." A few years later *Time* concluded an article on virtual reality with a promoter's prediction that couples wearing electronic data suits at opposite ends of the continent would someday "visit artificial tropical islands together." Their activities would "be limited only by their imagination—and the power of their computers"—a remark differing only in its coy salaciousness from former projections of chemical beefsteaks and one-piece injection-molded houses.[40]

Just as plastic had transformed the material world in the century and a quarter since Hyatt sought to win $10,000 with an improved billiard ball, it seemed likely that digitized artificiality would transform American culture in ways that could be assessed only after another century had passed. Far beyond plastic, at the outer edges of speculation about digital immateriality, were promoters of cyberspace, an ultimate realm of pristine artificiality prefigured by Pynchon's offhand reference to a "clean, honest, purified Electroworld" into which rebellious youth would escape by plugging in rather than by taking drugs (p. 699). The concept of cyberspace, as well as the word itself, came from William Gibson's novel *Neuromancer*, which acquired cult status after publication in 1984. Before long, *Neuromancer* and similar science fictions had precipitated a cyberpunk mentality—a hybrid of anti-establishment punk, industrial music, smart drugs, kinky New Age-ism, and a hacker's subversive attitude toward databases. Gibson, who claimed he wrote the novel on a manual typewriter, provided a definition of cyberspace often quoted by Silicon Valley engineers: "a consensual hallucination experienced daily by billions of legitimate operators" comprising "a graphic representation of data abstracted from the banks of every computer in the human system . . . lines of light ranged in the nonspace of the mind, clusters and constellations of data" arranged "like city lights, receding."[41] In Gibson's novel, freelance cyberspace cowboys use commercial decks to jack into the matrix and soar above astrally precise architectonic structures representing the databases of multinational conglomerates. The goal is to penetrate the protective "ice" of these structures to steal information for the highest bidder. None of it is physically real, yet all of it seems psychologically more alive than the "meat" world to which every hacker must return. *Neuromancer* crystallized a common consciousness for computer engineers engaged in experiments with data gloves, video goggles, and other devices for simulating physical experience in the interface of humans with computers. Gibson offered a vision of perfect synthetic artificiality that transcended their crude virtual reality gadgets—a vision powerful enough to attract future development.

Gibson's novel also appealed to a punk mentality because it contrasted the holographic projection of cyberspace with a compelling meat-world landscape of *film noir* intensity that owed much to the dystopian Los Angeles depicted in *Blade Runner* in 1982. Central to the effectiveness of these scenes is Gibson's awareness of plastic as just another element among the randomly layered scraps of an imperfect world that has never known the unnatural perfection of a utopian "world of tomorrow" or "house of the future." An intentionally instructive vista early in the novel extends from the "towering hologram logo of the Fuji Electric Company" to the "black expanse" of Tokyo Bay "where gulls wheeled above drifting shoals of white styrofoam." Down below, a down-and-out protagonist makes the rounds of the bars, one of which is furnished with "pale Milanese plastics" in a style "of the previous century," their "once glossy" surfaces "fogged with something that could never be wiped away." The "brown laminate" of a tabletop is "dull with a patina of tiny scratches" caused by "countless random impacts." In another place a bartender wears an artificial arm "cased in grubby pink plastic," outrageously ugly for "an age of affordable beauty." At another point, as the scene shifts to Istanbul, the protagonist wanders "along a broad concourse, beneath soot-stained sheets of plastic and green-painted ironwork out of the age of steam." Such scenes of plastic decay are grittily romanticized but somehow as real in their makeshift accretions as we can ever know the future will be. To compare them to Gibson's cyberspace, with its "bright lattices of logic unfolding across that colorless void," is to recognize his fundamental agreement with another of Pynchon's throwaway lines—a line that affords a corrective to the utopian dreams of synthesizers, both chemical and digital, all the way back to the beginning of the twentieth century. Speaking as a well-meaning father to a rebellious "electrofreak" son, Pynchon warns that "you always do come back to old Realityland, don't you."[42]

This fundamental issue of the material and the immaterial came up at a conference on virtual reality in 1990. Its organizer, architect Michael Benedikt, expanding on Gibson's definition of cyberspace as a new "consensual" reality, insisted on a set of organizing principles—natural laws, so to speak—to avoid a confusing welter of solipsisms. Designers of cyberspace would have to decide on an appropriate "level of malleability" for an "ongoing, real-time, consensual public realm." Breaking ranks with dreamers of artificiality from Slosson onward, Benedikt proposed a "principle of indifference" stating that resistance to the desires of users establishes the "felt realness" of a system or world. In other words, in a phrase that echoes James, "what is real always pushes back."[43] But it is hard to believe that cyberspace's actual (or virtual) development will follow this spartan agenda. In moving from traditional materials to plastic to electronic impulse as the medium of artificial environments, a casual malleability has triumphed. An apparent dissolving of limits makes anything seem possible—and perhaps of little worth. We question the value of flickering images just as an earlier generation questioned plastic's facile surfaces. To claim to "like anything that's fake" reveals fascination with the power of synthesis. But it also hides a fear that the surface ingenuity of artificial environments indicates profound cultural exhaustion. Fulfilling our dreams has become ever easier through the deformations and extrapolations of chemical and electronic synthesis. Proliferating layers of synthetic things and images suggest a vast expansion of creative freedom. But they may also conceal our culture's increasing lack of grounding in the resistant stuff of nature. This insoluble paradox, encapsulating the American experience of plastic, seems likely to persist into the next century and beyond.

ACKNOWLEDGMENTS

It is impossible to acknowledge everyone who has helped since the day I discovered, as a graduate student researching a dissertation on industrial design, that not much was published on the history of plastic. Scribbling the idea on a note card along with a dozen or so other projects, I hardly suspected it would expand to fill much of my life for ten years. I would like first to thank Robert Friedel, whose study of celluloid defined many of the issues essential to any consideration of plastic, and who has responded thoughtfully, often critically, to my ideas on the subject almost from the moment I began taking it seriously. Others who encouraged initial exploratory essays are Joe Corn, Bruce Sinclair, and Gerhard Hoffmann. Work on the project began in earnest in 1984, when an NEH Summer Stipend and a Summer Research Award from the University Research Institute of the University of Texas at Austin gave me time to examine the entire run of *Modern Plastics*, the industry's trade journal and the only source of a detailed chronology. I completed much of the research during 1986–87, supported by a Faculty Research Assignment from the University Research Institute and by a crucial NEH-Mellon Research Fellowship that allowed me to spend six months in residence at the Hagley Museum and Library in Wilmington—site of the Du Pont archive and an excellent jumping-off point for other resources in the northeast. I am equally indebted to the Getty Grant Program for a Senior Research Grant in 1989–90 that supported me while I wrote much of the first draft. A small grant from the University Research Institute in 1994 helped pay for illustrations. I also want to acknowledge the Institute of United States Studies at the University of London, which welcomed me as a Fulbright lecturer in 1992–93 and gave me an opportunity to gain a different perspective on the material.

The Hagley Museum and Library afforded a stimulating environment for the research phase of the project. Glenn Porter, Liz Kogen, Ed Lurie, and Dick Williams offered a warm welcome and frequent opportunities for informal discussions. Mike Nash, Heddy

Richter, and Marge McNinch guided me through archive and library holdings and fielded many requests. Other researchers, among them David Hounshell, John K. Smith, Jr., Olivier Zunz, and especially Ken Mernitz and Roland Marchand, helped focus my ideas. I am indebted to the latter for an actual fortune-cookie slogan—"make friendships more lasting, seal someone in plastic"—that until now has frustrated all attempts to work it into the text. During my stay at Hagley I often consulted materials at the Smithsonian Institution. Jon Eklund of the National Museum of American History, who has encouraged this project from the start, was helpful as always. John Fleckner and Robert Harding of the Smithsonian's Archives Center facilitated my work there. I'm especially grateful to Toby Appel and Robert Friedel for their hospitality during Washington research trips.

I would like to thank many people from the plastics industry who provided information and shared insights. Gordon Kline submitted patiently to a long interview at his home in Lake Worth, Florida. William Cruse discussed his career over take-out Chinese in his Manhattan apartment. Don Featherstone, whose signature appears on every pink flamingo molded by Union Products, Inc., gave me an enthusiastic tour of their plant in Leominster. Keith Lauer, John Keville, and Evelyn and Ed Hachey revealed much about that city's celluloid industry. I learned about the trials of custom molders through a discussion with Robert Davidson and Bud Kasch of Kurz-Kasch in Dayton. Sidney Gross, who interpreted the industry as an editor for many years, contributed an admirably frank interview. I am also indebted in various ways to Albert Dietz, Joan Goody, Jerome Heckman, Harper Landell, and Jean Reinecke, and to many organizations that answered queries, sent publications, or, in the cases of E. I. du Pont de Nemours & Company, the Rohm and Haas Company, and The Society of the Plastics Industry, Inc., granted permission to consult and quote from archives.

Projects that are long under way often benefit from responses to preliminary publication. I would like to thank the MIT Press, the American Philosophical Society, and the journal *Amerikastudien* for enabling my first thoughts on plastic to reach print as articles. I am also indebted to Rizzoli, Oxford University Press, the Royal Society of Chemistry, the Center for American Studies at Odense University, and the journal *Textile Chemist and Colorist* for permission to reproduce portions of the text that first appeared in somewhat different form under their auspices.

Many colleagues have influenced my thinking, often directly but just as often by osmosis. I benefited from careful readings of the entire text by Robert Friedel and Michael Schiffer. John Clarke and John K. Smith, Jr., read and commented helpfully on portions of the text. I learned much from comments on conference papers by Paul Boyer, Cecelia Tichi, and especially Michael L. Smith, who alerted me early on to some of the pitfalls. Jeff Sturchio broadened my knowledge of the history of polymer chemistry by inviting me to the *conversazione* of the Center for History of Chemistry in Philadelphia. Steve Spivak was an enthusiastic collaborator on an article about the naming of nylon. I am indebted to Tony Fry, John Heskett, Wendy Kaplan, Mel McCombie, Victor Margolin, Penny Sparke, Teal Triggs, and Fredrik Wildhagen for sharing their perspectives on design and material artifacts. The interdisciplinary side of the project was stimulated by discussions with Bob Bednar, Emily Cutrer, Tim Davis, Dennis Doordan, Peter Hales, Brian Horrigan, Cynthia Meyers, Terry Smith, and Ed Tenner. The year in London gave me opportunities to discuss the project with colleagues and students throughout the U.K., and with Sylvia Katz, Peter Morris, Susan Mossman, Percy Reboul, and Tony

Walker of the Plastics Historical Society, a lively organization that has no counterpart in the United States. At the University of Texas I am indebted to Janice Bradley for overseeing many administrative details. My colleagues in the American Studies Program offered frequent suggestions. Among them I especially want to thank Mark C. Smith for his friendship and encouragement over the years, and for giving the manuscript a careful reading at a critical point.

My family has shown great good humor throughout, tolerating a near obsession and adjusting to erratic schedules and odd vacation detours. Vanessa and Jason, who both grew up while this project was under way, discovered apt quotations and made whimsical contributions to an expanding "plastics museum" in my study. I'm especially proud that Vanessa took several of the photographs. Above all I'm grateful to Alice, who endured with me through good times and bad. She listened with enthusiasm to initial discoveries, made tough-minded comments on early drafts, and gave the final manuscript a perceptive reading filled with common sense. As always she remains my best critic and strongest support.

SOURCES

Materials for this study include individual and corporate archives; interviews conducted by me or made available to me; the complete run of *Modern Plastics* and its predecessor journals; books, trade catalogues, pamphlets, and articles published by and for the plastics industry; articles on plastic in the business press and in general-circulation magazines and newspapers; a wide range of secondary historical works; and an ever-expanding collection of references to plastic from the larger culture—with sources ranging from novels and films to advertisements and other ephemera. The most tangible (and elusive) sources are the plastic objects and surfaces that have surrounded me since long before I consciously thought about them. Given the scope of the reference notes, I mention here only key works.

The idea of writing about plastic first occurred to me when the only history of the subject I could find was J. Harry DuBois's *Plastics History U.S.A.* (Boston: Cahners, 1972). Written by a retired plastics engineer, it seemed uninformed by any larger issues, but I have since come to appreciate its succinct accounts of materials, processes, and applications—based on interviews with members of the Plastics Pioneers Association and a collection of early trade catalogues. A similar history, more culturally oriented and mostly focused on developments in Great Britain, is Morris Kaufman, *The First Century of Plastics* (London: Plastics Institute, 1963). The process of defining and commercializing new materials was first analyzed by Robert Friedel in *Pioneer Plastic: The Making and Selling of Celluloid* (Madison: University of Wisconsin Press, 1983), a work whose scope extends beyond its announced subject. David A. Hounshell and John Kenly Smith, Jr., *Science and Corporate Strategy: Du Pont R&D, 1902–1980* (Cambridge: Cambridge University Press, 1988) is essential for an understanding of twentieth-century developments in the chemical industry. Williams Haynes's massive six-volume *American Chemical Industry* (New York: D. Van Nostrand, 1945–54) is useful for information on specific companies and chemical industry trends. Three postwar surveys of synthetic materials and their applications are especially helpful: E. G. Couzens and V. E. Yarsley, *Plastics in the Modern World*, 3rd ed. (Baltimore: Penguin, 1963); Sylvia Katz, *Plastics: Designs and Materials* (London: Studio Vista, 1978); and Ezio Manzini, *The Material of Invention* (Cambridge: MIT Press, 1989). Andrea DiNoto's *Art Plastic: Designed for Living* (New York: Abbeville Press, 1984) and Sylvia Katz's *Plastics: Common Objects, Classic Designs* (New York: Harry N. Abrams, 1984) are well-illustrated, intelligently presented surveys of design in plastic. Two useful anthologies of historical articles, one emphasizing issues of design and marketing, the other issues of invention and manufacturing, are *The Plastics Age from Modernity to Post-Modernity*, ed. Penny Sparke (London: Victoria & Albert Museum, 1990; Woodstock, N.Y.: Overlook Press, 1993), and *The Development of Plastics*, ed. S.T.I. Mossman and P.J.T. Morris (Cambridge: Royal Society of Chemistry, 1994). Finally, Peter J. T. Morris's brief *Polymer Pioneers: A Popular History of the Science and Technology of Large Molecules* (Philadelphia: Center for the History of Chemistry, 1986) is a good source for quick reference.

The following abbreviations are used throughout:

AISI   American Iron and Steel Institute Archive, Hagley Museum and Library, Wilmington, Delaware

Adams   Roger Adams Papers, Archives of the University of Illinois at Urbana-Champaign

CCR   Celluloid Corporation Records, Archives Center, National Museum of American History, Smithsonian Institution, Washington, D.C.

DDC   Donald Deskey Archive, Cooper-Hewitt, National Design Museum, Smithsonian Institution, New York

DPA   Du Pont Archive, Hagley Museum and Library, Wilmington, Delaware

DuBois   J. Harry DuBois Collection of the History of Plastics, ca. 1900–1975, Archives Center, National Museum of American History, Smithsonian Institution, Washington, D.C.

Eklund   Jon Eklund interview tapes, National Museum of American History, Smithsonian Institution, Washington, D.C.

Kurz   Kurz-Kasch, Inc., corporate records, Dayton, Ohio

LHB   Leo H. Baekeland Papers, 1881–1968, Archives Center, National Museum of American History, Smithsonian Institution, Washington, D.C.

MOMA   Museum of Modern Art Library, New York

MP   *Modern Plastics*

NBG   Norman Bel Geddes Collection, Hoblitzelle Theatre Arts Library, Harry Ransom Humanities Research Center, University of Texas at Austin

PPA Tapes   Plastics Pioneers Association interview tapes, Division of Physical Sciences, National Museum of American History, Smithsonian Institution, Washington, D.C.

R&H   Rohm and Haas Company Archive, Philadelphia
SPIA   The Society of the Plastics Industry, Inc., Archive, Hagley Museum and
         Library, Wilmington, Delaware
SPIF   The Society of the Plastics Industry, Inc., Files, Washington, D.C.
WCBA   Warshaw Collection of Business Americana, Archives Center, National
         Museum of American History, Smithsonian Institution, Washington,
         D.C.

PREFACE

1   Mark Helprin, *Winter's Tale* (1983; rpt., New York: Pocket Books, 1984), 373–378.

INTRODUCTION

1   For production figures see Joel Frados, *The Story of the Plastics Industry*, 13th ed. (New York: Society of the Plastics Industry, 1977), 6; for a chart comparing plastic production with GNP see Dominick V. Rosato, William K. Fallon, and Donald V. Rosato, *Markets for Plastic* (New York: Van Nostrand Reinhold, 1969), 2.

2   "New Products: Prometheus Unbound," *Time* 76 (September 19, 1960): 94; Brooke Hindle, "How Much Is a Piece of the True Cross Worth?," in *Material Culture and the Study of American Life*, ed. Ian M. G. Quimby (New York: W. W. Norton, 1978), 18. See also David M. Potter, *People of Plenty: Economic Abundance and the American Character* (Chicago: University of Chicago Press, 1954).

3   "What Man Has Joined Together . . . ," *Fortune* 13 (March 1936): 69.

4   "Editorial Impressions: We Believe in Dreams!," *Plastics* 3 (February 1927); Haynes as quoted in "Industry Prepares for War Role," "special insert" in *MP* 19 (May 1942): unpaginated.

5   Marcia Seligson, "Hollywood's Hottest Writer—Buck Henry," *New York Times Magazine*, July 19, 1970, reprinted in *The New York Times Encyclopedia of Film 1969–1971*, ed. Gene Brown (New York: Times Books, 1984), unpaginated; Bob Martino, "Human Factors," *MP* 63 (November 1986): 43. *The Graduate* (Embassy Pictures) was directed by Mike Nichols and released at the end of 1967. A popular and critical success, the film enjoyed an initial gross that placed it behind only *Gone With the Wind* and *The Sound of Music*, according to Seth Cagin and Philip Dray, *Hollywood Films of the Seventies* (New York: Harper & Row, 1984), 32. For a cultural assessment of the scene see Landon Y. Jones, *Great Expectations: America and the Baby Boom Generation* (1980; rpt., New York: Ballantine, 1986), 136. Although Calder Willingham shared screenwriting credit with Buck Henry, the latter wrote the final script, according to Seligson, including the plastics scene, which did not appear in the original novel by Charles Webb. Interviewed in 1992, Henry recalled that at Dartmouth College in the 1950s he had heard *plastic* used, presumably as an adjective, by philosophy professor Eugen Rosenstock-Huessy to refer (in the interviewer's paraphrase) to "a civilization that abandons its values." See Jay Boyar, "When 'Plastics' Became a Bad Word," *Washington Post* (August 30, 1992): G4.

6   Roland Barthes, "Plastic," in *Mythologies*, trans. Annette Lavers (New York: Hill and Wang, 1972), 97–99; originally written between 1954 and 1956 and reprinted as "Le Plastique" in *Mythologies* (Paris: Editions du Seuil, 1957), 192–194.

7   Henry George Liddell and Robert Scott, *A Greek-English Lexicon* (Oxford: Oxford

University Press, 1968), 1412; G.W.H. Lampe, *A Patristic Greek Lexicon* (Oxford: Oxford University Press, 1961), 1089; and Charlton T. Lewis and Charles Short, *A Latin Dictionary* (Oxford: Oxford University Press, 1969), 1385.

8   Samuel Johnson, *A Dictionary of the English Language* (London: W. Strahan, 1755), vol. 2: unpaginated; Emma C. Embury, "Female Education," *Ladies' Companion* 8 (January 1838): 18, as quoted by Barbara Welter in "The Cult of True Womanhood, 1820–1860," reprinted in *The National Temper: Readings in American Culture and Society*, 2nd ed., ed. Lawrence W. Levine and Robert Middlekauff (New York: Harcourt Brace Jovanovich, 1972), 161; William Dwight Whitney, ed., *The Century Dictionary* (New York: The Century Co., 1890), 4:4535; Benjamin E. Smith, ed., *The Century Dictionary Supplement* (New York: The Century Co., 1910), 2:1016; and William A. Neilson, ed., *Webster's New International Dictionary of the English Language*, 2nd ed. (Springfield, Mass.: G. & C. Merriam, 1934), 1883. See also the *Oxford English Dictionary* entry for "Plastic."

9   Ernest J. Parry, *Shellac: Its Production, Manufacture, Chemistry, Analysis, Commerce and Uses* (London: Sir Isaac Pitman & Sons, 1935). On natural plastics see Robert Friedel, *Pioneer Plastic: The Making and Selling of Celluloid* (Madison: University of Wisconsin Press, 1983), 24–28.

10   United States Patent Office, *Classification Bulletin*, no. 10 (Washington: Government Printing Office, 1903), 7–11; L. H. Baekeland, "The Synthesis, Constitution, and Uses of Bakelite," *The Journal of Industrial and Engineering Chemistry* 1 (March 1909): 156–157; Alan S. Cole interview, 1952, PPA Tapes, reel 1, side 1; and "Editorial Impressions," *Plastics* 1 (October 1925): 20.

11   "A Little Cotton and a Little Camphor Make You This Finer Fountain Pen!," *Du Pont Magazine* 32 (Midsummer 1938): advertisement inside front cover; Morris Sanders, "Plastics and Architecture," *Architectural Record* 88 (July 1940): 66; and *Webster's New Collegiate Dictionary* (Springfield, Mass.: G. & C. Merriam, 1945), 646. As late as 1961 *Webster's New Collegiate Dictionary* stated that rubber was "sometimes included among plastics" (646). Rubber did not disappear from the definition, bringing it in line with current usage, until 1963, in *Webster's Seventh New Collegiate Dictionary* (648).

12   Stephen Bass, *Plastics and You* (New York: Eastwood-Steli, 1947), 24; "The Plastics Industry Has Come of Age," *MP* 23 (April 1946): 132; and S. I. Hayakawa, *Funk & Wagnalls Modern Guide to Synonyms and Related Words* (New York: Funk & Wagnalls, 1968), 358.

13   Ken Kesey, *One Flew Over the Cuckoo's Nest* (1962; rpt., New York: Signet, n.d.), 101, 220; George Bernard Shaw as quoted by *OED Supplement* (Oxford: Oxford University Press, 1982), 3:554 (italics supplied); D. H. Lawrence as quoted by *OED Supplement* (Oxford: Oxford University Press, 1972), 1:466 (italics supplied); and Russell Lynes, "After Hours: Highbrow, Lowbrow, Middlebrow Reconsidered," *Harper's* 235 (August 1967): 19. Lynes was cited by Clarence L. Barnhart, Sol Steinmetz, and Robert K. Barnhart, *The Barnhart Dictionary of New English Since 1963* (Bronxville, N.Y.: Barnhart/Harper & Row, 1973), 366, and by *OED Supplement* (Oxford: Oxford University Press, 1982), 3:556.

14   As quoted in interviews edited by Leonard Wolf with Deborah Wolf, *Voices from the Love Generation* (Boston: Little, Brown, 1968), 90, 260; Charles A. Reich, *The Greening of America* (1970; rpt., New York: Bantam, 1971), 408–409.

15   Eugene E. Landy, *The Underground Dictionary* (New York: Simon and Schuster, 1971), 151. See also "Glossary" in Wolf, *Voices from the Love Generation*, 280; William

and Mary Morris, *Dictionary of Word and Phrase Origins* (New York: Harper & Row, 1971), 3:217; *A Dictionary of Contemporary and Colloquial Usage* (Chicago: The English Language Institute of America, 1972), CC-22; and Robert L. Chapman, ed., *Roget's International Thesaurus*, 4th ed. (New York: Thomas Y. Crowell, 1977), 47, 522.

[16]   *Webster's New Collegiate Dictionary* (Springfield, Mass.: G. & C. Merriam, 1973), 879; Sidney I. Landau and Ronald J. Bogus, eds., *The Doubleday Roget's Thesaurus* (Garden City, N.Y.: Doubleday, 1977), 508.

[17]   Phrases from *Newsweek*, June 30, 1975, and the *New York Times*, September 11, 1976, cited by Clarence L. Barnhart, Sol Steinmetz, and Robert K. Barnhart, *The Second Barnhart Dictionary of New English* (Bronxville, N.Y.: Barnhart/Harper & Row, 1980), 360; *Time*, September 29, 1980, cited by *The Barnhart Dictionary Companion*, vol. 2, no. 2 (Summer 1983): 50. This use was defined in Stuart Berg Flexner, ed., *The Random House Dictionary of the English Language*, 2nd ed. (New York: Random House, 1987), 1482. See also "Credit Cards in Plastics," *MP* 35 (November 1957): 105.

[18]   Paul Durbin, Department of Philosophy, University of Delaware, supplied this information.

[19]   Wolfgang Schivelbusch, *The Railway Journey: Trains and Travel in the 19th Century*, trans. Anselm Hollo (New York: Urizen, 1979), 17–18. He was referring to a rapid shift among Europeans during the mid-nineteenth century from fear and distrust of rail travel to a mood of acceptance, even celebration.

[20]   Forrest Wilson, "Plastics, Past and Future," *Architecture* 77 (April 1988): 108.

[21]   J. Harry DuBois and Frederick W. John, *Plastics*, 6th ed. (New York: Van Nostrand Reinhold, 1981), v. "Volume" referred to measurement of spatial volume rather than of material density (or weight), an entirely different matter.

[22]   For an economic discussion see Eric D. Larson, Marc H. Ross, and Robert H. Williams, "Beyond the Era of Materials," *Scientific American* 254 (June 1986): 34–41.

[23]   Ralph Waldo Emerson, "The American Scholar" (1837), reprinted in *Selections from Ralph Waldo Emerson: An Organic Anthology*, ed. Stephen E. Whicher (Boston: Houghton Mifflin, 1957), 75; William James, *Pragmatism* (New York: Longmans, Green, 1907), 65, 201, 257, 243. The meaning of *plasticity* as used by James was ambiguous. He seemed unable or unwilling to clarify whether he regarded it as an effect of the fallibility and hence constant revision of human knowledge, or as an attribute of the cosmos. His opinion was in transition, moving from the more pessimistic position to the more optimistic.

[24]   Julian Hill, an associate of Wallace H. Carothers in the development of nylon during the 1930s, as quoted by Henry Allen, "Their Stocking Feat: Nylon at 50 & the Age of Plastic," *Washington Post* (January 13, 1988): D10.

[25]   Headline of an advertisement for Wemco, a laminating firm in Austin, Texas, *Third Coast* (July 1985): 49.

CHAPTER 1: CELLULOID

[1]   For trade names see Edwin E. Slosson, *Creative Chemistry* (New York: The Century Co., 1919), 132.

[2]   Hyatt told his version of the story to Leo Baekeland on March 24, 1909, to Edward Chauncey Worden in 1910, and to chemists gathered to hear him accept the Perkin Medal on January 23, 1914. See Baekeland, Diary 3, March 24, 1909, LHB; Edward Chauncey Worden, *Nitrocellulose Industry* (New York: D. Van Nostrand, 1911), 2:576–

579; and John Wesley Hyatt, "Address of Acceptance," *The Journal of Industrial and Engineering Chemistry* 6 (February 1914): 158–160. See also Charles F. Chandler, "Presentation Address," *The Journal of Industrial and Engineering Chemistry* 6 (February 1914): 156–158; "John Wesley Hyatt," *The National Cyclopaedia of American Biography* (New York: James T. White, 1904), 12:148; Philip B. McDonald, *Dictionary of American Biography* (New York: Charles Scribner's Sons, 1961), 5:447–449; Morris Kaufman, *The First Century of Plastics* (London: The Plastics Institute, 1963), 33–38; and Peter J. T. Morris, *Polymer Pioneers: A Popular History of the Science and Technology of Large Molecules* (Philadelphia: Center for History of Chemistry, 1986), 33–35. My discussion of celluloid relies throughout on Robert Friedel, *Pioneer Plastic: The Making and Selling of Celluloid* (Madison: University of Wisconsin Press, 1983), here esp. 3–6, 12–17.

[3] For the opinion that Parkes preceded Hyatt in recognizing the importance of camphor see Susan Mossman, "Parkesine and Celluloid," in *The Development of Plastics*, ed. S.T.I. Mossman and P.J.T. Morris (Cambridge: Royal Society of Chemistry, 1994), 10–25. This account should be compared with Friedel, *Pioneer Plastic*, 8–16.

[4] Friedel, *Pioneer Plastic*, 35–36.

[5] *Celluloid as a Base for Artificial Teeth* (New York: Celluloid Manufacturing Company, 1878), 3–4. For similar statements see untitled article, *New York Times* (January 7, 1880): 4.

[6] Brian Moore, *The Great Victorian Collection* (New York: Farrar, Straus and Giroux, 1975), 173, 203.

[7] On imitation in mass production see Herwin Schaefer, *Nineteenth Century Modern: The Functional Tradition in Victorian Design* (New York: Praeger, 1970), 68, 70. For general background see Reed Benhamou, "Imitation in the Decorative Arts of the Eighteenth Century," *Journal of Design History* 4 (1991): 1–13. For cultural analysis of imitation see Miles Orvell, *The Real Thing: Imitation and Authenticity in American Culture, 1880–1940* (Chapel Hill: University of North Carolina Press, 1989), esp. xv–xxvi, 31–72.

[8] A. Welby Pugin, *The True Principles of Pointed or Christian Architecture* (London: John Weale, 1841), 27, 34; John Ruskin, *The Seven Lamps of Architecture* (1849; rpt., New York: Farrar, Straus and Giroux, 1986), 38–39, 50–59 (quotation from 51); idem, *The Stones of Venice* (1853; rpt., New York: Merrill and Baker, n.d.), 3:26–31; and William Morris, "Art and the Beauty of the Earth" (lecture delivered October 13, 1881), *The Collected Works of William Morris* (London: Longmans Green, 1914), 22:169.

[9] Charles L. Eastlake, *Hints on Household Taste in Furniture, Upholstery and Other Details* (New York: Dover, 1969; reprint of 4th London ed., 1878), 27, 42–49. See also Friedel, *Pioneer Plastic*, 88–89. On the Arts and Crafts movement see T. J. Jackson Lears, *No Place of Grace: Antimodernism and the Transformation of American Culture 1880–1920* (New York: Pantheon, 1981), 59–96; Eileen Boris, *Art and Labor: Ruskin, Morris, and the Craftsman Ideal in America* (Philadelphia: Temple University Press, 1986).

[10] John Ruskin, *The True and the Beautiful in Nature, Art, Morals & Religion*, ed. Mrs. L. C. Tuthill (New York: ca. 1859), 245–246. This popular anthology of undocumented selections from Ruskin's work was frequently reprinted. The copy consulted lacked a title page.

[11] Robert Friedel, "The First Plastic," *American Heritage of Invention and Technology* 3 (Summer 1987): 19. This comment departed from *Pioneer Plastic*, in which Friedel focused more on imitative celluloid as the source of an image of plastic as cheap and shoddy (see especially xv, xix, 88–89). Friedel attributed his changed perspective to a

recognition of the artistry and inventiveness displayed in a collection of seven thousand celluloid artifacts that came to his attention.

[12]   Celluloid advertising brochure bound as *Lady on Sleigh* (title derived from its first illustration), ca. 1890, 25–26, CCR, box 1, folder 2.

[13]   "Relics and Razors," *Du Pont Magazine* 14 (April 1921): 6. This description quoted and paraphrased an earlier, more complete characterization in *Salesmen's Instructions: The Manufacture of Pyroxylin Plastics: P-1* (Wilmington, Del.: E. I. du Pont de Nemours & Co., October 1919), 4.

[14]   "Practical Results in the Preparation of Plastic Material," *Scientific American Supplement* 62 (November 10, 1906): 25, 790; Worden, *Nitrocellulose Industry*, 2:567.

[15]   Robert Kennedy Duncan, "The Wonders of Cellulose," *Harper's Monthly Magazine* 113 (September 1906): 573. On Duncan's career see Edward R. Weidlein and William A. Hamor, *Glances at Industrial Research During Walks and Talks in Mellon Institute* (New York: Reinhold, 1936), 46–47.

[16]   The value of celluloid produced in the U.S. in 1914 was $8.9 million, representing an increase of nearly 500 percent over the $1.5 million production of 1899. The increase in value of annual production from 1899 to 1904 was a modest 40 percent, compared to increases of 171 percent from 1904 to 1909 and 56 percent from 1909 to 1914. See Table 108, "Chemicals and Allied Products," in U.S. Department of Commerce, Bureau of the Census, *Abstract of the Census of Manufactures 1914* (Washington, D.C.: Government Printing Office, 1917), 175.

[17]   John W. Hyatt, "Improvement in the Manufacture of Combs from Celluloid &c.," U.S. Patent 199,909 (February 5, 1878); *Lady on Sleigh*, CCR; *Py-ra-lin Combs* (Wilmington, Del.: E. I. du Pont de Nemours Export Co., ca. 1918), 2; George Frederick Kunz, *Ivory and the Elephant in Art, in Archaeology, and in Science* (Garden City, N.Y.: Doubleday, Page, 1916), 261, 290–291; *Salesmen's Course: Pyralin Sales Policies: P-4* (Wilmington, Del.: E. I. du Pont de Nemours & Co., 1921), 16–24; and J. A. Burckel, assistant general manager, Du Pont Plastics Department, to Charles Copeland, May 14, 1937, DPA, series II, part 2, box 74, folder 6.

[18]   On the camphor problem see "The Camphor of Commerce," *Du Pont Magazine* 27 (April 1933): 1–3, 15–16; Jasper E. Crane, *A Short History of The Arlington Company* (Wilmington, Del. [?]: E. I. du Pont de Nemours & Co., 1945), 10.

[19]   "Tortoise" and "Tortoiseshell," *Encyclopaedia Britannica*, 11th ed. (1911), 27:69–71; *Salesmen's Course: Pyralin Industrial Uses and Sale: P-2* (Wilmington, Del.: E. I. du Pont de Nemours & Co., 1921), 56; and "Tortoise-Shell and Shell Pyralin," *Du Pont Magazine* 14 (April 1921): 13.

[20]   On horn imitations of tortoiseshell see "Tortoiseshell," *Encyclopaedia Britannica*, 71; [Harriet E. O'Brien], *Comb Making in America* (Boston: Perry Walton, 1925), 23; and Benhamou, "Imitation in the Decorative Arts of the Eighteenth Century," 7. On scarcity of horn see "Viscoloid Gets a Good Start," *Leominster Enterprise* (December 20, 1901): 1 (typed copy in Leominster Public Library); O'Brien, *Comb Making in America*, 114; and Wallace W. Cameron, "But One Horn Company Remains in Leominster," *Worcester Sunday Telegram* (February 3, 1935), clipping, Leominster Public Library.

[21]   *Salesmen's Instructions: The Manufacture of Pyroxylin Plastics*, 6; Friedel, *Pioneer Plastic*, 32 (in general see 29–34). On import volume and prices see Kunz, *Ivory and the Elephant*, 448.

[22]   For 1879 production figures see Friedel, *Pioneer Plastic*, 52. Approximations for

1913 come from a detailed study of the industry conducted by Du Pont prior to its purchase of the Arlington Company (producer of Pyralin). See Development Department to Executive Committee, "Progress Report on Celluloid Investigation," February 16, 1914, reprinted in *United States v. E. I. du Pont de Nemours and Company, General Motors Corporation, et al.*, Civil Action No. 49 C-1071, Printed Documents, vols. 1–2 (1909–1917), 88–89, 106.

23 *Ivory Py-ra-lin Toilet Ware de Luxe* (New York: The Arlington Co., 1917), 3; *Fiberloid Toilet Articles* (Indian Orchard, Mass.: The Fiberloid Corp., 1923), 34.

24 Edwin E. Slosson, "Plastics and Elastics," *The Independent* 92 (December 22, 1917): 558.

25 *Ivory Pyralin Toiletware* (Wilmington, Del. [?]: E. I. du Pont de Nemours & Co., 1920), 3.

26 On competitors see Crane, *A Short History of The Arlington Company*, 3–6; Friedel, *Pioneer Plastic*, 68–69. On the origins of Viscoloid see Bernard W. Doyle, "Doyle Works of the Du Pont Viscoloid Company," *Leominster Daily Enterprise* (December 5, 1935), typed copy, Leominster Historical Society; "Bernard Wendell Doyle," *The National Cyclopaedia of American Biography* (New York: James T. White, 1946), G:268–269. On the history of the Arlington Company see Arlington Company, bound minute books, board of directors and stockholders meetings (1886–1917), DPA, series II, part 1, boxes 79, 80, 81; Worden, *Nitrocellulose Industry*, 2:580; "The Arlington Plant," *Du Pont Magazine* 9 (August 1918): 8–9; *Salesmen's Instructions: History of the Du Pont Company*, 2nd ed. (Wilmington, Del.: E. I. du Pont de Nemours & Co., 1920), 21–23; W. E. Price, "Up the Trail with Pyralin," *Du Pont Magazine* 24 (January 1930): 8–9; Jasper E. Crane, "Talk at Dinner of Twenty-Five Year Club, Arlington Works," June 3, 1943, DPA, series II, part 2, box 1037; and Crane, *A Short History of The Arlington Company*.

27 Unless otherwise noted, description of celluloid manufacture is based on Worden, *Nitrocellulose Industry*, 1:107–114; idem, *Nitrocellulose Industry*, 2:582–614, 668–670; Masselon, Roberts, and Cillard, *Celluloid: Its Manufacture, Applications and Substitutes*, trans. Herbert H. Hodgson (London: Charles Griffin, 1912); O. W. Pickering of Du Pont Experimental Station, report on "Celluloid: Description of Processes Used in Experimental Plant," October 21, 1915, and report on "Description of Experimental Machinery and Equipment at the Experimental Station for the Manufacture of Celluloid," November 2, 1915, both DPA, accession 1784, box 5; *Salesmen's Instructions: The Manufacture of Pyroxylin Plastics*, 9–31; Sumner Kean, *Mold of Fortune: Lionel B. Kavanagh and the First Half Century of Plastics* (Leominster: The Standard Tool Co., 1959), 12; John Merriam, *Pioneering in Plastics* (Ipswich: East Anglian Magazine, 1976), 9–11, 77–85; and Friedel, *Pioneer Plastic*, 17–18. Pickering's reports are especially useful because they were intended for a technical audience unfamiliar with celluloid production.

28 Crane, "Talk at Dinner of Twenty-Five Year Club." Three German chemists, W. Schmidt, Hermann Goetter, and Robert C. Schüpphaus, worked for Arlington during the 1890s. The Cellonite Manufacturing Company was organized in May 1886, and plans were laid to erect a factory at Arlington, New Jersey; a suit pressed by Celluloid for infringement of trade name led in September 1887 to adoption of the name Arlington Manufacturing Company (shortened to the Arlington Company in February 1899 when the company was reorganized). This chronology of events, which conflicts with that of prior accounts, even Worden's of 1911, is taken directly from the Arlington Company directors minutes, DPA, series II, part 1, boxes 79, 80.

29   See "Camphors," *Encyclopaedia Britannica*, 11th ed. (1910) 5: 135–136; Worden, *Nitrocellulose Industry*, 1:235–248; Arlington Company directors minutes, December 18, 1913, and February 19, 1914, DPA, series II, part 1, box 80; "Progress Report on Celluloid Investigation," February 16, 1914, *United States v. E. I. du Pont de Nemours*, Printed Documents, 87; Francis A. Gudger, "Camphor," *Du Pont Magazine* 9 (July 1918): 4–7; "The Camphor of Commerce," 1–3, 15–16; Crane, *A Short History of The Arlington Company*, 10; Williams Haynes, *American Chemical Industry: The World War I Period: 1912–1922* (New York: D. Van Nostrand, 1945), 2:270–272; and Merriam, *Pioneering in Plastics*, 75.

30   *The Autobiography of an Ivory Py-ra-lin Brush* (Wilmington, Del. [?]: E. I. du Pont de Nemours & Co., 1917), 3.

31   Uncertainty of supply became acute during the Russo-Japanese War of 1904–05, when the Arlington Company eliminated two cheap brands of celluloid collars and in agreement with Celluloid and Fiberloid raised and fixed prices of more expensive brands. Three years later Arlington built a plant to recover camphor from collar and cuff scraps. See Arlington Company directors minutes, March 17, April 19, and May 19, 1904, and stockholders minutes, February 20, 1908, DPA, series II, part 1, box 80.

32   As late as 1935, however, Du Pont bought at least 600,000 pounds of natural camphor from Mitsui & Company and an equal amount of synthetic camphor from the Schering Company of Germany; the following year Du Pont's own manufacturing capacity for synthetic camphor reached 2.5 million pounds. See Du Pont Viscoloid Co. board minutes, September 24, 1935, DPA, series II, part 4, box 141; Williams Haynes, *American Chemical Industry: Decade of New Products* (New York: D. Van Nostrand, 1954), 5: 230–231.

33   Pickering, "Celluloid: Description of Processes Used in Experimental Plant," 9–10, DPA, accession 1784, box 5. The "wet" method of Parkes thus ironically replaced Hyatt's "dry" method.

34   *Salesmen's Instructions: The Manufacture of Pyroxylin Plastics*, 14.

35   Pickering, "Celluloid: Description of Processes Used in Experimental Plant," 12, DPA, accession 1784, box 5.

36   "Tortoise-Shell and Shell Pyralin," 13.

37   Pickering, "Celluloid: Description of Processes Used in Experimental Plant," 15, DPA, accession 1784, box 5.

38   *Salesmen's Instructions: The Manufacture of Pyroxylin Plastics*, 21.

39   John Hix, "Strange As It Seems" (McNaught Syndicate), reprinted by *Du Pont Magazine* 31 (June 1937): 21.

40   Worden, *Nitrocellulose Industry*, 2:682–683. See also *Salesmen's Instructions: The Manufacture of Pyroxylin Plastics*, 29–30.

41   According to a randomly surviving "Report of Operations, The Arlington Co., May 1917," DPA, accession 1410, box 58. In 1917 the company employed 1,300 workers, including those at a paper mill, the Florida camphor plantation, and a comb plant at Poughkeepsie. Roughly one in ten workers was injured in a random month.

42   Arlington Company stockholders minutes, June 22, 1888, May 19, 1893, and board minutes, May 3, 1898, December 14, 1899, DPA, series II, part 1, boxes 79, 80; Crane, *A Short History of The Arlington Company*, 4, 11.

43   Crane, "Talk at Dinner of Twenty-Five Year Club," DPA, series II, part 2, box 1037; *Salesmen's Course: Pyralin Sales Policies*, 16–24; and "Celluloid Visor Blazed," *New York World*, December 10, 1902.

44   On fabricating methods see Masselon, *Celluloid*, 199ff.; *Salesmen's Course: Pyralin Articles: P-3* (Wilmington, Del.: E. I. du Pont de Nemours & Co., 1921), 4–11; and Merriam, *Pioneering in Plastics*, 88–92ff. I also benefited from a discussion with Ed Hachey, Leominster Historical Society, October 8, 1986.

45   Untitled article, *New York Times* (January 7, 1880): 4; *Challenge Cleanable Collars* (Wilmington, Del.: E. I. Du Pont de Nemours & Co., 1918), Hagley Imprints Collection; George H. Boehmer, *Celluloid—Grand-Daddy of 'Em All* (Celluloid Corp., 1935), CCR, box 1, folder 1; and Worden, *Nitrocellulose Industry*, 2:700.

46   The best accounts of manufacture of imitation linen are Masselon, *Celluloid*, 259–265; Worden, *Nitrocellulose Industry*, 2: 699–711. See also Friedel, *Pioneer Plastic*, 78–84.

47   "A Successful Toy Industry," *Du Pont Magazine* 13 (November 1920): 8–9.

48   "Py-ra-lin," *Du Pont Magazine* 8 (January 1918): 7. As part of the same strategy Du Pont obtained controlling interest in General Motors. As John J. Raskob argued in 1917, "Our interest in General Motors will undoubtedly secure for us the entire Fabrikoid [artificial leather], Pyralin, paint and varnish business of these companies." As quoted by Alfred D. Chandler, Jr., and Stephen Salsbury, with Adeline Cook Strange, *Pierre S. du Pont and the Making of the Modern Corporation* (New York: Harper & Row, 1971), 454; see also 247–250, 382–386. According to a company-sponsored history by William S. Dutton, *Du Pont: One Hundred and Forty Years* (New York: Charles Scribner's Sons, 1949), 262, between 1913 and 1939 Du Pont's assets increased from $75 million to $850 million while its income went from being 97 percent derived from explosives to being 90 percent derived from non-explosives. See also Alfred D. Chandler, Jr., *Strategy and Structure: Chapters in the History of the American Industrial Enterprise* (Cambridge: MIT Press, 1962), 78–86; David A. Hounshell and John Kenly Smith, Jr., *Science and Corporate Strategy: Du Pont R & D, 1902–1980* (Cambridge: Cambridge University Press, 1988), 56–75.

49   *Du Pont Pyralin: Its Manufacture and Use* (New York: Du Pont Viscoloid Co., [1928]), 3–4, 12–13, Hagley Imprint Collection.

50   *Du Pont Toiletware: Lucite & Pyralin* (New York: Du Pont Viscoloid Co., 1928), brochure in the Leominster Historical Society; *Your Reference Guide of Du Pont Accessories for the Boudoir* (New York: Du Pont Viscoloid Co., [ca. 1934]), Hagley Imprint Collection. For an account of the styling of the initial Lucite offering see Clarence F. Brown, "Why and How Du Pont 'Lucite' Was Created," February 11, 1928, typescript, Hagley Pictorials Collection.

51   *Py-ra-lin Transparent Sheeting* (New York: Arlington Works, E. I. du Pont de Nemours & Co., 1918; photographic image copyrighted 1912), advertising card, Hagley Imprint Collection (vertical file, flat box).

52   "Taking the Peril Out of Glass," *Du Pont Magazine* 15 (July-August 1921): 6–7, 9; "Duplate—A New Improved Non-Shatterable Glass," *Du Pont Magazine* 22 (September 1928): 1–3; Gordon M. Kline, "History of Plastics and Their Uses in the Automotive Industry," *MP* 17 (July 1940): 49–51; and, on the technical history of safety glass, P. Willard Crane, 1968, PPA Tapes, reel 2, side 2.

53   Walter O. Locke, production manager, Ryan Airlines, Inc., San Diego, to Du Pont Viscoloid Co., San Francisco office, June 16, 1927, DPA, series II, part 4, box 146; *What Next Will Be Made of Pyralin?*, leaflet, ca. 1925, Hagley Imprint Collection.

54   "Colorful Bathroom Furnishings," *Du Pont Magazine* 23 (October 1929): 17; "How It Started," *Plastics* 2 (August 1926): 268.

55   Friedel, *Pioneer Plastic*, 95–96; in general 90–96. See also Reese V. Jenkins, *Images and Enterprise: Technology and the American Photographic Industry 1839–1925* (Baltimore: Johns Hopkins University Press, 1975), 122–133, 288–290, *passim*.

56   J. O. Reinecke, "Design Dates Your Product," *MP* 15 (October 1937): 123.

57   Leonard Cohen, "Queen Victoria and Me," *Selected Poems 1956–1968* (New York: Viking, 1968), 143.

58   Development Department to Executive Committee, "Progress Report on Celluloid Investigation," February 16, 1914, 108; R.R.M. Carpenter to Executive Committee, "Purchase of Arlington Company," September 12, 1915, 215; both in *United States v. E. I. du Pont de Nemours and Company, General Motors Corporation, et al.*, Civil Action No. 49 C-1071, Printed Documents, 1909–1917, vols. 1–2.

59   Photographs of Charles Sheehan and son John (with captions) are in the Hagley Pictorial Collection.

60   For examples of these designations see Wallace W. Cameron, "But One Horn Company Remains in Leominster," *Worcester Sunday Telegram* (February 3, 1935); Davis Bushnell, "Pots from the 'Plastic City,' " *Boston Globe* (May 30, 1983); both clippings in the Leominster Public Library. Friedel discusses the transformation of Leominster in *Pioneer Plastic*, 71–75.

61   "Du Pont Will Close New Jersey Plant," *Wilmington Journal* (June 5, 1957): 1, 4, in bound volume *Du Pont Plant Histories*, Hagley Imprint Collection; Du Pont Public Relations Department, "Press Analysis for November 19, 1958," DPA, series II, part 2, box 910; interview with Keith Lauer, Leominster, October 7, 1986; and Friedel, "The First Plastic," 22.

CHAPTER 2: BAKELITE

1   Quotations are from *The Material of a Thousand Uses* (New York: Bakelite Corporation, 1926), title page and 3, DuBois, box 1.

2   *Fact Versus Fancy* (New York: Bakelite Corporation, 1924), pamphlet, DuBois, box 1; Leo H. Baekeland, U.S. Patent 957,137 (May 3, 1910); and Allan Brown, "Bakelite—What It Is," *Plastics* 1 (October 1925): 17.

3   Diary 16, October 8, 1914, LHB; *Time* 4 (September 22, 1924): cover; "Father of Plastics," *Time* 35 (May 20, 1940): 50; Walter Clemons, "Macabre Americana," *Newsweek* (August 5, 1985): 66B; and Robert Friedel, "The Plastics Man," *Science 84* 5 (November 1984): 49.

4   I have benefited from Wiebe E. Bijker's analysis of the social construction of technology through differing frames of reference. My account reinforces the theoretical outlines of his brief Bakelite case study but expands the discussion of the material's invention beyond chemists to include social and cultural perspectives not defined by the scientific literature he surveyed. See Bijker, "The Social Construction of Bakelite: Toward a Theory of Invention," in *The Social Construction of Technological Systems: New Directions in the Sociology and History of Technology*, ed. Wiebe E. Bijker, Thomas P. Hughes, and Trevor J. Pinch (Cambridge: MIT Press, 1987), 159–187.

5   Baekeland pronounced his Flemish name in two syllables ("bake-land"), as his secretary G. S. Bidwell reported in a letter to Charles E. Funk of *The Literary Digest* (April 22, 1935), LHB, division I, box 5, folder I-88. Unless otherwise noted, biographical information is from "Leo Hendrik Baekeland," *The National Cyclopaedia of American Biography* (New York: James T. White, 1945), 32:20–21; Charles F. Kettering, "Bio-

graphical Memoir of Leo Hendrik Baekeland," *National Academy of Sciences Biographical Memoirs* 24 (1947): 281–302; Aaron J. Ihde, "Leo Hendrik Baekeland," *The Dictionary of American Biography*, supplement 3 (New York: Charles Scribner's Sons, 1973), 25–27; and Carl B. Kaufmann, "Grand Duke, Wizard, and Bohemian: A Biographical Profile of Leo Hendrik Baekeland (1863–1944)," M.A. thesis, University of Delaware, 1968. The *DAB* sketch is flawed by errors of fact and interpretation. Only Kaufmann benefited from access to Baekeland's diaries, sixty-three volumes covering the years 1907 to 1941, discovered in a private household in Scarsdale, N.Y., in 1966 and restored to the family (according to a letter from son George Baekeland's secretary Gertrude B. Gould to Carl B. Kaufmann, January 20, 1966, LHB, division V, box 10, folder V-6). By 1994, forty-seven diaries (into 1930) had been deposited with the National Museum of American History, with the rest to follow.

6   According to grandson Brooks Baekeland as transcribed by Natalie Robins and Steven M. L. Aronson, *Savage Grace* (1985; rpt., New York: Dell, 1986), 71. The book, an example of the true-life crime genre profiling a psychopathic great-grandson, contains useful interviews with and quotations from relatives and acquaintances of Leo Baekeland, interspersed among pp. 56–120. The authors imply by means of diary quotations taken out of context that the chemist lacked human feeling and warped two generations of descendants; the diaries reveal otherwise.

7   Reese V. Jenkins, *Images and Enterprise: Technology and the American Photographic Industry 1839 to 1925* (Baltimore: Johns Hopkins University Press, 1975), 124–125, 131–132.

8   L. H. Baekeland, "Address of Acceptance of Perkin Medal," *Metallurgical and Chemical Engineering* 14 (February 1, 1916): 153.

9   Baekeland's letterhead in a photograph of a letter to M. Mayer of the Camillus Cutlery Co., October 16, 1908, DuBois, box 7; Diary 1, July 22, 1907, LHB; and Baekeland, "Address of Acceptance of Perkin Medal," 156. Williams Haynes claimed he heard the million-dollar figure directly from Baekeland. More conservative accounts place the figure at $750,000; all ignore the fact that Jacobi presumably received a share. See Haynes, "Leo Hendrik Baekeland," in *Great Chemists*, ed. Eduard Farber (New York: Interscience, 1961), 1186.

10   See Charles F. Chandler, "Presentation Address," *Metallurgical and Chemical Engineering* 14 (February 1, 1916): 149; Baekeland, "Address of Acceptance of Perkin Medal," 154; "Some Facts Concerning the Late Clinton Paul Townsend," 1931, typescript, LHB, division I, box 7, folder I-127; Williams Haynes, *American Chemical Industry: Background and Beginnings* (New York: D. Van Nostrand, 1954), 1:278–279; idem, *American Chemical Industry: The Chemical Companies* (New York: D. Van Nostrand, 1949), 6:210–211; and Martha Moore Trescott, *The Rise of the American Electrochemicals Industry, 1880–1910: Studies in the American Technological Environment* (Westport, Conn.: Greenwood Press, 1981), 81–82. On Baekeland's professional milieu see 226–311.

11   Diary 12, March 20, 1913, LHB; May Sarton on Celine Baekeland as quoted by Robins and Aronson, *Savage Grace*, 74; and Diary 6, May 29, 1910, LHB. Baekeland devoted one of his longest diary entries (ten pages) to a dinner conversation with Orville Wright (Diary 17, April 7, 1915).

12   See especially sections of the following articles by Baekeland, all published in *The Journal of Industrial and Engineering Chemistry*: "The Synthesis, Constitution, and Uses of Bakelite," 1 (March 1909): 150–155; "On Soluble, Fusible, Resinous Condensation

Products of Phenols and Formaldehyde," 1 (August 1909): 545–547; "The Chemical Constitution of Resinous Phenolic Condensation Products," 5 (June 1913): 506–507; and (with H. L. Bender) "Phenol Resins and Resinoids," 17 (March 1925): 225–237. For a somewhat jumbled historical overview whose virtue lies in its neutrality and completeness, see Carleton Ellis, *Synthetic Resins and Their Plastics* (New York: The Chemical Catalog Co., 1923), 75–92.

[13]   Baekeland noted the high cost of shellac in "The Chemical Constitution of Resinous Phenolic Condensation Products," 508. An otherwise comprehensive work on shellac, Ernest J. Parry's *Shellac: Its Production, Manufacture, Chemistry, Analysis, Commerce and Uses* (London: Sir Isaac Pitman & Sons, 1935), provides no production or sales figures.

[14]   A phrase from the folklore of American chemistry cited by Patrick D. Ritchie, *A Chemistry of Plastics and High Polymers* (London: Cleaver-Hume, 1949), 94. On efforts in Great Britain, where James Swinburne organized the Damard ["Damn Hard"] Lacquer Company to produce phenolic lacquers and coatings, see Percy Reboul, "Britain and the Bakelite Revolution," in *The Development of Plastics*, ed. S.T.I. Mossman and P.J.T. Morris (Cambridge: Royal Society of Chemistry, 1994), 26–37.

[15]   The term is from Bijker, "The Social Construction of Bakelite"; I have extended its significance.

[16]   L. H. Baekeland, "The Invention of Celluloid," *Journal of the Society of Chemical Industry* 32 (June 16, 1913): 90; "Silver Anniversary Number," *Bakelite Review*, vol. 7, no. 3 (1935), 11.

[17]   Thurlow to Baekeland, November 30, 1904, LHB, division II, box 7, folder II-1; U.S. Patent 698,761 (April 29, 1902). Biographers seem to agree that it took five years to achieve Bakelite but are unclear whether that five-year period ended with its discovery in 1907 or its public announcement in 1909. Kaufmann, usually most accurate, stated in 1968 that Baekeland began in 1902 (87). Kettering, offering no date, stated in 1947 that the project took five years (287). Haynes, who knew Baekeland, stated in 1961 that the work began five years before the 1909 announcement—that is, in 1904 (1189). To complicate matters, the "Silver Anniversary Number" of the *Bakelite Review* stated in 1935 that Baekeland carried out the project between 1905 and 1909. 1904 or 1905 seems a more likely starting date than 1902 for two reasons: because Hooker's pilot plant was well under way by 1904 and because Thurlow, whose hiring was discussed in November 1904, was crucial to the project.

[18]   Baekeland, "The Synthesis, Constitution, and Uses of Bakelite," 151.

[19]   Diary 1, June 18-19-20-21 [sic], 1907; Diary 2, March 18, 1908; both LHB.

[20]   "BKL I" (Bakelite Lab Notebook I), June 20, 1907, 12–13, LHB, series I, box 2; Baekeland, "The Synthesis, Constitution, and Uses of Bakelite," 155–156.

[21]   U.S. Patent 949,671 (February 15, 1910), filed February 18, 1907, original application filed as docket 358,156 at National Archives Center, Suitland, Maryland; H. Lebach, "Bakelite and Its Applications," *Journal of the Society of Chemical Industry* 32 (June 16, 1913): 561, 559. See also Parry, *Shellac*, 152–154; Philip Schidrowitz, *Rubber* (London: Methuen, 1911), 183–184, 197–205.

[22]   "BKL I," 1, LHB, series I, box 2. See also U.S. Patent 949,671 (February 15, 1910), filed February 18, 1907.

[23]   "BKL I," June 20, 1907, 16–18, LHB, series I, box 20.

[24]   "BKL I," June 20, 1907, 32–33.

25   Diary 1, July 11, 18–20, 23, 1907, LHB.

26   Baekeland, "The Chemical Constitution of Resinous Phenolic Condensation Products," 507, 509.

27   Diary 2, March 18, 1908, LHB.

28   The chemical patents were U.S. Patent 942,700 (December 7, 1909) and U.S. Patent 942,809 (December 7, 1909). The application patents were issued as U.S. Patent 954,666 (April 12, 1910), U.S. Patent 942,808 (December 7, 1909), and U.S. Patent 1,213,726 (January 23, 1917).

29   Eventually issued as U.S. Patent 1,054,265 (February 25, 1913), U.S. Patent 939,966 (November 16, 1909), and U.S. Patent 941,605 (November 30, 1909).

30   Baekeland, "Address of Acceptance of Perkin Medal," 155.

31   Diary 1, August 1, 1907, LHB; Aaron J. Ihde, "Charles Frederick Burgess," in *American Chemists and Chemical Engineers*, ed. Wyndham D. Miles (Washington, D.C.: American Chemical Society, 1976), 54–55.

32   Diary 1, October 17–19, 24, 1907, LHB; George P. Meade, "Ferdinand Gerhard Wiechmann," in *American Chemists and Chemical Engineers*, 509–510; and Haynes, *American Chemical Industry*, 6:66–68.

33   Diary 1, November 15, 8, 1907, LHB.

34   Diary 1, January 15, 1908; Diary 2, February 10, March 1, 1908; LHB.

35   General information on Boonton comes from Richard W. Seabury, "Historical Report of Boonton Rubber Manufacturing Company," April 15, 1916, typescript, DuBois, box 3; Richard W. Seabury, "Historical Report of Boonton Rubber Manufacturing Company," March 27, 1923, typescript, DuBois, box 3; "The Pioneer of Phenolic Molding," *SPI Voice* (December 11, 1940): 6–7, SPIA; George K. Scribner interview, 1952, PPA Tapes, reel 2, side 1; and an account by Richard W. Seabury quoted verbatim in J. Harry DuBois, *Plastics History U.S.A.* (Boston: Cahners, 1972), 82, 85.

36   Diary 2, February 4, 1908; see also Diary 1, January 7, 11, 1908; Diary 2, January 31, 1908; all LHB.

37   The percentage is based on a comparison of charts from "Actual Sales Shipments," 1921, and Seabury, "Historical Report of Boonton Rubber Manufacturing Company," March 27, 1923, both DuBois, box 3. See also Diary 2, March 28, June 6, August 5, 1908; Diary 3, December 5, 1908; all LHB.

38   Lawrence V. Redman and A.V.H. Mory, "The Bakelite Corporation," *Industrial and Engineering Chemistry* 23 (May 1931): 595. Descriptions of this molding differ slightly. See E. F. Weston, president, Weston Electrical Instrument Corporation, to Allan Brown, advertising manager, Bakelite Corporation, May 20, 1935, DuBois, box 3; "The Pioneer of Phenolic Molding," 6; George K. Scribner interview, 1952, PPA Tapes, reel 2, side 1; Richard W. Seabury, "Note," November 12, 1963, DuBois, box 3; and DuBois, *Plastics History U.S.A.*, 85–86, 156.

39   *Boonton Bakelite: Moulded Insulation* (Boonton, N.J.: Boonton Rubber Co., October 1911), DuBois, box 3; Scribner interview, PPA Tapes, reel 2, side 1.

40   Diary 8, March 9, 1911, LHB.

41   Testimony of Joseph Rockhill, *General Bakelite Company v. General Insulate Company* (1919), 2:940, LHB, division VIII, box 15.

42   Baekeland to Ferdinand G. Wiechmann, April 20, 1910, LHB, division II, box 8, folder II-9.

43   Baekeland, "Address of Acceptance of Perkin Medal," 155.

[44] On these changes see Diary 2, March 18, April 19, May 6, 12, 14, 18, 1908; Diary 3, December 7, 10, 1908; LHB. The autoclave mentioned may be the "Bakelizer" long known in company lore as "Old Faithful," which is in the possession of the National Museum of American History and was designated a National Historic Chemical Landmark by the American Chemical Society on November 9, 1993.

[45] Baekeland to M. Mayer, Camillus Cutlery Co., October 16, 1908, photograph of letter, DuBois, box 7.

[46] Unless otherwise noted, the source is Diary 3, February 4, 5, 1909, LHB.

[47] Lawrence P. Eblin, "Charles Baskerville," and Wyndham D. Miles, "Wilder Dwight Bancroft," both in *American Chemists and Chemical Engineers*, 25–26, 22–23.

[48] Although Baekeland extemporized, I must rely on the published version: Baekeland, "The Synthesis, Constitution, and Uses of Bakelite," 149–161; quotations from 155–156.

[49] "New Chemical Substance, *New York Times* (February 6, 1909): 4; *New York Daily Tribune* (February 6, 1909): 2.

[50] Diary 3, February 8, 10, 1909, LHB.

[51] Diary 3, February 24, March 18, 24, 1909, LHB; Baekeland to Ferdinand G. Wiechmann, September 2, 1909, LHB, division II, box 7, folder II-4; and "Hylton Swan," *SPI Bulletin*, no. 788 (December 28, 1960): 3, SPIA.

[52] Baekeland, "Address of Acceptance of Perkin Medal," 155.

[53] "Orders" ledger, February 12, 1908, to March 15, 1911, LHB. See also Diary 3, March 22, 19, 1909; Diary 5, August 31, 1909; Diary 6, April 11, 18, May 11, 1910; LHB.

[54] Diary 5, September 20, 1909, January 27, 1910, LHB; Baekeland to Harry T. Peters, April 23, 1926, LHB, division I, box 5, folder I-109.

[55] Diary 4, June 11, 1909, LHB. See also Baekeland to Wiechmann, March 22, 1909, LHB, division II, box 7, folder II-2; Diary 4, April 28, 1909, LHB; Baekeland to Wiechmann, May 5, 1909, LHB, division II, box 7, folder II-3; Baekeland to Wiechmann, September 2, 1909, LHB, division II, box 7, folder II-4; and Baekeland to Wiechmann, January 20–21, 1910, LHB, division II, box 8, folder II-7. On vegetable ivory see George Frederick Kunz, *Ivory and the Elephant in Art, in Archaeology, and in Science* (Garden City, N.Y.: Doubleday, Page, 1916), 279–281, 286; George C. Williamson, *The Book of Ivory* (London: Frederick Muller, 1938), 16.

[56] Diary 6, April 1, 1910, LHB. See also Protal Co. press release, ca. January 1910; Wiechmann to board of directors, Protal Co., January 3, 1910; both LHB, division II, box 8, folder II-7; and Diary 6, March 29–30, 1910, LHB.

[57] Hylton Swan memorandum to Baekeland, April 28, 1926, LHB, division I, box 5, folder I-109; Baekeland to Mr. van Sinderen, September 15, 1910, LHB, division II, box 8, folder II-13; and Diary 7, August 8, 1910, LHB. On the Protal collapse see also Wiechmann to Baekeland, September 20, 1910, division II, box 8, folder II-13; Baekeland to Wiechmann, September 23, 1910 (marked "*not sent*"), division II, box 8, folder II-13; Baekeland to Walter Damrosch, November 25, 1910 (10 pp.), division I, box 5, folder I-109; Baekeland to Wiechmann, March 3, 1911, division I, box 5, folder I-109; Wiechmann to Baekeland, March 18, 1911, division VII, box 14, folder VII-5; and Baekeland to Harry T. Peters, April 23, 1926, division I, box 5, folder I-109; all LHB.

[58] Diary 15, June 19, 1914; Diary 26, February 13, 1919; Diary 27, July 27, 1919;

Diary 30, October 10, 1920; LHB. See also "Roessler & Hasslacher—Partners," *Industrial and Engineering Chemistry* 21 (October 1929): 989–991; "They Founded an Industry," *Du Pont Magazine* 26 (Fall 1932): 1–2, 24.

[59] Diary 8, September 24, 1910, LHB. General Bakelite became organizationally independent during the First World War owing to R & H's affiliation with German companies. See Diary 21, November 11, 13, December 14, 1916, January 3, 1917, LHB.

[60] Figures are from Baekeland's account of the incorporation (Diary 8, September 29, 1910) and from an estimate of personal worth (Diary 8, October 30, 1910). Neither entry reflects agreement with two stock plans proposed in Baekeland's typed "Project for the Formation of a Bakelite Company and Possible Sources of Income," 8–9, LHB, division VII, box 14, folder VII-6. Records of funds subscribed and stock issued are not available. It is possible the capital attributed to Hasslacher actually came from an informal consortium of R & H's directors. According to George Roll, one-time son-in-law and aide to Baekeland, "the Doctor" became majority stockholder in 1930 when Du Pont bought R & H; a bylaw of General Bakelite Co. required the offer of Hasslacher's stock in such an event and Baekeland bought it. Roll interview, October 10, 1976, Eklund.

[61] Diary 8, October 4, 19, 1910, LHB.

[62] *Bakelite: Information Number One* (New York: General Bakelite Co., November 1910), 3, DuBois, box 1.

[63] Redman and Mory, "The Bakelite Corporation," 595; Rupert B. Lowe interview, November 18, 1976, Eklund.

[64] Diary 9, February 11, 1911, LHB. Compare with Lawrence C. Byck, "The Bakelite Story and Dr. Baekeland," typescript, 23–24, LHB, division VI, box 12, folder VI-C-6.

[65] L. C. Byck, "A Survey of the Bakelite Thermosetting Business 1910 Through 1951," November 3, 1952, typescript, 9, LHB, division VI, box 12, folder VI-B-6.

[66] Seabury, Laboratory Notebook #1, from August 30, 1911, to May 6, 1919, DuBois, box 3.

[67] Diary 11, April 13, 19, 1912, LHB.

[68] Baekeland to Remy Electric Co., April 20, 1912, DuBois, box 3.

[69] Charles Brock, president of Boonton Rubber Manufacturing Co., to William Hull Wickham, September 26, 1912; Seabury to Baekeland, August 31, 1912; both DuBois, box 3.

[70] As reported by Brock in his letter of September 26, 1912. For Baekeland's account of their meeting see Diary 11, September 25, 1912, LHB.

[71] Diary 12, January 17, 1913, LHB.

[72] Seabury to Baekeland, August 31, 1912, and Brock to Wickham, September 26, 1912, both DuBois, box 3.

[73] Unless otherwise noted, information about Condensite comes from Redman and Mory, "The Bakelite Corporation," 595; Byck, "The Bakelite Story and Dr. Baekeland," 31–35; "Kirk Brown," *SPI Bulletin*, no. 313 (February 18, 1949): 6, SPIA; Williams Haynes, *American Chemical Industry: The World War I Period: 1912–1922* (New York: D. Van Nostrand, 1945), 3:380; Allan Brown interview, December 26, 1974 (tape courtesy of William T. Cruse); and Rupert B. Lowe interview, November 18, 1976, Eklund.

[74] U.S. Patent 1,033,044 (July 16, 1912); U.S. Patent 1,146,384 (July 13, 1915); and U.S. Patent 1,146,385 (July 13, 1915); all filed May 1, 1909.

75   U.S. Patent 1,020,594 (March 19, 1912), filed May 14, 1909. Compare with U.S. Patent 1,020,593 (March 19, 1912), filed February 11, 1910. See also Ellis, *Synthetic Resins and Their Plastics*, 115–116.

76   Quotations from Diary 8, November 17, 1910; Diary 8, March 9, 1911, LHB. See also Diary 9, January 24, March 16, June 29, July 11, 1911; Diary 11, June 12, October 21, 1912; all LHB.

77   J. W. Aylsworth, letter to the editor, *The Journal of Industrial and Engineering Chemistry* 3 (August 1911): 615–616. See also "Condensite, A New Acid- and Alkali-Proof Plastic Material," 3 (June 1911): 439; Leo H. Baekeland, "Condensation Products of Phenols and Formaldehyde," 3 (July 1911): 518–520; idem, "Recent Developments in Bakelite," 3 (December 1911): 936; and idem, "Phenol-Formaldehyde Condensation Products," 4 (October 1912): 737–739.

78   Diary 10, November 2, 1911; Diary 11, June 28, October 30, 1912; Diary 13, May 14, 1913; Diary 14, July 10, 1913; and Diary 11, October 1, 1912; LHB. Unless otherwise noted, information about Redmanol comes from *General Bakelite Company v. General Insulate Company*, 1:236–238; Ellis, *Synthetic Resins and Their Plastics*, 109–118; Lawrence V. Redman to Baekeland, May 26, 1924, LHB, division I, box 6, folder I-115; Redman and Mory, "The Bakelite Corporation," 595; Edward R. Weidlein and William A. Hamor, *Glances at Industrial Research During Walks and Talks in Mellon Institute* (New York: Reinhold, 1936), 46–47; Byck, "The Bakelite Story and Dr. Baekeland," 31–35; Haynes, *American Chemical Industry*, 3: 380–381; "Lawrence Vincent Redman," *The National Cyclopaedia of American Biography* (New York: James T. White, 1953), 38:632–633; George Roll interview, October 10, 1976, Eklund; and Rupert B. Lowe interview, November 18, 1976, Eklund.

79   U.S. Patent 1,188,014 (June 20, 1916), filed February 24, 1914.

80   U.S. Patent 1,107,703 (August 18, 1914), filed February 17, 1911.

81   Diary 14, September 11, 1913, LHB.

82   Leo H. Baekeland, "Synthetic Resins," *The Journal of Industrial and Engineering Chemistry* 6 (February 1914): 167–170.

83   L. V. Redman, A. J. Weith, and F. P. Brock, "On Synthetic Phenol Resins," *The Journal of Industrial and Engineering Chemistry* 8 (May 1916): 473–474; Leo H. Baekeland, "Synthetic Phenol Resins," *The Journal of Industrial and Engineering Chemistry* 8 (June 1916): 568–570 (esp. 569). Owing to unpaid fees Redman had not actually received a Ph.D. from the University of Toronto.

84   Baekeland, "Synthetic Phenol Resins," 569; L. V. Redman, A. J. Weith, and F. P. Brock, "Synthetic Phenol Resins," *The Journal of Industrial and Engineering Chemistry* 8 (November 1916): 1,077.

85   According to Byck, "A Survey of the Bakelite Thermosetting Business 1910 Through 1951," 9, poundage of "laminating varnish" sold by General Bakelite Co. exceeded "molding material" every year from 1911 to 1921. They approached parity in 1915, but in 1916 varnish increased by 99 percent to 1,099,336 pounds while molding material increased by only 42 percent to 630,986 pounds. Discussion of phenolic laminates is based in general on R.W.E. Moore, "Properties and Uses of Bakelite Micarta," *The* [Westinghouse] *Electric Journal* 10 (July 1913): 645–650; Ward Moore, "Big Industry Built On Vision and Few Thousand Dollars," *Cincinnati Post*, August 15, 1940; "Formica Is On Top," *Fortune* 44 (October 1951): 116–117; Formica Foremen's Business Club, "This Is Your Life: Mr. D. J. O'Conor," 75 pp. typescript, June 1, 1955 (provided by

Formica Corporation); Daniel J. O'Conor interview, February 22 [1968?], PPA Tapes, reel 2, side 2; and a pamphlet on *Historical Growth of High Pressure Decorative Laminates* (Wayne, N.J.: Formica Corporation, ca. 1980). See also essays in Susan Grant Lewin, ed., *Formica & Design: From the Counter Top to High Art* (New York: Rizzoli, 1991).

[86] Diary 1, December 9, 1907; Diary 2, January 23, 1908; LHB.

[87] See "Bakelite-Micarta-D Gears and Pinions," *Electrical Review and Western Electrician* 69 (October 14, 1916): 693–694; "New Material for Gears," *Industrial Management* 54 (October 1917): 118–121.

[88] Diary 16, November 4, 1914, LHB. See also Diary 24, April 12, 1918; Diary 25, December 2, 1918; Diary 32, September 19, 1921; LHB.

[89] Diary 14, October 20, 1913; Diary 17, April 10, 1915; Diary 14, February 2, 1914; Diary 12, January 16, 1913; Diary 19, January 3, 1916; Diary 11, July 30, 1912; all LHB; and interview with Garson Meyer of Eastman Kodak, 1952, PPA Tapes, reel 1, side 2.

[90] Diary 25, August 10, 1918, LHB. Fred C. Meacham, Northern Industrial Chemical Co., Boston, discussed wartime applications in 1952, PPA Tapes, reel 1, side 2; Sandford Brown recalled the shaving cream caps in 1952, PPA Tapes, reel 2, side 1.

[91] Diary 17, April 7, 1915, LHB. For typical presentations of Bakelite to the automotive industry see "Bakelite—A Synthetic Resin," *The Automobile* 34 (January 6, 1916): 36, 50; "Making Bakelite Parts," *The Automobile* 36 (May 24, 1917): 1018–1019; and Ray P. Jackson, "Fabricated Bakelite Materials," *The Automobile and Automotive Industries* 37 (September 27, 1917): 544–545, 563.

[92] W. D. Richardson, "Presentation Address," *The Journal of Industrial and Engineering Chemistry* 5 (June 1913): 505; Diary 25, August 26, 1918, LHB.

[93] Baekeland testimony in *General Bakelite Company v. General Insulate Company*, March 31, 1919, 1:34, LHB, division VIII, box 15.

[94] Diary 27, April 22, 1919, LHB. On the trial see the typed transcript, *General Bakelite Company v. General Insulate Company*, two volumes totaling 1,216 pp., LHB, division VIII, box 15; and Diary 27, March 31–May 2, 1919, LHB.

[95] Diary 32, July 30, August 2, 5, 9, 1921, LHB. General Bakelite issued a two-page leaflet for its customers, "Bakelite Patents Declared Valid and Infringed," August 4, 1921, LHB, division V, box 10, folder V-11. See also "Decision in Bakelite Patent Suit," *Chemical and Metallurgical Engineering* 25 (August 17, 1921): 279–280; "Bakelite Company Wins Decree," *The Journal of Industrial and Engineering Chemistry* 13 (September 1921): 769.

[96] Byck, "A Survey of the Bakelite Thermosetting Business 1910 Through 1951," 3. See also Diary 31, November 3, 1920, February 10, 1921, LHB.

[97] Diary 31, March 2, 9, April 13, May 5, 12, 18, 1921; Diary 32, August 11, 18, 1921; LHB.

[98] Diary 32, November 14, 1921, LHB. My account of the consolidation is based on diary entries running from Diary 32, November 11, 1921, through Diary 33, December 23, 1921. Karpen's maneuver entered the folklore of the Bakelite Corporation and was discussed by Byck, "The Bakelite Story and Dr. Baekeland," 31–35; George Roll interview, October 10, 1976, Eklund; and Rupert B. Lowe interview, November 18, 1976, Eklund.

[99] Diary 34, April 8, 1922; Diary 33, December 10, 1921; Diary 34, November 23, May 8, 1922; LHB.

100 Diary 33, December 23, 1921; Diary 34, May 5, 1922; LHB. See also "Combination in Condensation Products Manufacture," *Chemical Age* 30 (June 1922): 246; "Bakelite Corporation Merges Condensite and Redmanol," *Oil, Paint and Drug Reporter*, June 19, 1922, reprint, DuBois, box 3; and "A Significant Chemical Merger," *Chemical and Metallurgical Engineering*, June 21, 1922, clipping, DuBois, box 3.

101 *Moulded Insulation* (Boonton, N.J.: Boonton Rubber Manufacturing Co., ca. 1915), inside back cover, DuBois, box 3; Diary 33, February 16, 1922; Diary 34, December 9, 1922; both LHB.

102 The issue of "phenol dumping" has led to misunderstanding of the Bakelite Corporation's expansion during the 1920s. Attempting to demonstrate the uncertainties associated with any technological development whose success later wrongly appears inevitable, Trevor J. Pinch and Wiebe E. Bijker have suggested that without immense postwar surpluses of phenol, Bakelite might never have enjoyed success in the marketplace and gained a reputation as a "marvellous synthetic resin." While agreeing with their general historiographic point, I believe they have misstated this particular case. Their argument is based on Haynes, who described phenol producers as desperate after the war because they had expanded to an annual capacity of 100 million pounds (for use in explosives and antiseptics) while the federal government held a stockpile of 35 million pounds. When prices fell from a wartime high of $1.00 per pound to eight cents in 1919 (or from forty-five cents to six cents as later reported by the government), the government engaged in stabilization by placing its surplus with Monsanto Chemical Works to be sold at a fixed twelve cents a pound. Haynes stated that release of the surplus stimulated expansion in synthetic plastics—no doubt true to a certain extent, but he failed to mention the even lower prewar price of eight cents a pound. Christy Borth, a science and technology popularizer, in 1939 took the more extreme position that surplus phenol served as "the driving force of two new industries," plastics and radio. Pinch and Bijker in 1984 expressed doubt "over whether Bakelite would have acquired its prominence if it had not profited from that phenol dumping." They incorrectly stated that market conditions for synthetic plastics had worsened during the war and that General Bakelite and other phenolic producers were in serious trouble prior to the "dumping"; in fact Bakelite production and sales prices both increased each year from 1914 to 1918, and the company dealt with high prices on raw materials by taking smaller profits. Pinch and Bijker also incorrectly argued that low postwar prices enabled Bakelite to compete effectively for market share with celluloid; in fact the two plastics shared few applications and were rarely in competition. The phenol surplus no doubt enabled the Bakelite Corporation to expand its postwar profit margin, but the independent impetus of the expanding radio industry genuinely taxed the company's productive capacity in other ways, contributed to transforming its image, and in fact somewhat ironically forced it to build its own synthetic phenol plant at Painesville, Ohio, in 1923, which remained in operation as late as 1927, long after the phenol "dumping," when a second period of oversupply forced its closing. A federal study published in 1938 reported that "the increased demand for synthetic resins used up the accumulated [wartime] stocks of phenol sooner than was expected"; by 1923 production of synthetic (as opposed to "natural") phenol was in full swing. See Pinch and Bijker, "The Social Construction of Facts and Artefacts: or How the Sociology of Science and the Sociology of Technology Might Benefit Each Other," *Social Studies of Science* 14 (1984): 406 (reprinted in *The Social Construction of Technological Systems*, 24); Haynes, *American*

Chemical Industry, 2:137–138; Christy Borth, *Pioneers of Plenty: The Story of Chemurgy* (Indianapolis: Bobbs-Merrill, 1939), 197; Redman and Mory, "The Bakelite Corporation," 596; Byck, "A Survey of the Bakelite Thermosetting Business 1910 Through 1951," 10; idem, "The Bakelite Story and Dr. Baekeland," 36; and United States Tariff Commission, *Synthetic Resins and Their Raw Materials*, Report no. 131, Second Series (Washington, D.C.: Government Printing Office, 1938), 111.

[103]   Diary 34, April 10, 8, 1922, LHB; Formica salesman Edwin M. Wolcott in "This Is Your Life: Mr. D. J. O'Conor," 15; Diary 36, October 10, 1923, LHB; and "Silver Anniversary Number," *Bakelite Review* 7, no. 3 (1935): 30. For sales figures see Byck, "A Survey of the Bakelite Thermosetting Business 1910 Through 1951," 17; "Formica Is On Top," 150.

[104]   Diary 32, August 11, 1921, LHB.

[105]   Direct quotation from Conrad Schrimpe interview transcribed in research notes of Carl B. Kaufmann, ca. 1965, LHB, division IX, box 16; C. William Cleworth interview, 1968, PPA Tapes, reel 3, side 1.

[106]   *Bakelite: The Material of a Thousand Uses* (New York: Bakelite Corporation, March 1924), DuBois, box 1; Leo H. Baekeland, "Bakelite, a Condensation Product of Phenols and Formaldehyde," *The Journal of the Franklin Institute* 169 (January 1910): 58; *Keeping Pyralin Sheeting Before the Trade* (Wilmington, Del.: E. I. du Pont de Nemours & Co., June 1920), a six-page pamphlet illustrating advertisements then appearing in *Novelty News, Playthings*, and *Automotive Industries*; and *Condensite: A Phenolic Condensation Product* (Bloomfield, N.J.: Condensite Company of America, March 1923), DuBois, box 2.

[107]   A. C. Tate, "The Protective Power of a Good Trade Mark," *Plastics* 1 (November 1925): 53. To compare versions of the trademark see *Bakelite: The Material of a Thousand Uses* (March 1924) and *The Material of a Thousand Uses* (1926), both DuBois, box 1. For an expression of concern about loss of the patent monopoly see Diary 36, July 30, 1923, LHB. On patent expiration see also "Editorial Impressions: New Era in Field of Phenol Resins," *Plastics* 2 (December 1926): 442. The role of the Brown brothers in recasting the Bakelite Corporation might make a fascinating psychological study. It is hard to accept Baekeland's assertion that they despised their father's apparent machinations as much as he did. Late in life Allan Brown recalled his father Kirk with great affection (interview, December 26, 1974, tape courtesy of William T. Cruse). In working to unify the three components of the Bakelite Corporation the Brown brothers might have symbolically compensated for the rift with their father. The B in the trademark might also hide a clever double meaning, standing for "Brown" as well as "Bakelite," the three half circles representing the three Brown brothers.

[108]   *Fact Versus Fancy: Why 95% of All Radio Manufacturers Use Bakelite* (New York: Bakelite Corporation, 1924), DuBois, box 1.

[109]   *Plastics* 1 (October 1925): front cover. See also Allan Brown, "Bakelite—What It Is," 17, 28–29, in the same issue; C. William Cleworth interview, 1968, PPA Tapes, reel 3, side 1.

[110]   Allan Brown interview, December 26, 1974, tape courtesy of William T. Cruse. See also A.V.H. Mory and Leon V. Quigley, "How the Bakelite Corporation Effectively Manages Organized Research," *Manufacturing Industries* 11 (January 1926): 13–16; "Bakelite's Selling and Advertising Discussed at Tariff Hearings," *Printers' Ink* 135 (June 10, 1926): 183–184. The first of these seemingly objective articles was planted to gain business, Mory being a Bakelite employee.

111   John Allen Murphy, "The Bakelite Caravan—A New Idea in Industrial Selling," *Advertising and Selling* 8 (April 20, 1927): 23–24. See also "Promoting Molded Products Through Novel Exhibition," *Plastics and Molded Products* 3 (May 1927): 234, 244.

112   *A Romance of Industry* (New York: Bakelite Corporation, 1924), LHB, division IX, box 16. See also Charles G. Muller, "Bakelite's Round-Robin Advertising Comes Home to Roost," *Printers' Ink* 131 (May 7, 1925): 133–134, 136, 139; Allan Brown interview, December 26, 1974, tape courtesy of William T. Cruse.

113   Ralph Abercrombie, *The Renaissance of Art in American Business*, General Management Series no. 99 (New York: American Management Association, 1929), 3–6.

114   Diary 36, July 8, September 3, 12, 1923, LHB. In a letter to the author, December 12, 1986, William T. Cruse volunteered the information that Allan Brown employed Mumford. He had recently finished a similar promotional survey for the Roebling company, *Outspinning the Spider: The Story of Wire and Wire Rope* (New York: Robert L. Stillson, 1921), and soon published a volume on *Anthracite* in the "Romance of Industry Series" (New York: Industries Publishing Co., 1925).

115   Diary 37, March 28, 1924, LHB; Slosson to Baekeland, May 13, 1924, LHB, division I, box 6, folder I-125.

116   Little to Baekeland, October 25, 1924; Baekeland to Little, October 27, 1924; both LHB, division I, box 5, folder I-87. See also George W. Baker, "Arthur Dehon Little," *American Chemists and Chemical Engineers*, 300–301. For publication figures see "How Bakelite Uncovered New Uses for the Product," *Sales Management* 13 (December 24, 1927): 1118, 1160.

117   John Kimberly Mumford, *The Story of Bakelite* (New York: Robert L. Stillson, 1924), 20, 7, 46, 8, 51, 20, 22, 80, 8.

118   Ibid., 32–33.

119   *Bakelite for Radio and Why* (New York: Bakelite Corporation, February 1924), 11, DuBois, box 1.

CHAPTER 3: VISION AND REALITY

1   Paul T. Frankl, *Form and Re-form: A Practical Handbook of Modern Interiors* (New York: Harper, 1930), 163.

2   Production of pyroxylin plastics fell from 27.5 million pounds in 1923 to 18 million pounds in 1925, according to U.S. Department of Commerce, Bureau of the Census, *Biennial Census of Manufactures, 1927* (Washington, D.C.: Government Printing Office, 1930), 673.

3   For examples see "Approach of the 'Plastic Age,'" *The Literary Digest* 112 (January 2, 1932): 42; Elinor Hillyer, "The Synthetics Become the Real," *Arts and Decoration* 42 (January 1935): 28; William Engle, "This Plastic Age," six-part series, *New York World-Telegram*, December 12–17, 1938; Julian P. Leggett, "The Era of Plastics," *Popular Mechanics Magazine* 73 (May 1940): 130A; and "The Shape of Things to Come," *Design* 42 (April 1941): 26. For production statistics see U.S. Tariff Commission, *Synthetic Resins and Their Raw Materials*, Second Series, Report no. 131 (Washington, D.C.: Government Printing Office, 1938), 8; U.S. Department of Commerce, Bureau of the Census, *Biennial Census of Manufactures, 1935* (Washington, D.C.: Government Printing Office, 1938), 634; and U.S. Department of Commerce, Bureau of the Census, *Biennial Census of Manufactures: 1937*, Part I (Washington, D.C.: Government Printing Office, 1939), 657.

4   "Plastics in 1940," *Fortune* 22 (October 1940): 92–93, 88–89.

5   Nearly thirty years later Oscar Gold, who had worked for Ortho Plastic Novelties, considered the map significant enough for mention in a brief written memoir that J. Harry DuBois read into the PPA Tapes, 1968, reel 2, side 2.

6   Mention of surrealism suggests that Salvador Dalí's Dream of Venus, an underwater attraction in the Amusement Zone at the New York World's Fair of 1939, inspired "An American Dream of Venus." See Rem Koolhaas, *Delirious New York: A Retroactive Manifesto for Manhattan* (New York: Oxford University Press, 1978), 227; Terry Smith, *Making the Modern: Industry, Art, and Design in America* (Chicago: University of Chicago Press, 1993), 421.

7   V. E. Yarsley and E. G. Couzens, "The Expanding Age of Plastics," *Science Digest* 10 (December 1941): 57–59. This article in an American magazine was excerpted from the conclusion of their *Plastics* (Harmondsworth: Penguin, 1941), 149–152.

8   Henry Smith Williams, "The Miracle-Workers: Modern Science in the Industrial World," *Everybody's Magazine* 17 (October 1907): 498; Edwin E. Slosson, "Chemistry in Everyday Life," *The Mentor* 10 (April 1922): 3. Quotations on the chemical crusade are from an unpublished paper by David J. Rhees, "'Making the Nation Chemically Conscious': The Popularization of Chemistry in America, 1914–1940," History of Science Society meeting, Chicago, December 28, 1985. See also Rhees, "The Chemists' Crusade: The Rise of an Industrial Science in Modern America, 1907–1922," Ph.D. dissertation, University of Pennsylvania, 1987.

9   Slosson to Baekeland, March 12, 1923, LHB, division I, box 6, folder I-125; Slosson as quoted by David J. Rhees, "A New Voice for Science: Science Service under Edwin E. Slosson," M.A. thesis, University of North Carolina at Chapel Hill, 1979, 38–39; and E. E. Slosson, *Creative Chemistry* (New York: Century, 1919), 9–10, 3–7, 13 (Stieglitz's introduction is unpaginated). On the emergence of a definition of chemistry as an active synthetic process of remaking the world, rather than as a passive analytic process of cataloguing its matter, see Robert Friedel, "Defining Chemistry: Origins of the Heroic Chemist," in *Chemical Sciences in the Modern World*, ed. Seymour H. Mauskopf (Philadelphia: University of Pennsylvania Press, 1993), 216–233.

10   Slosson, *Creative Chemistry*, 132–135.

11   Slosson, "Chemistry in Everyday Life," 4. This slightly rewritten version from 1922 corresponds to material in *Creative Chemistry*, 61–62.

12   Slosson, *Creative Chemistry*, 135–136.

13   Slosson, "Chemistry in Everyday Life," 12.

14   John A. Craig, "Leo Hendrik Baekeland," *The World's Work* 31 (April 1916): 651, 652; "Millions for Leadership: Du Pont Chemists Keep Great Chemical Company in Van of Progress," *Barron's* 7 (January 31, 1927), rpt., Pictorial Division, Hagley Museum and Library; Floyd L. Darrow, "The New Age of Chemistry," *The Saint Nicholas Magazine* 55 (May 1928): 551; "Plastics Industry Serves Many Needs," *New York Times* (May 8, 1932): sections 11 and 12: 15; "Synthetic Age . . . Era of Make-Believe," *The Review of Reviews* 86 (November 1932): 62; John E. Lodge, "Chemistry Gives Us Amazing New Substances for Art and Industry," *Popular Science Monthly* 127 (October 1935): 20; Alden P. Armagnac, "New Feats of Chemical Wizards Remake the World We Live In," *Popular Science Monthly* 129 (July 1936): 9, 11; and "Plastics' Progress," *Business Week* (December 21, 1935): 17.

15   "Millions for Leadership"; Darrow, "The New Age of Chemistry," 552; and "Modern Plastics: A Rapidly Expanding Industry," *The Index* 16 (May 1936): 88, 94.

[16] Pauline G. Beery, *Stuff: The Story of Materials in the Service of Man* (New York: D. Appleton, 1930), vii–viii, 169, 491.

[17] C. C. Furnas, *The Next Hundred Years: The Unfinished Business of Science* (New York: Reynal & Hitchcock, 1936), 159; "What Man Has Joined Together . . . ," *Fortune* 13 (March 1936): 69; and Lewis Mumford, *Technics and Civilization* (1934; rpt., New York: Harcourt, Brace & World, 1963), 52.

[18] Furnas, *The Next Hundred Years*, 161; Beery, *Stuff*, 481; du Pont as quoted by Leonard Mosley, *Blood Relations: The Rise & Fall of the du Ponts of Delaware* (New York: Atheneum, 1980), 373; and "What Man Has Joined Together . . . ," 71.

[19] "What Man Has Joined Together . . . ," 69, 150; "And Now, In Closing," *Plastics and Molded Products* 6 (February 1930): 126; E. F. Lougee, "Editorial Comment," *MP* 14 (December 1936): 24; and idem, "The Rise of Synthetic Plastics," *Dun's Review* 47 (December 1939): 23.

[20] "Editorial Impressions," *Plastics and Molded Products* 3 (July 1927): 336.

[21] "Editorial Impressions: We Believe in Dreams!," *Plastics* 3 (February 1927).

[22] "Gordon M. Kline: Transcript of an Interview Conducted by Jeffrey L. Meikle in Lake Worth, Florida, on 15 and 16 May 1987" (Philadelphia: Beckman Center for the History of Chemistry, 1989), 49. The books were Edward Chauncey Worden, *Nitrocellulose Industry*, 2 vols. (New York: D. Van Nostrand, 1911); Emile Hemming, *Plastics and Molded Electrical Insulation* (New York: Chemical Catalog Co., 1923); and Carleton Ellis, *Synthetic Resins and Their Plastics* (New York: Chemical Catalog Co., 1923). The journals were *Plastics and Molded Products* (earlier known as *Plastics*), *Kunststoffe*, and *Revue Générale des Matières Plastiques*.

[23] Marshall McLuhan, *Understanding Media: The Extensions of Man* (1964; rpt., New York: Signet, 1966), 51–56.

[24] Letter from Mr. A. Munro, manufacturer of the Research fountain pen, October 3, 1928, as quoted by Donald Murray, "Action of Light on Coloured Bakelite," *Nature* 122 (December 1, 1928): 845.

[25] "Color in Industry," *Fortune* 1 (February 1930): 85. On color as a marketing tool see Roland Marchand, *Advertising the American Dream: Making Way for Modernity, 1920–1940* (Berkeley and Los Angeles: University of California Press, 1985), 120–127.

[26] "Trade Name Index," in *Plastics: Directory, Index & Buyers' Guide* (New York: Plastics Publications, 1929), 99–101; "Plastic Products and Producers," in Louis F. Rahm, *Plastic Molding* (New York: McGraw-Hill, 1933), 205–232; and Williams Haynes, *American Chemical Industry: The Merger Era* (New York: D. Van Nostrand, 1948), 4:348.

[27] Diary 37, May 22, 1924, LHB; *New Age Manufacturers Use Durez* (North Tonawanda, N.Y.: General Plastics, May 1931). Information on Dent and Durez is from B.D.B. to Lammot du Pont, ca. December 15, 1936, and Lammot du Pont to C. G. Shannon, December 18, 1936, both DPA, accession 1662, box 62; Haynes, *American Chemical Industry*, 4:348–349; Williams Haynes, *American Chemical Industry: The Chemical Companies* (New York: D. Van Nostrand, 1949), 6:138; Herbert S. Spencer interview, 1952, PPA Tapes, reel 1, side 2; George Roll interview, October 10, 1976, and Rupert B. Lowe interview, November 18, 1976, both Eklund; Bud Kasch interviewed by the author, July 22, 1985; and a photocopy of an unidentified brochure, ca. 1970, provided by Bruce F. Battaglia, Hooker Chemical Co.

[28] "A New Plastic Material," *Plastics and Molded Products* 4 (July 1928): 397; *The Gem of Modern Industry* (New York: American Catalin Corp., 1932). Unless otherwise

noted, information on cast phenolic comes from R. L. Kramer of Du Pont's London office to J. A. Burckel, Du Pont Viscoloid Co., DPA, series II, part 2, box 544; Ruth Lampland, "American Catalin," *MP* 12 (June 1935): 16–24; U.S. Tariff Commission, *Synthetic Resins and Their Raw Materials*, Second Series, Report no. 131 (Washington, D.C.: Government Printing Office, 1938), 20–21; Haynes, *American Chemical Industry*, 4:349; idem, *American Chemical Industry*, 6: 70–71; and R. Norris Shreve, *Selected Process Industries* (New York: McGraw-Hill, 1950), 687.

29   "Colorful New Bakelite Resinoid," *Bakelite Information*, no. 36 (November 1933).

30   "A Cloud No Bigger Than a Lady's Wrist," *MP* 12 (September 1934): 28–29, 76, 78. See also Eleanor Gordon and Jean Nerenberg, "Everywoman's Jewelry: Early Plastics and Equality in Fashion," *Journal of Popular Culture* 13 (Spring 1980): 629–644; Corinne Davidov and Ginny Redington Dawes, *The Bakelite Jewelry Book* (New York: Abbeville, 1988).

31   Ellis, *Synthetic Resins and Their Plastics*, 238, 244. See also Cyril S. Dingley, *The Story of B.I.P. (1894–1962)* (Oldbury, Birmingham: British Industrial Plastics Ltd., [1963]), 26–33; Raymond B. Seymour, "History of the Development and Growth of Thermosetting Polymers," in *History of Polymer Science and Technology*, ed. Raymond B. Seymour (New York: Marcel Dekker, 1982), 104.

32   A. C. Blackall, "Molded Dishes From New Synthetic Resin," *Plastics & Molded Products* 3 (November 1927): 606–607; "By-Product Becomes Unbreakable Tableware," *Business Week* (October 29, 1930): 15. On Kurz-Kasch's involvement see "Molded Dishes and Household Fixtures," *Plastics and Molded Products* 5 (February 1929): 95; William L. Sanders, "Dayton Industries: Kurz-Kasch First Molder of Plastic Dinnerware," *Dayton Daily News* (October 23, 1958): 14.

33   From Synthetic Plastics Co. (American Cyanamid) advertisements in *Plastic Products* 9 (September 1933): 266, and *Plastic Products* 9 (March 1933): 5.

34   From Synthetic Plastics Co. (American Cyanamid) advertisements in *Plastic Products* 9 (May 1933): 87, and *Plastic Products* 10 (January 1934): 8. On premiums see also Nathaniel Williamson, "Premiums from the Plastics Point-of-View," *Plastic Products* 10 (February 1934): 50–52; Bernard Wolfe, *Plastics: What Everyone Should Know* (Indianapolis: Bobbs-Merrill, 1945), 24–26; and Arthur J. Pulos, *American Design Ethic: A History of Industrial Design to 1940* (Cambridge: MIT Press, 1983), 337.

35   The slogan appeared in an advertisement for American Cyanamid's Beetleware Division, *MP* 13 (February 1936): 10.

36   On development of cellulose acetate see George H. Boehmer, *Celluloid—Grand-Daddy of 'Em All* (Celluloid Corporation, 1935), pamphlet, CCR, box 1, folder 1; Gordon M. Kline, "History of Plastics and Their Uses in the Automotive Industry," *MP* 17 (August 1940): 58–59; V. E. Yarsley and E. G. Couzens, *Plastics* (Harmondsworth, Middlesex: Penguin, 1941), 49–53; Haynes, *American Chemical Industry*, 3:375–376; 4:350–351, 382–383; 5:202–205, 330–332; 6:72–73, 139–140; and E. G. Couzens and V. E. Yarsley, *Plastics in the Modern World* (Baltimore: Penguin, 1968), 65–67.

37   John J. Keville, "Third Annual Salute to the Plastics Industry, January 12–February 20, 1981," scripts for broadcasts for radio station WLMS, Leominster, Massachusetts, 11; provided by John J. Keville.

38   Polyvinyl butyral in 1936 replaced cellulose acetate for the same reason. See J. M. DeBell and J. Dahle, "Safety Glass," *MP* 15 (January 1938): 37; J. Harry DuBois, *Plastics History U.S.A.* (Boston: Cahners, 1972), 284.

[39]   J. Harry DuBois interview, 1968, PPA Tapes, reel 2, side 2. On techniques of injection molding see Maurice L. Macht, Walter E. Rahm, and Harold W. Paine, "Injection Molding," *Industrial and Engineering Chemistry* 33 (May 1941): 563–567; Yarsley and Couzens, *Plastics*, 78–81; Herbert R. Simonds, Carleton Ellis, and M. H. Bigelow, *Handbook of Plastics* (New York: D. Van Nostrand, 1943), 579–584; and DuBois, *Plastics History U.S.A.*, 213–236. An accurate account of the development of injection molding is hard to come by. That of DuBois is most complete, but I have tried to collate sometimes conflicting information from the following: Du Pont Viscoloid Co., directors minutes, August 22, 1933, October 29, November 26, 1935, DPA, series II, part 4, box 141; Ernest W. Halbach, "Injection Molding and Extrusion," in *Fundamentals of Plastics*, ed. Henry M. Richardson and J. Watson Wilson (New York: McGraw-Hill, 1946), 369–389; John F. Geers, Index Machinery Corp., to Walter E. Rahm, Du Pont, February 17, 1951, DuBois, box 7; Stuart Landes of Celanese, 1952, PPA Tapes, reel 1, side 1; Tennessee Eastman, 1952, written statement read into PPA Tapes, reel 2, side 1; Joseph F. Geers to N. A. Backscheider, Plastics Pioneers Association, March 16, 1953, SPIA, miscellaneous carton; Walter F. Grote to Nicholas Backscheider, June 3, 1953, SPIA, miscellaneous carton; Walter E. Rahm, "The Historical Background and Development of Modern Injection Machines," talk delivered to the Society of Plastics Engineers, Toronto, January 28, 1954, DuBois, box 7; Joseph C. Foster, "Foster Grant Co. and Leominster, Mass.," *Plastics Industry* (March 1956): 16–18; John J. Keville, paper presented to Leominster Historical Society, May 10, 1957, read into PPA Tapes, 1968, reel 3, side 1; Sumner Kean, *Mold of Fortune: Lionel B. Kavanagh and the First Half Century of Plastics* (Leominster: Standard Tool Co., 1959), 23–27; J. Harry DuBois, recorded statement, 1968, PPA Tapes, reel 3, side 1; Walter F. Grote, 1968, PPA Tapes, reel 3, side 1; John G. Slater of Tennessee Eastman, 1968, written statement read into PPA Tapes, reel 3, side 1; Glenn D. Kittler, *"More Than Meets the Eye": The Foster Grant Story* (New York: Coronet Books, 1972), 68–81; Edward van Vlaanderen, *Pronounced Success: America and Hoechst, 1953–1978* (n.p.: American Hoechst, 1979), 140–141; and Keville, "Third Annual Salute to the Plastics Industry," 9–13.

[40]   Geers to Backscheider, March 16, 1953, SPIA.

[41]   *Molding with Lumarith* (New York: Celluloid Corp., 1936), 1–8, 11.

[42]   *Tenite: A Thermoplastic Molding Material Made from Eastman Cellulose Esters*, edition 4A (Kingsport, Tenn.: Tennessee Eastman Corp., 1940), 3, 12; Keville, "Third Annual Salute to the Plastics Industry," 11, 13. Simonds and Ellis, *Handbook of Plastics,* 10, reported 12,500 compression machines and 3,250 injection machines in 1941. For production figures and prices see Haynes, *American Chemical Industry*, 4:455–456; 5:415, 465–466, 531.

[43]   "Plastics in 1940," 92–93.

[44]   Morris Kaufman, *The Chemistry and Industrial Production of Polyvinyl Chloride: The History of PVC* (New York: Gordon and Breach, 1969), esp. 1–32, 80, 103–104, 123–124; DuBois, *Plastics History U.S.A.*, 280–288.

[45]   Quotation is from "Shower Curtains," *MP* 18 (January 1941): 40. On B. F. Goodrich and Koroseal see Haynes, *American Chemical Industry*, 6:193; Waldo L. Semon and G. Allan Stahl, "History of Vinyl Chloride Polymers," in Seymour, *History of Polymer Science and Technology*, 206–207.

[46]   "Vinyl Ester Resins Now Commercially Available," *Plastics & Molded Products* 6 (February 1930): 76, 81, 92–93; J. G. Davidson and H. B. McClure, "Applications of

Vinyl Resins," *Plastic Products* 9 (June 1933): 143–150; L. K. Merrill, "New Vinyl Molding Materials," *MP* 17 (April 1940): 64–65, 90, 92, 94; Haynes, *American Chemical Industry*, 3:167–168; idem, *American Chemical Industry*, 5:338–339; idem, *American Chemical Industry*, 6:435–437; and Joseph G. Davidson, "Petrochemical Survey: An Anecdotal Reminiscence," *Chemistry and Industry* 34 (May 19, 1956): 392–398.

47   A. E. Buchanan, Jr., "Synthetic Houses," *Scientific American* 149 (October 1933): 180. See also "Carbide's 'Age of Plastics,' " *Plastic Products* 9 (May 1933): 110; Davidson and McClure, "Applications of Vinyl Resins," 145–149; Joseph B. Singer, *Plastics in Building* (London: Architectural Press, 1952), 25–27; and Betty Pepsis, "Plastics Pictured as Building Items," *New York Times* (October 28, 1954): 44.

48   Sheldon Hochheiser, *Rohm and Haas: History of a Chemical Company* (Philadelphia: University of Pennsylvania Press, 1986), esp. 36, 54–74. See also *Synthetic Resins and Their Materials*, 35–38; Haynes, *American Chemical Industry*, 5:337–338; idem, *American Chemical Industry*, 6:357–358; *Chemicals for Industry* (Philadelphia: Rohm and Haas Co., 1959), 20–23, 67–87; and DuBois, *Plastics History U.S.A.*, 289–290.

49   Memo by W. R. Gawthrop of Du Pont Ammonia Department prepared for meeting with Fin Sparre, with cover letter from Gawthrop to Wendell R. Swint, May 9, 1933; A. F. Randolph, reports on "Methyl Methacrylate" submitted by H. W. Paine to A. F. Odell, May 7, July 18, October 23, 1934; January 14, 1935; all DPA, series II, part 2, box 544. On Du Pont and ICI see David A. Hounshell and John Kenly Smith, Jr., *Science and Corporate Society: Du Pont R&D, 1902–1980* (Cambridge: Cambridge University Press, 1988), 190–205.

50   F. A. Wardenburg of Du Pont's Ammonia Department to the Du Pont Executive Committee, March 3, 1936, DPA, series II, part 2, box 544; Hochheiser, *Rohm and Haas*, 76–78; and Hounshell and Smith, *Science and Corporate Society*, 204.

51   "New Plastic," *Du Pont Magazine* 30 (October 1936): 16; W. Hamilton Gordon, "Transparent Resins in England," *MP* 13 (June 1936): 13–15. See also a letter from Arnold E. Pitcher, general manager of Du Pont's new Plastics Department (formerly Du Pont Viscoloid Co.), to the Executive Committee, accompanying a booklet on *Pontalite: A New du Pont Plastic Material for Industrial Use*, October 5, 1936, DPA, accession 1662, box 62.

52   A. E. Pitcher, form letter addressed "To Our Trade," January 4, 1937, DPA, accession 1662, box 62.

53   W. & J. Sloane advertisement, *The New Yorker*, tearsheet, ca. 1938–1941; ellipses in original. See also *Plastics by Swedlow* (Glendale, Calif.: Swedlow, Fall-Winter 1941) (both MOMA, vertical file, "Plastics").

54   Swedlow as quoted by Ray McGlew, "The Shoot and Scoot Helicopters," *Du Pont Magazine* 62 (September-October 1968): 21. See also "Giant Retort Created in 'Lucite,' " press release from Du Pont Style News Service, September 22, 1939, DPA, accession 1410, box 49.

55   Unless otherwise noted, this paragraph is based on Donald S. Frederick, "Plastics Aloft," *MP* 15 (October 1937): 302–303, 318; "'Lucite' Cast Sheeting," anonymous Du Pont report, ca. 1945, DPA, accession 1410, boxes 29 and 30 (combined); Donald S. Frederick interview, 1968, PPA Tapes, reel 2, side 1; Donald S. Frederick interviewed by Sheldon Hochheiser, September 27, 1983, transcript, 14–18, 20, R&H; and Hochheiser, *Rohm and Haas*, 59–65.

56   On other applications see "New Plastic Exhibited at New York Electrical Society,"

MP 14 (January 1937): 42–43; H. W. Magee, "Things Are Not What They Seem," *Popular Mechanics Magazine* 68 (July 1937): 130A-131A; *Synthetic Resins and Their Uses*, 36–37; "Cold Light Surgical Instruments Developed with 'Lucite' Plastic," press release from Du Pont Public Relations Department, November 21, 1938, DPA, accession 1410, box 49; "Use of 'Lucite' Plastic Display Fixtures Increasing," press release from Du Pont Public Relations Department, March 28, 1939, DPA, accession 1410, box 49; Henry Pearson, "Piping Light with Acrylic Materials," *MP* 23 (August 1946): 123–127; and transcript of Frederick interviewed by Hochheiser, 19.

[57] On August 10, 1942, Rohm and Haas and Du Pont and eight of their executives were indicted under the Sherman Anti-Trust Act for conspiring to fix prices and control the market for acrylic resins; ICI, Röhm and Haas AG, and IG Farben were named unindicted co-conspirators. Despite acquittal on criminal charges and an eventual negotiated settlement of related civil suits, it remains an article of folklore that both companies not only sold acrylic denture compounds at forty-five dollars per pound when the going rate for acrylic molding compounds was eighty-five cents but also laced the regular compounds with arsenic to prevent their unauthorized use in dentures. In its defense Rohm and Haas argued somewhat convincingly that it sold denture material not in hundred-pound drums but, according to the terms of a licensing agreement with Du Pont, as a specially formulated compound in preformed blanks or "units," about thirteen to a pound. The company sold each unit for a dollar to two regular distributors who in turn received $2.40 from dental supply houses; an individual dentist had to pay $3.60 per unit (or a bit more than $45 a pound). Supply houses seeking to bypass the major distributors could purchase regular molding compound in bulk directly from Rohm & Haas for $1.25 a pound and formulate it themselves. To prevent this end run, and supposedly to protect consumers from manufacturers who used contaminated scrap (it was claimed that dentures containing radium had been found), one of the distributors suggested placing denture materials under the Pure Food and Drug Act and incorporating in regular molding compounds a substance (arsenic?) that would have disqualified them by FDA standards. The most telling argument mustered by Rohm and Haas in its defense was the observation that ordinary baking soda was sold in drugstores in four-ounce packages priced at 10 cents (or 40 cents a pound) but cost less than two cents a pound when sold in bulk on the industrial market. For typical expressions of these charges see Corwin D. Edwards, *Economic and Political Aspects of International Cartels: A Study Made for the Subcommittee on War Mobilization of the Committee on Military Affairs, United States Senate*, 78th Congress, 2nd Session, Monograph No. 1 (Washington, D.C.: Government Printing Office, 1944), 13, 18–19; Wendell Berge, *Cartels: Challenge to a Free World* (Washington, D.C.: Public Affairs Press, 1944), 28–30; "Du Pont Lie Nailed," *In Fact* 13 (July 8, 1946): 1–2; Vernon A. Mund, *Government and Business* (New York: Harper, 1950), 127–128; and Corwin D. Edwards, "Standards and Product Differentiation," in *National Standards in a Modern Economy*, ed. Dickson Reck (New York: Harper, 1956), 328. For defenses against the charges see "Anti-Trust Indictments Involving Rohm and Haas Company," September 29, 1943, R&H, FJR 12–300; Donald S. Frederick, "Confidential Memorandum: Information Regarding Rohm and Haas Co. Activity in the Denture Field," June 27, 1944, R&H, FJR 12–300; and Du Pont Public Relations Department, "Press Analysis for June 1946," DPA, series II, part 2, box 872. See also transcript of Frederick interviewed by Hochheiser, 29–30; Hochheiser, *Rohm and Haas*, 75–79.

[58] Unless otherwise noted, information on polystyrene is from *Synthetic Resins and*

Their Raw Materials, 41–43; L. E. Humphrey, "Styrene Resins," *MP* 17 (October 1939): 100, 102, 104; Simonds and Ellis, *Handbook of Plastics*, 325–327; Haynes, *American Chemical Industry*, 5:339; idem, *American Chemical Industry*, 6:121–122; Don Whitehead, *The Dow Story: The History of the Dow Chemical Company* (New York: McGraw-Hill, 1968), 144–146; DuBois, *Plastics History U.S.A.*, 290–292; and Peter J. T. Morris, *Polymer Pioneers: A Popular History of the Science and Technology of Large Molecules* (Philadelphia: Center for History of Chemistry, 1986), 52.

[59]   The price was recalled by Ray Boyer, who began at Dow in 1935. See Robert Leaversuch, "The SPI at 50: Looking into an Exciting, Uncertain Future," *MP* 64 (September 1987): 49.

[60]   Monsanto Lustron advertisement, *MP* 18 (February 1941): 11.

[61]   "Trade Name Index," *Plastics: Directory, Index & Buyers' Guide* (New York: Plastics Publications, 1929), 99–101; "Appendix: Plastic Products and Producers," in Louis F. Rahm, *Plastic Molding* (New York: McGraw-Hill, 1933), 205–232; and "Directory of Trade Names," *MP* 17 (October 1939): 447–448.

CHAPTER 4: AN INDUSTRY TAKES SHAPE

[1]   On consolidation see Williams Haynes, *American Chemical Industry: Decade of New Products* (New York: D. Van Nostrand, 1954), 5:341–342, and brief company histories published in Haynes, *American Chemical Industry: The Chemical Companies*, 6 (New York: D. Van Nostrand, 1949). Allan Brown explained the sale of Bakelite in an interview, December 26, 1974, tape provided by William T. Cruse.

[2]   *An Engineering Interpretation of the Economic and Financial Aspects of American Industry: Volume VII: Plastics and the Plastics Industry* (New York: George S. Armstrong, 1946), 8, 12–13.

[3]   Molders and fabricators are singled out as the end-users of plastic resins because we typically think of plastic as a material from which solid objects, laminates, or films are made—all artifacts with discrete identities. However, in 1946 only about 43 percent of the resins produced went into such applications. The rest were used in paints, coatings, and adhesives. Although classed by the federal government and by business as intermediate plastic products, these often reached the market in forms that kept people from recognizing them as "plastic." See *An Engineering Interpretation*, 68.

[4]   "New Plastics Plant," *Business Week* (January 1, 1938): 33; John S. Egan, "Plastics Link Pittsfield, GE," *GE Jubilee* (October 7, 1978): 30, provided by E. J. Clark of GE Public Relations.

[5]   George S. Scribner as quoted by William Engle, "This Plastic Age: In Business For 21 Years, Still Learning," *New York World-Telegram* (December 15, 1938): 25. PPA Tapes reveal the world of the plastic molder, especially interviews with Clinton W. Blount, 1952, reel 2, side 1; J. Harry DuBois, 1968, reel 2, side 2; Oscar Gold, 1968, reel 2, side 2; Alfred C. Manovill, 1968, reel 2, side 2; Elmer E. Mills, 1968, reel 3, side 1; Walter E. Rahm, 1952, reel 1, side 1; Frank H. Shaw, 1952, reel 1, side 1; Hans H. Wanders, 1952, reel 1, side 2; and James S. Wilson, 1952, reel 2, side 1.

[6]   Cameron Hawley, *Cash McCall* (Boston: Houghton Mifflin, 1955), 117, 166; *A Brief Description of the Commonly Used Plastics Compiled for the Guidance of Engineers and Buyers* (Boonton, N.J.: Boonton Molding Co., 1933), 8, DuBois, box 3; and Elmer E. Mills interview, 1968, PPA Tapes, reel 3, side 1. Scribner began acknowledging authorship of the *Brief Description* by at least the third revision (1950).

[7] *Plastics: Directory, Index & Buyers' Guide* (New York: Plastics Publications, 1929), 84–90; "Molders, Fabricators, Designers," *MP* 17 (October 1939): 441–443; and *An Engineering Interpretation of the Economic and Financial Aspects of American Industry*, 67–68.

[8] Richard W. Seabury, "Historical Report of Boonton Rubber Manufacturing Company," March 27, 1923, typescript, DuBois, box 3; "Confidential Report/ Bakelite Molding Industry/ November 19, 1926," typescript, DuBois, box 3; and Robert L. Davidson, "Kurz-Kasch, Inc.: A Short Background," November 4, 1971, typescript, Kurz. Although not signed, the "Confidential Report" expresses ideas similar to Seabury's, lists his company first among the molders, and is prepared on legal-sized paper with a typing style similar to those of documents signed by Seabury and located in the same part of the DuBois Collection.

[9] This paragraph is based on comparison of several sources that sometimes disagree on minor points (such as Kurz's middle initial). Quotations from "Confidential Report/ Bakelite Molding Industry/ November 19, 1926," DuBois, box 3; an undated ms. by Robert L. Davidson, "Kurz-Kasch Profile: Part I—the CEO's," Kurz. Other sources include "Christian S. Kurz, Jr." and "Henry J. Kasch," both photocopies from a history of *Dayton and Montgomery County*, 321, 332, Kurz; "The Kurz-Kasch Company," *Industrial History of Dayton*, 501, photocopy, Kurz; Kasch obituary, *MP* 25 (April 1948): 240; mimeographed program for "Business-Industry-Education Day" for Dayton school teachers, April 18, 1951, Kurz; Walter E. Rahm interview, 1952, PPA Tapes, reel 1, side 1; William L. Sanders, "Dayton Industries: Kurz-Kasch First Molder of Plastic Dinnerware," *Dayton Daily News* (October 23, 1958): 14; and Robert L. Davidson, "Kurz-Kasch, Inc.: A Short Background." I am also indebted to Robert L. Davidson and to Bud Kasch, son of the founder, for interviews on July 22, 1985.

[10] Quotations are from George K. Scribner interview, 1952, PPA Tapes, reel 2, side 1; *A Brief Description of the Commonly Used Plastics* (1933), 2. Other sources are Richard W. Seabury, "Historical Report of Boonton Rubber Manufacturing Company," April 15, 1916, typescript, DuBois, box 3; "The Pioneer of Phenolic Molding," *SPI Voice* (December 11, 1940): 7, SPIA, misc. carton; Charles A. Norris, Jr., interview, 1952, PPA Tapes, reel 1, side 2; "G. K. Scribner Dies; Plastics Expert," *New York Times* (August 4, 1963): 80; Scribner obituary, *MP* 41 (September 1963): 248, 250; "Boonton, At 60, Sees New Ways To Grow," *MP* 58 (February 1981): 16; and letter from William F. Glennon, *MP* 58 (April 1981): 12.

[11] *A Brief Description* (1933), 1–2; George K. Scribner, *A Brief Description*, 3rd revision (1940), 1–2; and Scribner as paraphrased by Engle, "This Plastic Age: In Business for 21 Years, Still Learning," 25.

[12] Carl Marx, untitled editorial, *Plastics* 1 (October 1925): 7.

[13] Williams Haynes, untitled editorial, *Plastics & Molded Products* 8 (January 1933): front cover.

[14] D.E.A. Charlton and Charles A. Breskin, "*Modern Plastics* Statement of Policy," *MP* 12 (September 1934): 22.

[15] E. F. Lougee, "Editorial Comment," *MP* 12 (August 1935): 24.

[16] See for example "Success Stories," *MP* 12 (October 1934): 18–20, 64.

[17] E. F. Lougee, "Raymond Loewy Tells Why," *MP* 12 (January 1935): 21–23, 54–55; idem, "From Old to New with Lurelle Guild," *MP* 12 (March 1935): 14–15, 66–68; Dock Curtis, "'Give Us Facts' Says Ely Jacques Kahn," *MP* 12 (May 1935): 9–13, 60–62; E. F. Lougee, "Looking Backward and Forward with Jay Ackerman," *MP* 12 (May

1935): 14–16, 59–60; idem, "Planning Ahead with Gilbert Rohde," *MP* 12 (July 1935): 13–15, 56–58; and Marcy Babbitt, "As a Woman Sees Design" [Belle Kogan], *MP* 13 (December 1935): 16–17, 49, 51.

18 "Modern Plastics Sponsors First Annual Plastics Exhibit," *MP* 12 (December 1934): 24–25, 43; "Plastic Exhibit Ends May 18th," *MP* 12 (May 1935): 33.

19 A single postwar competition was held in 1946. See "Announcing Modern Plastics Competition," *MP* 13 (April 1936): 22–23; "The Judges of the Modern Plastics Competition," *MP* 13 (August 1936): 12; "Judges of Modern Plastics Second Annual Competition," *MP* 14 (August 1937): 27; "News," *MP* 15 (May 1938): 54; "Awards in the 4th Annual Modern Plastics Competition," *MP* 17 (November 1939): 30; and "Plastics Progress," *MP* 19 (November 1941): 54–55.

20 Robert L. Davidson, chairman, Kurz-Kasch, Inc., to the author, August 23, 1985.

21 Unless otherwise noted, the account of SPI's formation is based on Minutes of Board and General Meetings, 1937–1940, SPIA, microfilm reel 2; "Fun at Buckwood," *MP* 13 (September 1935): 23; "Plastics Society to Organize," *MP* 14 (May 1937): 49; "Society of the Plastics Industry," *MP* 14 (June 1937): 22–23; "News," *MP* 15 (June 1938): 48; a file of correspondence to and from Herbert S. Spencer, chairman of the SPI program committee, 1938, SPIF; "News," *MP* 17 (November 1939): 112; "In the Limelight," *MP* 18 (November 1940): 76; Gordon Brown to William T. Cruse, September 20, 1950, SPIF; Gordon Brown interview, 1952, PPA Tapes, reel 2, side 1; Gordon Brown to [J.] Harry [DuBois], December 10, 1970, SPIF; Gordon Brown to William T. Cruse, November 12, 1972, SPIF; and Allan Brown interview, December 26, 1974, tape provided by William T. Cruse. Documents designated as coming from the SPI Files (SPIF) were examined at SPI's Washington office in November 1986. They are presumably all included in the SPI Archive (SPIA) later deposited with the Hagley Museum and Library. Designations SPIF and SPIA indicate where a document was located when I examined it.

22 A. W. Fox of Johns Pratt Co. to Sandford Brown, November 23, 1923, quoted in Brown to William T. Cruse, November 12, 1972, SPIF.

23 On this point see W. A. Freyer to H. S. Spencer, July 20, 1938; Prescott Huidekoper to H. S. Spencer, August 2, 1938; and Brown to Cruse, November 12, 1972; all SPIF.

24 Brown to Cruse, November 12, 1972, SPIF.

25 List of SPI charter members, May 24, 1937; SPI attendance record, September 28–29, 1937; and SPI attendance record, May 23–24, 1938; all SPIF. For typical complaints see W. F. Reibold to Herbert S. Spencer, July 27, 1938; Prescott Huidekoper, president of American Insulator Corp., to Spencer, August 2, 1938; both SPIF.

26 SPI Program Committee report, ca. autumn 1938, typescript, SPIF.

27 Spencer to G. F. Nordenholt, editor of *Product Engineering*, June 24, 1938, SPIF.

28 Minutes of SPI general meeting, May 23, 1938, SPIA, microfilm reel 2.

29 Quotation is from Cruse's Senate testimony, September 20, 1967, printed in U.S. Senate, Committee on the Judiciary, Subcommittee on Antitrust and Monopoly, "New Technologies and Concentration," in *Economic Concentration* (Washington, D.C.: Government Printing Office, 1968), 2641. See also E. F. Lougee, letter, *MP* 17 (June 1940): 39; letter from Cruse to Charles A. Breskin, *MP* 19 (October 1941): 63; biographical sketch of Cruse, ca. 1950, SPI Press Releases, 1946–1969, SPIA, microfilm reel 1; "The SPI's Cruse: Mr. Plastics Industry," *MP* 40 (September 1962): 48; and "Harding Succeeds Cruse in SPI," *MP* 45 (October 1967): 226.

30 Rexmond C. Cochrane, *Measures for Progress: A History of the National Bureau of*

Standards (Washington, D.C.: U.S. Department of Commerce, National Bureau of Standards, 1966); Gordon M. Kline, "A Narrative Vignette of My Years at NBS," December 1980, typescript, provided by Beckman Center for the History of Chemistry, University of Pennsylvania, Philadelphia; and "Gordon M. Kline: Transcript of an Interview Conducted by Jeffrey L. Meikle in Lake Worth, Florida, on 15 and 16 May 1987" (Philadelphia: Beckman Center for the History of Chemistry, 1989).

[31] "Kline . . . Interview," 5, 9.

[32] "Plastics: War Makes Gimcrack Industry into Sober Producer of Prime Materials," *Life* 14 (May 3, 1943): 65. See also "Plastic Fords," *Time* 36 (November 11, 1940): 65; "75 Awards Listed for Plastics Uses," *New York Times* (November 5, 1941): 40; and John Delmonte, "The Postwar Role of Plastics," *MP* 20 (April 1943): 59.

[33] For a related discussion of pastiche and film see Fredric Jameson, "Postmodernism, or The Cultural Logic of Late Capitalism," *New Left Review*, no. 146 (July-August 1984): 64–71.

[34] James M. Cain, *Mildred Pierce* (1941; rpt., New York: Vintage, 1978), 5–6.

[35] Anonymous design writer quoted by William Grimes, "Material Value: The Sound of Plastic," *Esquire* 105 (January 1986): 17. See also Sylvia Katz, *Plastics: Common Objects, Classic Designs* (New York: Harry N. Abrams, 1984), 60–61, 64; Andrea DiNoto, "Bakelite Envy," *Connoisseur* (July 1985): 68; Philip Collins, *Radios: The Golden Age* (New York: Chronicle Books, 1987); and William C. Ketchum, Jr., "Antiques and Collectibles: Classic Plastic!," *Better Homes and Gardens* (August 1989): 129.

[36] Andrea DiNoto, *Art Plastic: Designed for Living* (New York: Abbeville Press, 1984), 136–143. On savings gained by painting naturally dark molded phenolic instead of molding naturally light-colored urea formaldehyde, see Robert L. Davis and Ronald D. Beck, *Applied Plastic Product Design* (New York: Prentice-Hall, 1946), 151.

[37] Nathan George Horwitt, "Plans for Tomorrow: A Seminar in Creative Design," *Advertising Arts* (July 1934): 29; "What Man Has Joined Together . . . ," *Fortune* 13 (March 1936): 69; "A Plastic a Day Keeps Depression Away," *Chemical and Metallurgical Engineering* 40 (May 1933): 248; Guild as quoted by E. F. Lougee, "From Old to New with Lurelle Guild," *MP* 12 (March 1935): 14; Muller-Munk, "The Future of Product Design," *MP* 20 (June 1943): 77, 144; and "New Jobs for Plastics," *Business Week* (December 28, 1935): 17.

[38] On the Machine-Age Exposition see Dickran Tashjian, "Engineering a New Art," in Richard Guy Wilson, Dianne H. Pilgrim, and Dickran Tashjian, *The Machine Age in America 1918–1941* (New York: Harry N. Abrams, 1986), 231–235. Elizabeth Rankin documented the irony that many of Gabo's pieces, whose materials were "intended to embody the utopian paradox of being eternally modern," later discolored, warped, and sometimes disintegrated. See Rankin, "A Betrayal of Material: Problems of Conservation in the Constructivist Sculpture of Naum Gabo and Antoine Pevsner," *Leonardo* 21 (1988): 285–290.

[39] "Modern Art in Pyroxylin," *Plastics & Molded Products* 4 (December 1928): 695; Paul T. Frankl, *Form and Re-form: A Practical Handbook of Modern Interiors* (New York: Harper, 1930), 31, 163; and idem, *Machine-Made Leisure* (New York: Harper, 1932), 112, 115, 117, 122, 120.

[40] Memories of John C. Pitzer in "This Is Your Life: Mr. D. J. O'Conor," 25–26, a presentation by the Formica Foreman's Business Club, Cincinnati, June 1, 1955, typescript, provided by Formica Corporation; *Historical Growth of High Pressure Decorative*

Laminates (Wayne, N.J.: Formica Corp., ca. 1980); undated press releases provided by Lenora Cerrato of Formica Corporation, August 1980; and DiNoto, *Art Plastic*, 170.

[41] "Bakelite Desk Tops," *Bakelite Information*, no. 2 (April 1927): 1; "The Guest Room Up-to-Date," *Plastics & Molded Products* 3 (July 1927): 362.

[42] Walter Rendell Storey, "Plastics Enter the Home," *House Beautiful* 74 (December 1933): 276–278, 291–292; E. F. Lougee, "Furniture in the Modern Manner," *MP* 12 (December 1934): 18–20, 61–62; idem, "Planning Ahead with Gilbert Rohde," *MP* 12 (July 1935): 14; Karen Davies, *At Home in Manhattan: Modern Decorative Arts, 1925 to the Depression* (New Haven, Conn.: Yale University Art Gallery, 1983); Diane H. Pilgrim, "Design for the Machine," in Wilson, Pilgrim, and Tashjian, *The Machine Age in America 1918–1941*, esp. 276–303; and David A. Hanks with Jennifer Toher, *Donald Deskey: Decorative Designs and Interiors* (New York: E. P. Dutton, 1987).

[43] "Color in Electric Iron Handles," *Plastics & Molded Products* 4 (July 1928): 398. Other typical manifestations include "Modern Cases for Modern Timepieces," *Plastics & Molded Products* 4 (January 1928): 37; advertisement for Karolith, *Plastics & Molded Products* 4 (March 1928): front cover.

[44] The quoted phrase is a journalist's paraphrase of Raymond Loewy in "Streamlining—It's Changing the Look of Everything," *Creative Design* 1 (Spring 1935): 22. Information on laminates for interior design comes from Donald Deskey, "Modern Wall Coverings," *Creative Art* 9 (October 1931): 324; "Rockefeller City Awards Design Job," *New York Times* (June 27, 1932): 32; Frank Settele, "The Best for the Biggest: Plastics in Radio City," *Plastics & Molded Products* 8 (January 1933): 455–456, 477; Robert Heller, "The Architect Reviews the Case for Laminated Plastics," *MP* 12 (November 1934): 17–21; Ward Moore, "Big Industry Built On Vision and Few Thousand Dollars," *Cincinnati Post*, August 15, 1940; "Formica Is On Top," *Fortune* 44 (October 1951): 116–118, 150, 154, 156; memories of John D. Cochrane, Jr., in "This Is Your Life: Mr. D. J. O'Conor," 32–36; D. J. O'Conor interview, [1968?], PPA Tapes, reel 2, side 2; and *Historical Growth of High Pressure Laminates*, 3–4.

[45] Allan Brown interview, 1952, PPA Tapes, reel 1, side 1; Allan Brown interview, December 26, 1974, tapes provided by William T. Cruse. See also "Industry Is Having Its Face Lifted," *Bakelite Information*, no. 31, February 1933; "Plastics in Pictures," *Plastic Products* 9 (March 1933): between 22 and 23; and J. Harry DuBois, *Plastics History U.S.A.* (Boston: Cahners, 1972), 184–185.

[46] Advertisements quoted are from *Sales Management* (August 15, 1933): 171; *Sales Management* (March 15, 1934): 237. Ellipses in original.

[47] *Sales Management* (May 15, 1934): 485.

[48] "Letters: Our Portfolio on Product Design," *Bakelite Review* 6 (October 1934): 1; *Bakelite Review* throughout 1938; and "Designers," *Bakelite Review* 11 (October 1939): 14. Sales figures are from L. C. Byck, "A Survey of the Bakelite Thermosetting Business 1910 Through 1951," November 3, 1952, 22, LHB, division VI, box 12, folder VI-B-6. Byck did not report figures for 1932; they were almost certainly lower than those for 1931.

[49] *Plaskon: Technical Data* (Toledo: Toledo Synthetic Products Co., 1933), 2; "News," *MP* 16 (February 1939): 56; Robert L. Davidson of Kurz-Kasch interviewed by the author, July 22, 1985; and letter from Jean Reinecke to the author, June 7, 1985.

[50] "News," *MP* 15 (March 1938): 52. See also "Plastics Travelcade," *MP* 15 (April 1938): 36–39; *Bakelite Horizons: The Story of Modern Plastics* (New York: Bakelite Corp.,

1939), pamphlet distributed at the New York World's Fair, New York World's Fair Collection, box 1002, New York Public Library; "In the Limelight," *MP* 18 (October 1940): 148; and "In the Limelight," *MP* 19 (October 1941): 100. I am grateful to the Buhl Science Center, Pittsburgh, and to the Museum of Science and Industry, Chicago, for providing copies of archival material, and to the Science Museum, London, for a glimpse of the film.

[51] Henry Glade, "Future City on Earth," *Amazing Stories* (April 1942): 240, as quoted by Kathleen Church Plummer, "The Streamlined Moderne," *Art in America* 62 (January-February 1974): 49.

[52] "Christmas Gifts for All," *Plastics & Molded Products* 6 (December 1930): 696; advertisement for the Marblette Corp., *Plastic Products* 9 (October 1933): 305.

[53] Quotations are drawn from Franklin E. Brill, "Our Homesick Plastics," *Plastics & Molded Products* 8 (June 1932): 235–236; idem, "Some Hints on Molded Design," *Plastic Products* 9 (April 1933): 54–55; and idem, "Phenolics for '36 Clocks," *MP* 13 (November 1935): 27. See also Franklin E. Brill, "Midget Radios Versus Electric Clocks," *Plastic Products* 9 (July 1933): 182. For other statements denigrating imitation see "The Expanding Decade," *Fortune* 3 (June 1931): 148; Joseph Sinel, "Artistic Abuse of the Plastics," *Plastic Products* 9 (March 1933): 13; Walter Rendell Storey, "Plastics Enter the Home," *House Beautiful* 74 (December 1933): 278; "Plastics," *Advertising Arts* (March 1934): 40; Harold Van Doren, "A Designer Speaks His Mind," *MP* 12 (September 1934): 24; E. F. Lougee, "In the Field of Decorative Design: An Hour with Morris B. Sanders, Jr.," *MP* 12 (October 1934): 14; Robert Heller, "The Architect Reviews the Case for Laminated Plastics," *MP* 12 (November 1934): 17; and Herbert Chase, "Buttons and Buildings," *Review of Reviews* 93 (May 1936): 36.

[54] Van Doren, "A Designer Speaks His Mind," 24; Raymond P. Calt, "A New Design for Industry," *The Atlantic Monthly* 164 (October 1939): 541–542. See also Joseph Thorp, "The New Plastic Materials," *The Architectural Review* 73 (June 1933): 265. Illustrations in Van Doren's article traced the Air-King radio from rough sketch to finished product. The Electrolarm appeared, along with the "Cathedral" model ("Gothic design. Bakelite case with mottled walnut finish."), in a Telechron advertisement headlined "Time You Can Trust!," *Literary Digest*, June 14, 1930, tearsheet in N. W. Ayer Collection, box 461, Archive Center, National Museum of American History, Smithsonian Institution, Washington, D.C.

[55] Quotations from Franklin E. Brill, "What Shapes for Phenolics," *MP* 13 (September 1935): 21; Davis and Beck, *Applied Plastic Product Design*, 44; and Franklin E. Brill and Joseph Federico, "Decorative Treatments for Molded Plastics," *Product Engineering* 8 (January 1937): 23. On streamlining and technical aspects of molding see Montgomery Ferar and Carl W. Sundberg, "Three-in-One," *MP* 14 (January 1937): 55–56; Frank H. Johnson, "Designing Plastic Parts," *Product Engineering* 9 (February 1938): 61; W. H. McHale, "Let's Look at Radio Cabinets," *MP* 15 (February 1938): 23; Carl Sundberg and Montgomery Ferar, "What the Designer Should Know About Plastics," *Bakelite Review* 10 (July 1938): 4; "Small Radios—Today and Tomorrow," *MP* 17 (March 1940): 80; V. E. Yarsley and E. G. Couzens, *Plastics* (Harmondsworth: Penguin, 1941), 113; László Moholy-Nagy, "Design Potentialities" (1944), in *Moholy-Nagy*, ed. Richard Kostelanetz (New York: Praeger, 1970), 85; John A. Sasso and Michael A. Brown, Jr., *Plastics in Practice: A Handbook of Product Applications* (New York: McGraw-Hill, 1945), 23–24, 89; J. Harry DuBois, "Plastics Product Design," in *Fundamentals of Plastics*, ed. Henry M.

Richardson and J. Watson Wilson (New York: McGraw-Hill, 1946), 261; John Sasso, *Plastics Handbook for Product Engineers* (New York: McGraw-Hill, 1946), 345, 347, 351; Joseph F. Geers to N. A. Backscheider, March 16, 1953, SPIA, misc. carton; and Sylvia Katz, *Plastics: Designs and Materials* (London: Studio Vista, 1978), 64.

56 Peter Muller-Munk, "Vending Machine Glamour," *MP* 17 (February 1940): 66.

57 "What Man Has Joined Together . . . ," *Fortune* 13 (March 1936): 71.

58 On plastic at the New York World's Fair see "Design Takes a Holiday," *MP* 16 (August 1939): 32–39, 72, 74, 76. An exhibition called "Plasti-City," sponsored by *Modern Plastics* and the Chemical Division of the Department of Commerce, opened in the foyer of the Department of Commerce Building in December 1940. See "Plasti-City, Washington, D.C.," *MP* 18 (January 1941): 35.

59 One of the first to discuss plastics designed to order was Archie J. Weith, as quoted in "A Plastic a Day Keeps Depression Away," 248. Treatment of Plaskon and the Toledo Scale Co. is based on A. M. Howald, "Systematic Study Develops New Resin Molding Compound," *Chemical & Metallurgical Engineering* 38 (October 1931): 583–584; James L. Rodgers, "Plaskon, a New Molding Compound the Result of Planned Research," *Plastics & Molded Products* 7 (December 1931): 664–665, 687; Norman Bel Geddes, *Horizons* (New York: Little, Brown, 1932), 205–221, 223–232; "Redesigned Scale, with Molded Plastic Housing, Is Tribute to Research," *Steel* 97 (August 5, 1935): 75; "Research—on a New Scale," *Business Week* (August 10, 1935): 9–10; "Giant Plastic Molding Press Produces Large Weighing Scale Housings," *The Iron Age* 136 (August 29, 1935): 13–14; H. D. Bennett, "Pushing Back Frontiers," *MP* 13 (September 1935): 25–27, 30–32; S. A. Maxom, "Hail to the Scale," *Du Pont Magazine* 29 (October 1935): 8–9; N. S. Stoddard, "Molding the Toledo Scale Housing," *MP* 13 (October 1935): 11–13, 56–57; "Plaskon: A New Plastic Material," *Machinery* 42 (November 1935): 169–174; Edward R. Weidlein and William A. Hamor, *Glances at Industrial Research During Walks and Talks in Mellon Institute* (New York: Reinhold, 1936), 59–82; "What Man Has Joined Together . . . ," 74–75; "Modernized Chopper," *Business Week* (August 8, 1936): 16; Harold Van Doren, *Industrial Design: A Practical Guide* (New York: McGraw-Hill, 1940), 48, 171–172, plates 16–18; Haynes, *American Chemical Industry*, 6:246–247; "Hubert Bennett Stressed Research in Industry," *Toledo Blade*, September 9, 1951, clipping, Toledo–Lucas County Public Library; "Hubert D. Bennett: Industrialist Once Headed Toledo Scale," *Toledo Blade*, September 9, 1951, clipping, Toledo–Lucas County Public Library; "J. L. Rodgers Jr., Plastic Leader," *New York Times* (February 26, 1955): 15; "H. Van Doren: Industrial Designer Was Ex-Toledoan," *Toledo Blade*, February 4, 1957, clipping provided by Harper Landell; J. Harry DuBois interview, 1968, PPA Tapes, reel 2, side 2; and DuBois, *Plastics History U.S.A.*, 117–121, 159–160, 190–191. I also relied on photographs and documents in the possession of Peter W. Bressler, Philadelphia, and Harper Landell, Downingtown, Pennsylvania; and on an interview with Bud Kasch, July 22, 1985. An earlier account of the Toledo Scale story in *Art and Industry: A Century of Design in the Products We Use* (London: The Conran Foundation, 1982), 28–35, has many errors of fact and interpretation.

60 Typed job summary for Toledo Scale Co., NBG, file 152.

61 "Bel Geddes," *Fortune* 2 (July 1930): 56. Other information is from the contract between Bennett and Geddes, December 18, 1928; "Record Copy Book" for Toledo Scale Co., 1928; typed job summary for Toledo Scale Co.; all NBG, file 152; and Geddes, "The Horse Race Game," typescript, December 2, 1951, NBG, Autobiography Chapter 47 file.

[62] Haynes gave the date of the fellowship's inception as January 1, 1928. On page 64 of their in-house history of the Mellon Institute, Weidlein and Hamor gave it as January 1, 1929; written with Bennett's assistance at an earlier date, their account is likely more accurate. Queries addressed to several librarians and archivists at Carnegie Mellon University failed to uncover records of the industrial fellowships.

[63] Geddes, "The Horse Race Game"; typed transcript of telephone conversation of Geddes with a Mr. "Edgerter," an executive in the plastic industry, August 27, 1954, NBG, Autobiography Chapter 40 file.

[64] Weidlein and Hamor, *Glances at Industrial Research*, 63.

[65] Rodgers, "Plaskon, a New Molding Compound the Result of Planned Research," 687; *Plaskon Molded Color* (Toledo: Toledo Synthetic Products Co., 1934).

[66] Van Doren's *Industrial Design* remained the profession's most intelligent guide for practitioners into the 1960s. Biographical information is from a one-page curriculum vitae provided by Harper Landell, who continued Van Doren's practice after his death in 1957.

[67] Weidlein and Hamor, *Glances at Industrial Research*, 68.

[68] Ibid., 78.

[69] Reprinted in DuBois, *Plastics History U.S.A.*, 119. Cf. a photograph on 160.

[70] Stoddard, "Molding the Toledo Scale Casing," 56.

[71] Frederick Simpich, "Chemists Make a New World," *National Geographic Magazine* 76 (November 1939): 632; Dorothy Walker and Amy Schaeffer, "A $500,000,000 Wonder Child," *The American Magazine* 131 (February 1941): 40; Elinor Hillyer, "The Synthetics Become the Real," *Arts & Decoration* 42 (January 1935): 31; and Julian P. Leggett, "The Era of Plastics," *Popular Mechanics Magazine* 73 (May 1940): 658.

[72] Gerald Wendt, *Science for the World of Tomorrow* (New York: W. W. Norton, 1939), 166 (italics supplied); "What Man Has Joined Together . . . ," 73; "Plastics in 1940," *Fortune* 22 (October 1940): 90; and Wendt, *Science for the World of Tomorrow*, 266.

[73] Waldemar Kaempffert, "The World Has Just Begun," *The American Magazine* 129 (January 1940): 42, 129–131.

CHAPTER 5: NYLON

[1] "Gordon M. Kline: Transcript of an Interview Conducted by Jeffrey L. Meikle in Lake Worth, Florida, on 15 and 16 May 1987" (Philadelphia: Beckman Center for the History of Chemistry, 1989), 27–28. See also Peter H. Spitz, *Petrochemicals: The Rise of an Industry* (New York: John Wiley, 1988), 227–270.

[2] "75 Awards Listed for Plastic Uses," *New York Times* (November 5, 1941): 40; "Plastics: War Makes Gimcrack Industry into Sober Producer of Prime Materials," *Life* 14 (May 3, 1943): 65; Joseph L. Nicholson and George R. Leighton, "Plastics Come of Age," *Harper's Magazine* 185 (August 1942): 300–301; "Plastics' Progress," *Time* 41 (March 29, 1943): 30; and Kurz-Kasch, Inc., advertisement, *MP* 18 (May 1941): 99. Statistics are from Dominick V. Rosato, William K. Fallon, and Donald V. Rosato, *Markets for Plastics* (New York: Van Nostrand Reinhold, 1969), 15, 30. Poundage figures exclude cellulosics (celluloid and cellulose acetate) and include resins used as adhesives and coatings.

[3] *1 Plastics Avenue* (Pittsfield, Mass.: General Electric Plastics Division, ca. 1946), Hagley Museum and Library, Imprints Collection, Pamphlets; Nicholson and Leighton, "Plastics Come of Age," 301; "Plastics: War Makes Gimcrack Industry into Sober Pro-

ducer of Prime Materials," 65; and "Test-Tube Marvels of Wartime Promise a New Era in Plastics," *Newsweek* 21 (May 17, 1943): 42.

[4] Haynes as quoted in "Industry Prepares for War Role," a "special insert" on the SPI convention, May 3–5, 1942, *MP* 19 (May 1942).

[5] Ruth Carson, "Plastic Age," *Collier's* 120 (July 19, 1947): 22, 49–50.

[6] See Paul Boyer, *By the Bomb's Early Light: American Thought and Culture at the Dawn of the Atomic Age* (New York: Pantheon, 1985).

[7] Stine as quoted by David A. Hounshell and John Kenly Smith, Jr., *Science and Corporate Strategy: Du Pont R&D, 1902–1980* (Cambridge: Cambridge University Press, 1988), 223–225; on scientific and technical aspects of nylon's development see 221– 274. For an accurate, detailed contemporary account see E. K. Bolton, "Development of Nylon," *Industrial and Engineering Chemistry* 34 (January 1942): 53–58. On Carothers see Roger Adams, "Wallace Hume Carothers," *National Academy of Sciences Biographical Memoirs* 20 (1939): 293–309; Louis Cook, Jr., "Iowan Who Fathered Nylon Never Lived to Taste Glory of His Work," *Des Moines Register and Tribune*, December 1945, clipping, DPA, series II, part 2, box 598; "Wallace Hume Carothers," *The National Cyclopaedia of American Biography* 38 (New York: James T. White, 1953), 55–56; John R. Johnson, "Wallace Hume Carothers," *Dictionary of American Biography*, supplement 2 (New York: Charles Scribner's Sons, 1958), 96–97; Julian W. Hill, "Wallace Hume Carothers," *Proceedings of the Robert A. Welch Foundation Conferences on Chemical Research* 20 (November 8–10, 1976): 231–251; and John K. Smith and David A. Hounshell, "Wallace H. Carothers and Fundamental Research at Du Pont," *Science* 229 (August 2, 1985): 436–442.

[8] See essays in *Polymer Science Overview: A Tribute to Herman F. Mark*, ed. G. Allan Stahl, ACS Symposium Series 175 (Washington: American Chemical Society, 1981), esp. Rudolf Brill, "Reminiscences of the Early Twenties," 21–23; Stahl, "Herman F. Mark: The Early Years, 1895–1926," 5–19; idem, "Herman F. Mark: The Geheimrat," 61–88; and idem, "A Short History of Polymer Science," 25–44. See also Mark's autobiography, *From Small Organic Molecules to Large: A Century of Progress* (Washington, D.C.: American Chemical Society, 1993), and a biographical profile, Morton M. Hunt, "Polymers Everywhere," *New Yorker* 34 (September 13, 1958): 48–72; (September 20, 1958): 46–83. Also useful is Herman F. Mark, "The Development of Plastics," *American Scientist* 72 (March-April 1984): 156–162.

[9] Carothers to Hamilton Bradshaw, November 9, 1927, as quoted by Hounshell and Smith, *Science and Corporate Strategy*, 231; see also 236–237; Herbert Morawetz, *Polymers: The Origins and Growth of a Science* (New York: John Wiley, 1985), 119–120.

[10] As quoted by Hounshell and Smith, *Science and Corporate Strategy*, 232.

[11] Sometimes written alone, sometimes with co-author(s), eight variously titled articles appeared under the general heading "Studies of Polymerization and Ring Formation," *Journal of the American Chemical Society* 51 (1929): 2548–2559, 2560–2570; 52 (1930): 314–326, 711–721, 3292–3300, 3470–3471, 5279–5288, 5289–5291. Carothers reorganized the material in "Polymerization," *Chemical Reviews* 8 (1931): 353–426.

[12] Carothers to Adams, May 17, 1930, photocopy from Adams; Hill, "Wallace Hume Carothers," 239, 241.

[13] Irénée du Pont as paraphrased and interpreted in a letter from Walter S. Carpen-

ter, Jr., chairman, Finance Committee, to Leonard A. Yerkes, president, Du Pont Rayon Co., November 5, 1930, DPA, accession 542, box 818.

14 Stine to Adams, December 2, 1938, photocopy, Adams.

15 Hounshell and Smith, *Science and Corporate Strategy*, 237 – 243.

16 U.S. Patent 2,071,250, February 16, 1937.

17 Carothers and Hill, "Giant Molecules and Synthetic Silk," *Silk Journal and Rayon World* (October 1931).

18 "Chemists Produce Synthetic Silk," *New York Times*, September 2, 1931, clipping, DPA, series II, part 2, box 598, p. 10.

19 "Topics in Brief," *The Literary Digest*, September 26, 1931, clipping, DPA, series II, part 2, box 598, p. 10; Du Pont president Walter S. Carpenter, Jr., as quoted by Bolton, "Development of Nylon," 56.

20 Lammot du Pont to Nelson H. De Foe, A Century of Progress, August 3, 1931, DPA, accession 1662, box 3. For vastly different interpretations see Gerard Colby, *Du Pont Dynasty: Behind the Nylon Curtain* (Secaucus, N.J.: Lyle Stuart, 1984), 330 – 346, and Hounshell and Smith, *Science and Corporate Strategy*, 332 – 333. See also Wayne S. Cole, *Senator Gerald P. Nye and American Foreign Relations* (Minneapolis: University of Minnesota Press, 1962), 65 – 80. Figures on profits are from the company's own defense, a forty-two-page "letter" addressed to stockholders and employees: Lammot du Pont, *The du Pont Company and Munitions* (Wilmington, Del.: E. I. du Pont de Nemours & Co., 1934), 27. H. C. Engelbrecht and F. C. Hanighen's muckraking *Merchants of Death: A Study of the International Armament Industry* (New York: Dodd, Mead, 1934) did not help matters by mentioning Du Pont frequently.

21 Barton to Lammot du Pont, May 18, 1935, DPA, accession 1662, box 3. See also L.L.L. Golden, *Only By Public Consent: American Corporations [sic] Search for Favorable Opinion* (New York: Hawthorne, 1968), 235 – 246. I also benefited from David J. Rhees, "'Making the Nation Chemically Conscious': The Popularization of Chemistry in America, 1914 – 1940," unpublished paper delivered at a meeting of the History of Science Society, Chicago, December 28, 1985.

22 *The Forum* 92 (July 1934): 32 – 33. Public Affairs Department files regarding Du Pont's new slogan contain tear sheets of the cartoon, DPA, accession 1410, box 36.

23 On the origin of the slogan see anonymous handwritten notes, DPA, series II, part 2, box 75; "Slogan: Better Things for Better Living," a file in DPA, accession 1410, box 35; and Edwin R. Manchester, "A History of Du Pont Advertising," typescript, DPA, accession 1866, box 1. The Hagley Imprint Collection includes bound scripts of *The Cavalcade of America*.

24 *Du Pont at the Texas Centennial Exposition* (Wilmington, Del.: E. I. du Pont de Nemours & Co., 1936), 10, 2, DPA, series II, part 2, box 50. See also R. H. Coleman to William A. Hart, "Report of the Du Pont Exhibit Activity at the Texas Centennial Exposition," December 1936, DPA, series II, part 2, box 50.

25 Jarrott Harkey, "Semi-monthly Report," n.d.; Florence E. Allen to R. H. Coleman, reports dated August 1, September 15, 1936; and Harkey, "Semi-Monthly Report," October 1, 1936; all DPA, series II, part 2, box 51.

26 Menter B. Terrill, "Report on Lecturing Section No. 4/ From September 15 to 30," 1936; Wilbyrn McKee, report, August 1, 1936; Harkey, "Semi-Monthly Report," October 1, 1936; Allen to Coleman, report, September 15, 1936; and Harkey, "Semi-monthly Report," n.d.; all DPA, series II, part 2, box 51.

27   Bolton, "Annual Report—1933," January 2, 1934, as quoted by Hounshell and Smith, *Science and Corporate Strategy*, 244. On practical development of nylon and nylon stockings see Hounshell and Smith, 244–245, 259–268, and Bolton, "Development of Nylon," 55–57.

28   Williams Haynes, *American Chemical Industry: Decade of New Products* (New York: D. Van Nostrand, 1954), 5:366; Hounshell and Smith, *Science and Corporate Strategy*, 257.

29   For a calculation of expenses, including those charged to the Chemical Department for fundamental research, see A. B. King, assistant treasurer, to the Executive Committee, "Out-of-Pocket Cost of Nylon Development to Du Pont Co. (to January 1, 1940)," July 31, 1940, DPA, accession 1813, box 33, folder 12.

30   William R. Tyson to G. D. Graves, "Fiber 66 Service Test," October 27, 1937, DPA, series II, part 2, box 970.

31   B. M. May to L. A. Yerkes, January 15, 1936, DPA, accession 542, box 819; Charles H. Rutledge, "The Name Nylon and Some of Its Adventures," typescript, June 20, 1966, DPA, series II, part 2, box 598, between pp. 18 and 19; laboratory notebooks of A. W. Staudt, March 26, 1936, and F. C. King, April 1, 1936, both DPA, series II, part 2, box 973; memo from L. A. Yerkes, general manager, Rayon Department, October 4, 1937, DPA, accession 1813, box 28, folder 2; and G. P. Hoff to E. K. Gladding, "Status of Fiber 66," November 10, 1937, DPA, series II, part 2, box 963. Steven M. Spivak and I summarized the naming process in "Nylon: What's in a Name?," *Textile Chemist and Colorist* 20 (June 1988): 13–16. This account draws from and expands on that article.

32   B. M. May sent the ten-page list under the title "Fiber 66" to eighteen executives on March 21, 1938, DPA, accession 1813, box 28.

33   Carpenter to B. M. May, March 30, 1938, DPA, accession 542, box 819; du Pont to O. F. Benz, March 24, 1938, DPA, accession 1662, box 66.

34   G. Tipp and V. J. Watson, *Polymeric Surfaces for Sports and Recreation* (London: Applied Science Publishers, 1982), 97.

35   The company denied this geopolitical interpretation in "Comical Suppositions as to Real Meaning of Nylon Contradicted by Du Pont Head," *Japan Times & Advertiser*, February 12, 1941, clipping, DPA, accession 1662, box 59.

36   Gladding to du Pont, March 26, 1941, DPA, accession 1662, box 59.

37   L. A. Yerkes, memo, October 19, 1938, DPA, accession 1813, box 28. The memo may have been issued earlier and misdated. Willis F. Harrington's copy bears a rubber stamp recording its having been "noted" on October 13.

38   Hounshell and Smith, *Science and Corporate Strategy*, 266.

39   G. P. Holt, "Confidential Memorandum," May 7, 1937, DPA, series II, part 2, box 951.

40   "Silk Is Done," *American Wool and Cotton Reporter* 51 (July 8, 1937), clipping, DPA, series II, part 2, box 598, p. 52.

41   *The Cavalcade of America*, no. 115, December 29, 1937, Hagley Imprint Collection; Stephen S. Marks, "Entirely New Synthetic Yarn For Women's Hose Developed," *New York Daily News Record*, July 18, 1938; "Fibers, Just Fibers," *Business Week*, September 3, 1938; "New du Pont Fibre May Threaten Market for Silk," *Wall Street Journal*, August 26, 1938; clippings from DPA, series II, part 2, box 598, pp. 65–67.

42 "New Silk Made on Chemical Base Rivals Quality of Natural Product," *New York Times*, September 22, 1938; "Coal and Castor Oil Challenge Silkworm," *Washington News*, September 22, 1938; both clippings, DPA, series II, part 2, box 598, p. 67; and "Castor Oil, Coal Newest 'Silkworms' for Stockings," *Science News Letter* 34 (October 1, 1938): 211–212. Other articles mentioning cadaverine included "Ladies' Hose and History," *Chicago Tribune*, September 26, 1938, photocopy, Adams; "Synthetic Silk May Hurt Japan," *Buffalo Evening News*, September 26, 1938, clipping, DPA, series II, part 2, box 598, p. 67; "Silk's Last Battle-Ground?," *Textile World*, October 1938, clipping, DPA, series II, part 2, box 598, p. 60; and "No. 2,130,948," *Time* 32 (October 3, 1938): 47.

43 Joseph C. Matera, "Report of Week—September 29–October 5," October 6, 1940, DPA, series II, part 2, box 36.

44 "Marvels of Tomorrow Unveiled at Fair Grounds as Forum Ends" and "Forum Closes 3-Day Session with Glimpse of Tomorrow's World at Fair Grounds," *New York Herald Tribune*, October 28, 1938, clippings, DPA, series II, part 2, box 598, pp. 81, 85. For the full advance text of the speech see Charles M. A. Stine, "What Laboratories of Industry Are Doing for the World of Tomorrow: Chemicals and Textiles," October 27, 1938, typescript, DPA, series II, part 2, box 598, p. 83. Lacking any evidence other than Gladding's recollection of the naming of the nylon, Charles H. Rutledge dismissed the connection between *nylon* and *Trylon* as mere speculation; it seems likely, however, that the company hoped to suggest a link between the fiber and the fair. See Rutledge, "The Name Nylon and Some of Its Adventures."

45 Harold L. Ickes, *The Secret Diary of Harold L. Ickes* (New York: Simon and Schuster, 1954), 2:497.

46 "New Hosiery Held Strong as Steel," *New York Times*, October 28, 1938; "No More Runs," *Orlando Sentinel*, October 1938; "More Durable Stockings," *Champaign Evening Courier*, October 31, 1938; and H. I. Phillips, "The Once Over," in a Birmingham, Alabama, newspaper, November 28, 1938; all clippings, DPA, series II, part 2, box 598, pp. 86, 94, 95, 96.

47 E. K. Gladding to L. A. Yerkes, August 17, 1944, DPA, series II, part 2, box 599. For examples of promotional texts paralleling the original announcement but omitting reference to "strong as steel" see C.M.A. Stine, "The Magic of Industrial Chemistry Today," *Du Pont Magazine* 33 (May 1939): 1; commercial for *The Cavalcade of America*, no. 166, May 22, 1939, Hagley Imprint Collection.

48 William S. Dutton, "The Golden Gate Glitters as Host to the World," *Du Pont Magazine* 33 (April 1939): 2–3. See also an untitled press release on nylon at San Francisco, February 1, 1939, DPA, series II, part 2, box 598.

49 *Official Guide Book of the New York World's Fair 1939*, 3rd ed. (New York: Exposition Publications, 1939), 180, 184, 189; press release, Du Pont Wonder World of Chemistry, May 1, 1939, DPA, accession 1410, box 44; William S. Dutton, "New York Presents 'A Drama of Opportunity,'" *Du Pont Magazine* 33 (June 1939): 3–8; and Frank D. Morris, "Sheer Magic," *Collier's* 105 (April 13, 1940): 13.

50 Paul W. Sampson to William A. Hart, "Report for week of July 2–8," July 9, 1939; Sampson to Hart, "Report for week of June 18," June 25, 1939; Irwin Heimer, "Lecturers' Reports/ Shift 'A'—Week Ending 6/24/39"; Thomas W. Witherspoon to Sampson, "Supervisor's Report—July 1–8, 1939," July 8, 1939; Heimer, "Lecturers' Reports—Shift 'A,'" July 15, 1939; and N. M. Walling, "Lecturers' Reports—Shift 'A,'" July 8, 1939; all DPA, series II, part 2, box 35.

51 "'Miss' of '39 Finds '64 Fair Different," *Wilmington Evening Journal*, April 24, 1964, clipping, DPA, series II, part 2, box 607.

52 B. S. Nicholson to Sampson, "Supervisor's Report," June 24, 1939; Katharine Mitton, "Lecturers' Report—Shift 'A,'" August 19, 1939; C. E. Speakman, "Excerpts from lecturers' reports (shift 'A')," July 1, 1939; and H. S. Chason, "Lecturers' Reports—Shift 'B,'" July 22, 1939; all DPA, series II, part 2, box 35.

53 As quoted in "'Miss' of '39 Finds '64 Fair Different"; Speakman, "Shift 'A'/ 8/26/ 39"; Mitton, "Shift 'A,'" September 2, 1939. Percentages are from "Lecturers' Report," September 9, 1939. On Du Pont and munitions, see John Boyko, "Lecturers' Reports—Shift 'B,'" September 2, 1939; Boyko, "Lecturers' Report," September 9, 1939. All sources DPA, series II, part 2, box 35.

54 Untitled press release, April 8, 1940, DPA, accession 1410, box 44; J. P. Kelley to W. H. Uffelman, June 2, 1940, DPA, series II, part 2, box 35; and Mildred Lang, "Report of Week—June 2–June 8," June 9, 1940, DPA, series II, part 2, box 35.

55 "Dear Silk," *Time* 35 (February 19, 1940): 76, 78. See also untitled press release, October 23, 1939, DPA, series II, part 2, box 598; "Nylon Hosiery on Sale," *Business Week* (October 28, 1939): 42–43.

56 Untitled press release, May 14, 1940, DPA, series II, part 2, box 598.

57 From script written for Dr. G. P. Hoff to read on *The Cavalcade of America*, no. 187, May 14, 1940, Hagley Imprint Collection. See also untitled press release, May 15, 1940, DPA, series II, part 2, box 598.

58 "Nylon Customers Swamp Counters," *Department Store Economist*, May 25, 1940, clipping, DPA, series II, part 2, box 598. See also "Nylon Hose Go On Sale Wednesday," *Business Week* (May 11, 1940): 38, 40–41; "Nylon Goes to Town," *Business Week* (May 25, 1940): 45; and George Albee, "Nylon Success Story," *Du Pont Magazine* 38 (September-October 1944): 8.

59 As quoted by Leonard Mosley, *Blood Relations: The Rise & Fall of the du Ponts of Delaware* (New York: Atheneum, 1980), 369. He was actually quoting reactions of Du Pont employees to an in-house offering of experimental stockings in February and March 1939. See "Hosiery Sale—Wilmington Office Employees," March 20, 1939, DPA, series II, part 2, box 951.

60 Raymond Hower, "Report of Week—May 26–June 1," June 2, 1940, DPA, series II, part 2, box 35. See also Elizabeth Kotz and Leonard Waller, "Report of Week—May 19–25, 1940," May 26, 1940, DPA, series II, part 2, box 35.

61 Duane L. Greenfield, "Report of Week—May 19–25, 1940," May 26, 1940; Greenfield, "Report of Week—June 9–June 15," June 16, 1940; both DPA, series II, part 2, box 35.

62 Quotations are from Waller, Greenfield, and Hower, "Report of Week—June 16–June 22," June 23, 1940; see also Greenfield, "Report of Week—June 2–June 8," June 9, 1940; all DPA, series II, part 2, box 35.

63 R. H. Legg, "Report of Week—June 2–June 8," June 9, 1940, DPA, series II, part 2, box 35. See also comments by Eileen Leonard and Carl Richter in the same report; Greenfield, "Report of Week—May 19–25, 1940," May 26, 1940, and Walling, "Report of Week—June 9-June 15," June 16, 1940, both DPA, series II, part 2, box 35; and "Report of Week—August 18–24," August 25, 1940, DPA, series II, part 2, box 36.

64 Greenfield, "Report of Week—May 19–25, 1940," May 26, 1940, DPA, series II,

part 2, box 35; Waller, "Report of Week—October 6–12," October 13, 1940, DPA, series II, part 2, box 36.

65   Anonymous poem, *New York Herald Tribune*, February 5, 1940, clipping, DPA, accession 1813, box 33, folder 12. On allergies see Virginia Routh, "Report of Week—August 18–August 24," September 1, 1940, DPA, series II, part 2, box 36; "Dye Used in Nylon Hose Blamed in Skin Trouble," *Richmond Times-Dispatch*, August 2, 1940, clipping, DPA, series II, part 2, box 598. On nylons dissolved by polluted air see U.S. Department of Commerce press release, March 22, 1941, and "Disintegrating Nylon Hose Big Mystery to Shoppers," *Jacksonville Florida Times-Union*, February 15, 1949, clipping, both DPA, series II, part 2, box 599. The Jacksonville case was reported by two wire services and dozens of newspapers; see clippings in box 599.

66   Williams Haynes, *This Chemical Age: The Miracle of Man-Made Materials* (New York: Alfred A. Knopf, 1942), 315.

67   Frank C. Quintana, "Report of Week—September 15–September 21," 1940, DPA, series II, part 2, box 36. The rumor that nylon was being siphoned off for parachute fabric surfaced as early as June and became the dominant theme of visitor comments in September. See Routh, "Report of Week—June 23–June 29," June 30, 1940, DPA, series II, part 2, box 35; and the entire "Report of Week—September 1–September 7," September 8, 1940, DPA, series II, part 2, box 36.

68   "Army Encouraged With Use of Nylon Parachute Cloth," *New York Daily News Record*, October 10, 1940, and "Nylon for Consumer Uses," *American Dyestuff Reporter*, September 30, 1940, both clippings, DPA, series II, part 2, box 598; G. P. Hoff, "Nylon Sales Meeting" memorandum, January 28, 1941, DPA, series II, part 2, box 959; press release, February 11, 1942, DPA, series II, part 2, box 599; and "Nylon," a major report on military applications, ca. 1946, I-10, DPA, accession 1410, boxes 29–30 (combined as one).

69   On wartime nylon production and use see "Nylon," ca. 1946, DPA, accession 1410, boxes 29–30; "Nylon's Life-Saving War Jobs," press release, July 16, 1945, DPA, series II, part 2, box 599.

70   "Paratroops to Ride From Sky On Nylon From Old Stockings," *New York Times*, December 30, 1942; "From Shapely Limbs to Parachutes Go Used Nylon Stockings," *Cisco* (Texas) *Press*, January 15, 1943; B. F. Goodrich advertisement, *Life*, May 3, 1943; and B. F. Goodrich advertisement (ellipsis in original), *Collier's*, June 23, 1945, p. 3; all clippings, DPA, series II, part 2, box 599.

71   "A Report on 'Nylonizing,' " *Consumer Reports* 10 (March 1945): 64. During the war years *Consumer Reports* covered leg makeup more frequently than any other product: "Stocking Savers," *Consumer Reports* 6 (September 1941): 232–234; "Stockings from a Bottle," *Consumer Reports* 7 (August 1942): 201–203; "Bottled 'Stockings,' " *Consumer Reports* 8 (July 1943): 181–183; "Cosmetic Stockings," *Consumer Reports* 9 (July 1944): 172–174; and "Instead of Stockings," *Consumer Reports* 10 (July 1945): 175–177.

72   Press release on twentieth anniversary of nylon, February 1959, DPA, accession 1410, box 50.

73   Based on Jay Walz, "Nylon Opens Tempting Field for the Black Marketeers," *New York Times*, August 13, 1944, clipping, DPA, series II, part 2, box 599; George Albee, "Nylon Success Story," *Du Pont Magazine* 38 (September-October 1944): 8–10; "Nylon:

From Test Tube to Household Word in 20 Years," *New York Times*, October 26, 1958, clipping, DPA, series II, part 2, box 602; and Mosley, *Blood Relations*, 372.

[74]   *The Cavalcade of America*, no. 384, May 22, 1944, Hagley Imprint Collection. See also memorandum from R. A. Ramsdell to E. K. Gladding, "Nylon Civilian Business in Wartime," December 22, 1943, DPA, series II, part 2, box 971.

[75]   Untitled press release, August 22, 1945, DPA, accession 1410, box 49.

[76]   "Nylon Sale and No Casualties," *Cortland* (New York) *Standard*, September 24, 1945; "Women Jam Store, Fight For Nylons," *Columbia* (South Carolina) *Record*, November 5, 1945; "Lone Man Spearheads Push For First Nylons Here," *Indianapolis Times*, September 26, 1945; "Peace, It's Here! Nylons on Sale," *Chicago Sun*, November 30, 1945; "Women Risk Life and Limb in Bitter Battle for Nylons," *Augusta* (Georgia) *Chronicle*, December 16, 1945; "Lady Raiders Take Nylon Beachhead," *Los Angeles Daily News*, January 19, 1946; "Women Win Battle For Nylons Here," *Portsmouth* (Virginia) *Star*, January 27, 1946; and "Nylon Mob, 40,000 Strong, Shrieks and Sways for Mile," *Pittsburgh Press*, June 13, 1946 (ellipsis in original); all clippings, DPA, series II, part 2, box 599; and "Nylons Are Getting Too Much Free Space," *Editor and Publisher*, March 2, 1946, as quoted by Du Pont Public Relations Department, "Press Analysis for February, 1946," DPA, series II, part 2, box 871.

[77]   Harold Brayman to Herbert B. Nichols, February 19, 1946, DPA, accession 1410, box 9.

[78]   Samuel L. Feldman of Town Shops, Inc., as quoted in "Picketing Bears, Nylon Mobs Start Postcard Protest," *Advertising Age*, March 4, 1946, clipping, DPA, series II, part 2, box 871.

[79]   Morris S. Verner as quoted in "Housewives Are Accused As 'Pigs' in Nylon Hoarding," *Wilmington Morning News*, March 11, 1946, clipping, DPA, series II, part 2, box 871.

[80]   "News Is All Bad on the Nylon Front," *Hattiesburg* (Mississippi) *American*, from summary of news reports, April 1946, DPA, series II, part 2, box 871.

[81]   *The Cavalcade of America*, no. 632, November 29, 1949, pp. 31–32, Hagley Imprint Collection; Public Relations Department, "Press Analysis for December 15, 1950," DPA, series II, part 2, box 882; and untitled press release, February 10, 1953, DPA, series II, part 2, box 601. On marketing of less durable nylons see E. K. Gladding to L. A. Yerkes, "Nylon's Plans for Immediate Post-War Period," April 5, 1943, 7–8; R. A. Ramsdell to E. K. Gladding, "Postwar Sales Report No. 3," August 20, 1944; both DPA, series II, part 2, box 971; and W. E. Coughlin and Michael Drury, "The Truth About Nylon Stockings," *Good Housekeeping* (September 1950): 60–61, 242, clipping, DPA, series II, part 2, box 601.

[82]   "Plastics: A Way to a Better More Carefree Life," *House Beautiful* 89 (October 1947): 141; Christine Holbrook and Walter Adams, "Dogs, Kids, Husbands: How to Furnish a House So They Can't Hurt It," *Better Homes and Gardens* 27 (March 1949): 38–39.

CHAPTER 6: GROWING PAINS

[1]   Jack Cole, "Plastic Man," *Police Comics*, no. 1, August 1941, reprinted in *A Smithsonian Book of Comic-Book Comics*, ed. Michael Barrier and Martin Williams (New York: Smithsonian Institution Press and Harry N. Abrams, 1981), 64–69; quotation from 69

(ellipsis in original). See also Don Thompson, "The Rehabilitation of Eel O'Brian," in *The Comic-Book Book*, ed. Don Thompson and Dick Lupoff (New Rochelle, N.Y.: Arlington House, 1973), 18–35.

[2] "Making Magic with Plastics," General Electric advertisement, tearsheet, ca. 1944, from General Electric Archive, W 4369, photocopy provided by Roland Marchand.

[3] According to David R. Smith, Walt Disney Archives, in a letter to the author, December 21, 1987, "The Plastics Inventor" was released on September 1, 1944. I am indebted to Anne Alexander of The Disney Channel for providing the cartoon's broadcast schedule, to James Davenport for taping it, and to Michael L. Smith who told me about it in the first place.

[4] Forrest Davis, "Airplanes, Unlimited!," *Scientific American* 161 (July 1939): 17. On acrylic glazing in aircraft see Roderick M. Grant, "Molding the World of Tomorrow," *Popular Mechanics* 80 (July 1943): 8–10.

[5] "Propaganda . . . a Threat or a Boost," *MP* 22 (March 1945): 198; "Plastic Buttons Sent to Dry Cleaners Institute for Testing," *SPI Bulletin*, no. 248 (February 14, 1947): 4, SPIA.

[6] Unless otherwise noted, discussion of Ford and plastic is based on E. F. Lougee, "Industry and the Soy Bean," *MP* 13 (April 1936): 13–15, 54–57; Robert L. Taylor, "How Soybeans Help Build Fords," *Chemical & Metallurgical Engineering* 43 (April 1936): 172–176; "Soybean Plastic," *Science* 87 (January 14, 1938): supp. 8, 10; William T. Cruse, "Ford and Plastics," *MP* 17 (January 1940): 23–29; "Plastic Fords," *Time* 36 (November 11, 1940): 65; Herbert Chase, "What's the Past, Present and Future of Plastic Automobile Bodies?," *The Iron Age* 147 (February 27, 1941): 44–48; Schuyler Van Duyne, "Mr. Ford Tells of Plans for Stronger Cars," *Popular Science Monthly* 138 (March 1941): 127–131; Herbert Chase, "Plastic Car Is Studied," *New York Times* (March 30, 1941): sec. 10, 5; "Ford Shows Auto Built of Plastic," *New York Times* (August 14, 1941): 19; "Ford from the Farm," *Newsweek* 18 (August 25, 1941): 39–40; "Plastic Ford Unveiled," *Time* 38 (August 25, 1941): 63; "Ford Builds a Plastic Auto Body," *MP* 19 (September 1941): 39–40, 78; Christy Borth, *Modern Chemists and Their Work* (New York: New Home Library, 1943), 200–212, 359–374; E. F. Lougee, *Plastics from Farm and Forest* (Chicago: Plastics Institute, 1943); William L. Stidger, "Henry Ford on Plastics," *The Rotarian* 62 (February 1943): 15–18, 58; Reynold Millard Wik, "Henry Ford's Science and Technology for Rural America," *Technology and Culture* 3 (Summer 1962): 247–258; David L. Lewis, "A Bushel in Every Car!," *Ford Life* 2 (May-June 1972): 14–24 (also published as "Henry Ford's Plastic Car," *Michigan History* 56 [Winter 1972]: 319–330); and Michael W. R. Davis, "Plastic Fords," *Special-Interest Autos* (June-July 1972): 18–19.

[7] Ford as paraphrased in "Ford from the Farm," 39; see also Borth, *Modern Chemists and Their Work*, 205.

[8] "Plastic Fords," 65. Gordon M. Kline described the event in conversation with the author, May 15, 1987.

[9] "Plastic Ford Unveiled," 63.

[10] As quoted and, in the case of the *Cleveland Press*, paraphrased by Lewis, "Henry Ford's Plastic Car," 324, 322.

[11] Boyer as quoted in "Ford Shows Auto Built of Plastic," 19.

[12] *"Lucite" Aircraft Manual* (Arlington, N.J.: Du Pont Plastics Dept., November 1942), preface; Stanton Kelton, Jr., interviewed by Sheldon Hochheiser, June 24, 1983,

transcript, 6, R&H. Both Swedlow and Kline (for the National Bureau of Standards) took credit for multiaxial stretching. See Ray McGlew, "The Shoot and Scoot Helicopters," *Du Pont Magazine* 62 (September-October 1968): 22; "Gordon M. Kline: Transcript of an Interview Conducted by Jeffrey L. Meikle in Lake Worth, Florida, on 15 and 16 May 1987" (Philadelphia: Beckman Center for the History of Chemistry, 1989), 3–4. See also Donald Frederick interviewed by Sheldon Hochheiser, September 27, 1983, transcript, 20, R&H; Sheldon Hochheiser, *Rohm and Haas: History of a Chemical Company* (Philadelphia: University of Pennsylvania Press, 1986), 64–65.

[13]   For Kline's complaint see "Kline . . . Interview," 4–5. For use of the term "plastic plywood" see "In the Limelight," *MP* 18 (February 1941): 74; "Twin-Motored Plastic-Plywood Plane," *MP* 19 (October 1941): 52–53, 110, 112; *Plastic Wood*, R1277 (Madison, Wisc.: Forest Products Laboratory, USDA, October 1941), Hagley Museum and Library, Imprints Collection, Pamphlets; and Borth, *Modern Chemists and Their Work*, 371. On plywood in aviation see Blaine Stubblefield, "Plastics in Aviation," *MP* 13 (February 1936): 17–19, 58–61; "The Plastic Airplane," *MP* 16 (March 1939): 41, 66, 68, 70; "Molded Airplanes for Defense," *MP* 17 (July 1940): 25–31, 78, 80, 82; and Gordon M. Kline, "Conclusion to Summary of Properties, Uses, and Salient Features of Families of Plastics," in *Symposium on Plastics* (Philadelphia: American Society for Testing Materials, 1944), 199. On the inflated-bag method of forming plywood see Herbert R. Simonds and Carleton Ellis, *Handbook of Plastics* (New York: D. Van Nostrand, 1943), 434–436; Paul I. Smith, *Synthetic Adhesives* (Brooklyn: Chemical Publishing Co., 1943), 92–93; and Thomas D. Perry, *Modern Wood Adhesives* (New York: Pitman, 1944), 137–141. On the transformation of plywood's reputation see Perry, *Modern Wood Adhesives*, 58; J. E. Gordon, *The New Science of Strong Materials or Why You Don't Fall Through the Floor*, 2nd ed. (1968; rpt., Princeton: Princeton University Press, 1976), 161. Other sources on plywood include H. S. Spencer, "Resin Bonded Plywood," *MP* 14 (October 1936): 296–298; R. Koch, "Resin-Bonded Wood," *MP* 17 (October 1939): 416, 418, 420; and Arthur W. Baum, "Wonderful Wooden Sandwich," *Saturday Evening Post* 229 (April 13, 1957): 48–49, 119–120, 122.

[14]   "An All-Plastic Boat," *MP* 25 (September 1947): 166; R. L. Van Boskirk, "The Plastiscope," *MP* 25 (September 1947): 196, 198; idem, "The Plastiscope," *MP* 29 (July 1952): 204; *Plastics Engineering Handbook of the Society of the Plastics Industry* (New York: Reinhold, 1954), 203–213; and J. Harry DuBois, *Plastics History U.S.A.* (Boston: Cahners, 1972), 392–405.

[15]   "Tumblers That Make Light of a Heavy Foot," Du Pont advertisement, *MP* 23 (May 1946): 37; R. L. Van Boskirk, "The Plastiscope," *MP* 25 (October 1947): 234. On Tupper's early career see "Life Begins!," *SPI Voice*, undated issue, ca. autumn 1940, 3, SPIA.

[16]   Robert L. Davis and Ronald D. Beck, *Applied Plastic Product Design* (New York: Prentice-Hall, 1946), v; Clark N. Robinson, *Meet the Plastics* (New York: Macmillan, 1949), 130–131; and "The Yen for a Plastics Business," editorial, *MP* 23 (August 1946): 5.

[17]   "King-Size Moldings," *MP* 34 (October 1956): 123; "Loma's Barnett: A Flair for 'Firsts,' " *MP* 39 (May 1962): 58; and *Why Loma Is the Leader in Plastics Custom Molding* (Fort Worth, Tex.: Loma Industries, ca. 1966), brochure provided by Loma Industries.

[18]   "In the News," *MP* 20 (May 1943): 114. See also "The Army Helmet Liner," *MP* 19 (May 1942): 35–38, 104, 106, 108; "Tough Bayonet Scabbard," *MP* 20 (November 1942): 66.

[19]   R. L. Van Boskirk, "Washington Round-Up," *MP* 21 (June 1944): 152; Kline, untitled typescript on plastics at war delivered at the Western Plastics Conference, February 23, 1943, made available by Gordon M. Kline; *24 Case Histories: How Plastics Solved War Problems* (Pittsfield, Mass.: General Electric Chemical Department, Plastics Division, 1946), 56–57, 34–35; and "Taps for Brass," *MP* 20 (November 1942): 62–63, 134, 136.

[20]   "Kline . . . Interview," 11–12, 15, 18–26. See also Gordon M. Kline, *Investigation of German Plastics Plants* (London [?]: Combined Intelligence Objectives Sub-Committee [CIOS], 1945); Gordon M. Kline, "Plastics in Germany, 1939–1945," *MP* 23 (October 1945): 152A-152P (the first of many reports and translations extending into 1947); John M. DeBell, William C. Goggin, and Walter E. Gloor, *German Plastics Practice* (Springfield, Mass.: DeBell & Richardson, 1946); Rexmond C. Cochrane, *Measures for Progress: A History of the National Bureau of Standards* (Washington, D.C.: U.S. Department of Commerce, National Bureau of Standards, 1966), 371–372, 422, 477–478; and, for a good overview of the activities of CIOS, John Gimbel, *Science, Technology, and Reparations: Exploitation and Plunder in Postwar Germany* (Stanford, Calif.: Stanford University Press, 1990), esp. 3–20. Subsequent accounts have suggested that Americans actually borrowed little from German practices, but for an opposing opinion see Peter H. Spitz, *Petrochemicals: The Rise of an Industry* (New York: John Wiley, 1988), 1–62.

[21]   Quotations from board of directors minutes, January 17, 1941, microfilm reel 2; "National Defense," *SPI Voice* (February 8, 1941): 1; both SPIA. See also "Invitation to Plastics Industry," *SPI Voice* (February 8, 1941): 2; board of directors minutes, February 21, 1941, microfilm reel 2; and minutes of first meeting of Plastics Defense Committee, March 18, 1941, microfilm reel 2; all SPIA. On SPE's history see articles by Bill Bregar in "Vision: Celebrating SPE's Past, Present and Future," *Plastics News* 4 (May 4, 1992): 7–31.

[22]   "Princess Mary White," General Electric advertisement, tearsheet hand-dated September 1942, General Electric Archive, photocopy provided by Roland Marchand; advertisement for Durez Plastics & Chemicals, *MP* 20 (May 1943): inside front cover.

[23]   Cameron Hawley, *Cash McCall* (Boston: Houghton Mifflin, 1955), 84.

[24]   *MP* 19 (May 1942): 20 (ellipsis in original). On wartime activities of industrial designers see Arthur J. Pulos, *The American Design Adventure: 1940–1975* (Cambridge: MIT Press, 1988), 11–47. See also Robert Friedel, "Scarcity and Promise: Materials and American Domestic Culture during World War II," in *World War II and the American Dream: How Wartime Building Changed a Nation*, ed. Donald Albrecht (Washington, D.C.: National Building Museum, and Cambridge: MIT Press, 1995), 42–89.

[25]   *MP* 20 (February 1943): 32; *MP* 20 (August 1943): 50; Carl Sundberg and Montgomery Ferar, "Planning Postwar Applications," *MP* 19 (August 1942): 52–53, 116, 118; and Carl Sundberg, "The Realities of the Future," *MP* 22 (May 1945): 105–108.

[26]   "Why Discuss a *Sportshack* When We're At War?," Durez advertisement in *Fortune* 25 (April 1942): 139, and in *MP* 19 (April 1942): inside front cover. See also "Look! No Hands!," Durez advertisement, *MP* 20 (May 1943): inside front cover; "Donald Deskey Looks at a Telephone for the Future," Durez advertisement in *Fortune* 28 (August 1943): 70, and in *MP* 21 (September 1943): inside front cover; and Pulos, *The American Design Adventure*, 45.

[27]   Mary Madison, "In a Plastic World," *New York Times Magazine* (August 22,

1943): 20; "Plastics Tomorrow," *Scientific American* 170 (March 1944): 103, 105; and Roderick M. Grant, "Molding the World of Tomorrow," *Popular Mechanics* 80 (July 1943): 9, 8.

28   Lougee as paraphrased in "Society of the Plastics Industry: Pacific Coast Conference," *MP* 21 (April 1944): 88; E. F. Lougee, "Plastics Post-War," *Art & Industry* 37 (August 1944): 34–40.

29   Cruse as quoted in "Fight Glamorizing of Plastics Role," *New York Times* (September 1, 1943): 31; William T. Cruse, "SPI in War and Peace," p. 13 of unidentified publication, ca. 1944, SPIF; and "Realities or Reveries?," *MP* 21 (April 1944): 78. See also SPI board of directors minutes, August 19, 1943; January 24, 1944; June 23, 1944; all SPIA, microfilm reel 2; and "What Happened to the Dreamworld?," *Fortune* 35 (February 1947): 90–93, 214–216.

30   R. L. Van Boskirk, "Washington Round-Up," *MP* 21 (June 1944): 152; Scribner as quoted in "Plastic Auto Held 'Engineer's Dream,'" *New York Times* (November 12, 1944): sec. 5, 6.

31   Quotation from John Campbell, "Plastics as a Post-War Business," *Pic* 17 (October 1945): 75.

32   M. M. Makeever interview, 1952, PPA Tapes, reel 1, side 2. See also "USDA Research Develops New Plastic from Farm Wastes," Office of War Information press release, May 18, 1943, AISI, accession 1631, vertical file on "Plastics—Properties."

33   For characterizations of shoddy plastics during and immediately after the war see "'Information, Please,'" *MP* 21 (June 1944): 168; Josephine von Miklos, "By Their Lines Ye Shall Know Them," *MP* 22 (November 1944): 119–120; "Propaganda . . . a Threat or a Boost," *MP* 22 (March 1945): 93–94, 198, 200; James J. Pyle, "New Horizons in Plastics," *Science Digest* 18 (August 1945): 85–86; Campbell, "Plastics as a Post-War Business," 75; George K. Scribner, "Plastics and Metals: Competitors or Collaborators?," in *Annual Report of the Smithsonian Institution 1945* (Washington, D.C.: Government Printing Office, 1946), 171; "Plastics Cannot Work Miracles," *MP* 23 (May 1946): 7; "The Buyer Is Reaching for His Crown," *MP* 24 (February 1947): 5; "The Strength of an Industry," *MP* 27 (November 1949): 5; Charles Lichtenberg, "Misuses of Plastics," in *A Manual of Plastics and Resins in Encyclopedia Form*, ed. William Schack (Brooklyn: Chemical Publishing Co., 1950), 296–298; "Consumer Item Failures Stressed by Joel Goldblatt," *SPI Bulletin*, no. 348 (April 26, 1950): 2, SPIA; "A 1950 Guide to the Plastics," *Fortune* 41 (May 1950): 109, 111; and "Stand-Ins That Made Good on Their Own," *Business Week* (April 10, 1954): 112.

34   L. H. Woodman, "Miracles? . . . Maybe," *The Scientific Monthly* 58 (June 1944): 421; "What Does the Public Know of Plastics?," *MP* 24 (December 1946): 5.

35   "The Shower Curtain Problem," *MP* 24 (July 1947): 5. See also Morris Kaufman, *The Chemistry and Industrial Production of Polyvinyl Chloride: The History of PVC* (New York: Gordon and Breach, 1969), 166–167.

36   Quotations from Campbell, "Plastics as a Post-War Business," 75; Charles A. Breskin, "Where Do Plastics Go From Here?," *Scientific American* 176 (May 1947): 214.

37   Kinnear as quoted in "Host of New Uses in Plastics Shown," *New York Times* (April 23, 1946): 31; "First National Plastics Exhibit Attracts Record Attendance," *SPI Bulletin*, no. 213 (May 1, 1946): 1, SPIA. Other sources include *National Plastics Exposition*, April 22–27, 1946, leaflet, SPIF; "Glamour Child's Growing Pains," *Business Week* (April 27,

1946): 20–22; "S.P.I. Conference and Plastics Exhibit," *MP* 23 (May 1946): 109–116; "Report of 1946 National Plastics Exposition Committee," typescript, SPIA, microfilm reel 2; and "Bill Cruse—25 Years on the Road for SPI," *MP* 44 (January 1967): 262.

[38] "National Plastics Exposition," leaflet.

[39] "Glamour Child's Growing Pains," 21; "S.P.I. Conference and Plastics Exhibit," 109.

[40] On new materials and problems with plasticizers see "Plasticizers," *MP* 26 (March 1949): 55–60; "A 1950 Guide to the Plastics," *Fortune* 41 (May 1950): 109–118, 120.

[41] "Flammability Laws," *MP* 24 (April 1947): 5.

[42] "2nd National Plastics Exposition," *MP* 24 (June 1947): 90; "Plastic Industry Gets Some Advice," *New York Times* (May 8, 1947): 42; and "Recognition of Selling," *MP* 24 (June 1947): 5.

[43] Alfred Auerbach, "I Don't Know," *MP* 23 (October 1945): 99–106; direct quotations from clerks and buyers are on 99–100.

[44] Scribner as paraphrased in "Society of the Plastics Industry: Annual Conference," *MP* 21 (June 1944): 87; see also "George K. Scribner Advocates Labeling of Plastics," *SPI Bulletin*, no. 108 (May 20, 1944): 3–4, SPIA.

[45] Report of D. Gray Maxwell to SPI board of directors, August 21–22, 1947; see also board of directors minutes, October 19, 1944, and September 26, 1945; all SPIA, microfilm reel 2. The following discussion is based also on extensive "Information Labeling Committee Minutes 1948–1951" contained in a looseleaf binder, SPIA, miscellaneous carton.

[46] Discussed by William T. Cruse, Promotion Sub-Committee of the Informative Labeling Committee minutes, September 20, 1949, 7, in binder "Informative Labeling Committee Minutes 1948–1951," SPIA.

[47] Label in the author's possession.

[48] Letter from David S. Hopping of Celanese Corp. of America to Amos Ruddock of Dow Chemical Co., January 20, 1950, copied into Promotion Sub-Committee of the Informative Labeling Committee minutes, January 26, 1950, 4, in binder "Informative Labeling Committee Minutes 1948–1951," SPIA.

[49] "Watch for the Real Competition," *MP* 24 (November 1946): 5.

[50] Board of Directors minutes, January 27–28, July 21–22, September 22–23, November 17–18, 1949, SPIA, microfilm reel 2; "Plastics Christmas Ornament Business Jeopardized," *SPI Bulletin*, no. 311 (February 1, 1949): 1–2; "Illuminated Plastic Christmas Tree Ornaments," *SPI Bulletin*, no. 330 (August 30, 1949): 1–2; and "Fire Chiefs Hear Joseph H. Ward of Noma Electric Company Discuss Plastics Christmas Tree Light Ornaments," *SPI Bulletin*, no. 335 (October 7, 1949): 1; all SPIA.

[51] "Modern Plastics Bulletin," *MP* 29 (September 1951): between 74–75; see also "The Tableware War," *MP* 29 (November 1951): 73–76, 182.

[52] Board of directors minutes, September 19–20, 1951, SPIA, microfilm reel 2.

[53] "Public Service on Plastics Questions Announced by SPI," press release, March 18, 1947, SPIA, microfilm reel 1; Public Relations Committee minutes, March 11, 1949, 3, SPIA, "Public Relations Minutes of Meetings 1947–1954," miscellaneous carton.

[54] Mary Roche, "New Ideas and Inventions," *New York Times Magazine* (November 10, 1946): 42; Christine Holbrook and Walter Adams, "Dogs, Kids, Husbands:

How to Furnish a House So They Can't Hurt It," *Better Homes and Gardens* 27 (March 1949): 37.

[55]   Marion Gough, "The Truth About Plastics," *House Beautiful* 89 (October 1947): 121; "This New Era of Easy Upkeep," 122–124; Charlotte Eaton Conway, "Use Plastics Where the Going Is Rough," 126. The subject of design was treated by Elizabeth Gordon, "What's Wrong with Plastics?," 166; that of proper applications by Helen Markel Herrmann, "They Laughed When I Said 'Methyl Methacrylate,' " 125, 289–291. See also "SPI Aid to All Plastics Issue of House Beautiful Magazine," *SPI Bulletin*, no. 269 (September 25, 1947): 3, SPIA.

[56]   Sidney Gross interviewed by the author, December 9, 1986; "Indestructible Room: New Plastics Protect Walls, Furniture and Rugs from Ravages of Kids and Dogs," *Life* 20 (January 14, 1946): 91; Holbrook and Adams, "Dogs, Kids, Husbands," 37; "Plastics Make Your Housework Easier," *Better Homes and Gardens* 28 (May 1950): 112; and "Formica Is On Top," *Fortune* 44 (October 1951): 116.

[57]   Mary Davis Gillies, "Over 100 Plastics in This Room," *McCall's* 80 (September 1953): 34–37, 104 (first and third set of ellipses in original). On the promotion see "Plastics Home Furnishings Promoted Nationally by McCall's Magazine," *SPI Bulletin*, no. 519 (September 28, 1953): 5, SPIA.

[58]   *Plastics: Everything a Woman Could Ask For* (New York: McCall Corp., 1953), SPIA.

[59]   Sidney Gross interviewed by the author, December 9, 1986.

[60]   "This Year Rediscover Plastics," *House Beautiful* 97 (January 1955): 56.

[61]   "Plastics Output Rose 30 Per Cent," *New York Times* (January 3, 1956): 90.

[62]   *Plastics: Everything a Woman Could Ask For*, unpaginated; "A New Way of Life in One Word: Plastics," *Good Housekeeping* 150 (April 1960): 84–95, 237–238. See also SPI board of directors minutes, November 19–20, 1959, SPIA, microfilm reel 3.

[63]   J. W. McCoy, "The Job Ahead Presents a Direct Challenge to Sales and Advertising," August 2, 1945, typescript, 2–4, DPA, series II, part 2, box 15.

[64]   On the material and cultural inflations of the postwar era see William Leiss, *The Limits to Satisfaction: An Essay on the Problem of Needs and Commodities* (Toronto: University of Toronto Press, 1976), esp. 3–23; Charles Newman, *The Post-Modern Aura: The Act of Fiction in an Age of Inflation* (Evanston, Ill.: Northwestern University Press, 1985), esp. 5–34; and Landon Y. Jones, *Great Expectations: America and the Baby Boom Generation*, rev. ed. (1980; rpt., New York: Ballantine, 1986).

[65]   Hiram McCann, "Doubling—Tripling—Expanding: That's Plastics," *Monsanto Magazine* 26 (October 1947): 4; F. A. Abbiati, "The Horn of Plenty Is Mechanized," *Monsanto Magazine* 26 (October 1947): 26; and General Electric, *24 Case Studies*, 63.

[66]   N.N.T. Samaras, "Something Nature Could Not Supply," *Monsanto Magazine* 26 (October 1947): 18, 21; executive quoted in *Plastics: The Story of an Industry*, rev. ed. (New York: SPI, 1946), 29; and Norman Mailer, "The Big Bite," *Esquire*, May 1963, reprinted in Norman Mailer, *The Presidential Papers* (New York: Berkley Medallion, 1970), 178.

[67]   On chemical shortages and the shift to petroleum and natural gas see "Is Plastics' Future Tied to Coal?," *MP* 24 (March 1947): 89–93, 198; "The Quest for Coal," *MP* 25 (May 1948): 5; R. L. Van Boskirk, "Benzol—Key to Plastics," *MP* 27 (August 1950): 154, 156, 158; "Why Benzol?," *MP* 28 (January 1951): 61, 148; and R. L. Van Boskirk, "The Plastiscope," *MP* 28 (August 1951): 190.

[68]   David M. Potter, *People of Plenty: Economic Abundance and the American Character* (Chicago: University of Chicago Press, 1954). One of Potter's central points was made four years earlier by two economists reviewing the plastic industry. After declaring that it offered "striking proof that our economy has not yet entirely matured," they concluded by observing that "the closing of the land frontier near the end of the last century was barely in time to make way for the opening of the new frontier in science and technology." See E. H. Anderson and W. C. Thompson, "Plastics—A Debutante Industry," *The Southern Economic Journal* 17 (October 1950): 186.

[69]   William J. Hennessey, *Russel Wright: American Designer* (Cambridge: MIT Press, 1983), 36–44, 52–54; Thomas Hine, *Populuxe* (New York: Alfred A. Knopf, 1986), esp. 3–14.

[70]   "Frontiers of Technology," *Life* 43 (October 7, 1957): 83; E. F. Lougee, "Plastics Post-War," *Art & Industry* 37 (August 1944): 40; John Gloag, "The Influence of Plastics on Design," *Journal of the Royal Society of Arts* 91 (July 23, 1943): 466–467. Gloag revised these remarks for inclusion in a short book, *Plastics and Industrial Design* (London: George Allen & Unwin, 1945), 30–41. For similar thoughts expressed by an American designer see Donald A. Wallance, "Design in Plastics," *Everyday Art Quarterly*, no. 6 (Winter 1947–48): 3–4.

[71]   Edgar Kaufmann, Jr., "Borax, or the Chromium-Plated Calf," *The Architectural Review* 104 (August 1948): 88–93; Robert Venturi, Denise Scott Brown, and Steven Izenour, *Learning from Las Vegas* (Cambridge: MIT Press, 1972); Hine, *Populuxe*, 3; and Philip Langdon, *Orange Roofs, Golden Arches: The Architecture of American Chain Restaurants* (New York: Alfred A. Knopf, 1986), 119, 124.

[72]   Alan Hess, *Googie: Fifties Coffee Shop Architecture* (San Francisco: Chronicle Books, 1985), 34.

[73]   Hine, *Populuxe*, 66.

[74]   James W. Sullivan, general manager of Union Products, interviewed by the author, October 9, 1986.

[75]   Wallance, "Design in Plastics," 3.

[76]   Sidney Gross interviewed by the author, December 9, 1986.

[77]   "Plastics Supply and Demand," *MP* 26 (January 1949): 142. See also "Precocious Plastic," *MP* 25 (February 1948): 73–74; "Seventh Modern Plastics Competition Awards," *MP* 26 (September 1948): 134–135; R. L. Van Boskirk, "The Plastiscope," *MP* 31 (October 1953): 236; idem, "The Plastiscope," *MP* 31 (April 1954): 246; idem, "The Plastiscope," *MP* 31 (June 1954): 386; idem, "The Plastiscope," *MP* 36 (November 1958): 41; and "Tupperware Company's Ansley: Quality Control," *MP* 40 (April 1963): 50.

[78]   Gloag, "The Influence of Plastic on Design," 466.

CHAPTER 7: DESIGN IN PLASTIC

[1]   "Plastics: The Future Has Arrived," *Progressive Architecture* 51 (October 1970): 92, 88–89.

[2]   Ronald Cuddon, "Design Review: Plastics," *The Architectural Review* 141 (June 1967): 453.

[3]   "Admiral Video in Plastic," *New York Times* (May 12, 1949): 47; "One-Piece Cabinet Weighs 35 Pounds," *MP* 26 (July 1949): 74–75; "Large Molded Parts: The Big Promise in Plastics," *Business Week* (January 3, 1953): 48; and "All-Plastic Console

Shocked TV Industry in 1948," *Appliance Manufacturer* (April 1966): 71, photocopy provided by Armour H. Titus of Admiral Home Appliances.

[4] Quotations are from transcript of Bob Whitesell interviewed by Sheldon Hochheiser, September 19, 1983, 18, R&H; Alvin Lustig, "Landscape of Lettering," *Architectural Forum* 102 (January 1955): 128. See also "Automotive Plastics: 1956," *MP* 33 (April 1956): 115, 118; Donald S. Frederick interview, 1968, PPA Tapes, reel 2, side 1; transcript of Stanton Kelton, Jr., interviewed by Sheldon Hochheiser, June 24, 1983, 23–31, 33, R&H; and Sheldon Hochheiser, *Rohm and Haas: History of a Chemical Company* (Philadelphia: University of Pennsylvania Press, 1986), 87–93.

[5] "Trends in Toys," *MP* 29 (June 1952): 71. See also "Plastics in the Product Revolution: Dolls," *MP* 38 (October 1960): 100–103, 194, 196.

[6] For short articles on dozens of products using an array of plastics (including vinyl) see "Achievement in Plastics," *MP* 31 (June 1954): 121–183. See also lists of dozens of applications scattered throughout William Schack, ed., *A Manual of Plastics and Resins in Encyclopedia Form* (Brooklyn: Chemical Publishing Co., 1950). For particular vinyl applications see Glenn L. Martin Co. press release no. 1049, February 3, 1946, in "Glenn L. Martin" folder, WCBA, collection 60, box 1; "A New Material for Upholstery," *MP* 23 (August 1946): 91–93; "Laminated Vinyl Tile Flooring," *MP* 24 (April 1947): 123–125; "Stitchless Quilted Vinyl," *MP* 26 (November 1948): 98–99; Egmont Arens, "Disposable Rainwear," *MP* 29 (March 1952): 116–117; "New Vinyl Film That 'Breathes,'" *MP* 30 (July 1953): 90, 169; "New Vinyl Upholstery Ready," *New York Times* (February 4, 1954): 42; "Vinyl Outerwear," *MP* 31 (April 1954): 106–108, 223; and E. Raymond Corey, *The Development of Markets for New Materials: A Study of Building New End-Product Markets for Aluminum, Fibrous Glass, and the Plastics* (Boston: Division of Research, Harvard Graduate School of Business Administration, 1956), 16–35.

[7] "Formica Is On Top," *Fortune* 44 (October 1951): 116–117. See also "What's New in Decorative Laminates," *MP* 28 (October 1950): 73–75.

[8] Scribner interview, 1952, PPA Tapes, reel 2, side 1. See also George K. Scribner, *A Brief Description of the Commonly Used Plastics and Their Origin*, 9th revision (Boonton, N.J.: Boonton Molding Co., 1950), 2; "Plastic Dishes," *Consumer Reports* 16 (January 1951): 10.

[9] On polystyrene applications see "Why Polystyrene Tableware," *MP* 24 (July 1947): 86–87; "Plastic Products," *MP* 28 (June 1951): 112; "Plastic Products," *MP* 30 (February 1953): 104–107; and R. L. Van Boskirk, "The Plastiscope," *MP* 31 (September 1953): 210. On Foster Grant see Glenn D. Kittler, *"More Than Meets the Eye": The Foster Grant Story* (New York: Coronet, 1972), esp. 121–122, 132–144; Edward Van Vlaanderen, *Pronounced Success: America and Hoechst, 1953–1978* (n.p.: American Hoechst Corp., 1979), 138–151. I am indebted to Robert Shelton of Foster Grant for allowing me to examine the company's trade catalogues running back to 1952. Robert L. Davidson of Kurz-Kasch explained the shift to polystyrene radio cabinets in an interview on July 22, 1985.

[10] *The Fisher-Pricer* (50th anniversary issue), 37 (May 1980); booklet on *Fisher-Price . . . Our Company in Brief* (East Aurora, N.Y.: Fisher-Price, 1985); both provided by Louise M. Lavere, Fisher-Price Toys.

[11] Quoted from Plasticville packaging, for which I am indebted to Glenn Porter; see also a Catalin Corporation advertisement, *MP* 30 (May 1953): 1.

[12] An excellent technical history of polystyrene development at Dow is provided by

357

Raymond F. Boyer, "Anecdotal History of Styrene and Polystyrene," in *History of Polymer Science and Technology*, ed. Raymond B. Seymour (New York: Marcel Dekker, 1982), 347–370. Sources on expanded polystyrene foam include D. W. McCuaig and O. R. McIntire, "An E-X-P-A-N-D-E-D Polystyrene," *MP* 22 (March 1945): 106–109, 202; Charles A. Breskin, "Expanding Fields for Expanded Plastics," *Scientific American* 177 (September 1947): 119–121; "Foam and the Future," *MP* 28 (October 1950): 83–86, 166–167; "For Portable Refrigerators, Molded Styrene Foam Takes Over From Metal," *MP* 36 (May 1959): 93–95; R. L. Van Boskirk, "The Plastiscope," *MP* 38 (March 1961): 212; idem, "The Plastiscope," *MP* 41 (June 1964): 45; and "Bright with Foam," *Du Pont Magazine* 68 (March-April 1974): 28–31. On tape cartridge and cassette housings see "New Concept in Recording," *MP* 36 (February 1959): 96–97; Mel Jacolow, "Biggest Hit in Tape Packs: Polystyrene, to the Tune of 19 Million Lb. in '68," *MP* 46 (June 1969): 68–69.

[13]   Eric Fawcett as quoted by Martin Sherwood, "Polythene and Its Origins," *Chemistry and Industry* (March 21, 1983): 239; anonymous molder as quoted in "1954— Fastest Growing Plastic," *MP* 32 (October 1954): 93; J. Harry DuBois interview, 1968, PPA Tapes, reel 2, side 2; and Sidney Gross interviewed by the author, December 9, 1986. Production information is from "Plastics Volume Up for 6th Year," *New York Times* (January 12, 1959): 114. See also M. W. Perrin, "The Story of Polythene," *Research* 6 (March 1953): 111–118; "The Polyethylene Gamble," *Fortune* 49 (February 1954): 134–137, 166, 170, 172, 174; J. A. Allen, *Studies in Innovation in the Steel and Chemical Industries* (Manchester: Manchester University Press, 1967), 7–52; Sherwood, "Polythene and Its Origins," 237–242; and David A. Hounshell and John Kenly Smith, Jr., *Science and Corporate Strategy: Du Pont R&D, 1902–1980* (Cambridge: Cambridge University Press, 1988), 474, 480–482, 494–495.

[14]   John Campbell, "Plastics as a Post-War Business," *Pic* 17 (October 1945): 77.

[15]   "Plastics for Disposables," *MP* 33 (April 1956): 5; "Plastics in Disposables and Expendables," *MP* 34 (April 1957): 206. See also "A Blowing Process for Thermoplastics," *MP* 20 (December 1942): 46; "Powder Dispensed from Squeezable Bottle," *MP* 25 (November 1947): 158; "Moral in a Plastic Bottle," *MP* 27 (October 1949): 5; Bedford Berry, "Now It's Bottles That Bounce," *Du Pont Magazine* 44 (June-July 1950): 32–33; R. P. Vuillemenot of Donald Deskey Associates, "Investigation of Packaging Materials and Method Suitable for Packing of Crisco Shortening," report to Procter & Gamble, February 26, 1951, typescript, DDC; " . . . And Now Blow Molding," *MP* 37 (November 1959): 83–88, 179, 181, 187–188; and "Jules Bernard Montenier," *National Cyclopaedia of American Biography* (New York: James T. White, 1970), 52:180.

[16]   "The Golden-Egg Goose," *MP* 36 (November 1958): 5. This editorial pegged consumption of resin for hula hoops during 1958 at 7.5 million pounds, but an article in the same issue ("Bonanza for Extruders," 146) claimed ten million pounds; nine years later, presumably with more complete statistics, the figure of fifteen million pounds appeared in an article on "Hula Hoops Back in the Spin," *MP* 45 (November 1967): 210. See also "Plastics Products," *MP* 36 (October 1958): 112–113. On pop beads see "Success in Accessories," *MP* 33 (April 1956): 100, 106, 235.

[17]   On development of linear polyethylene and polypropylene see Frank M. McMillan, *The Chain Straighteners: Fruitful Innovation: The Discovery of Linear and Stereoregular Synthetic Polymers* (London: Macmillan, 1979); Hounshell and Smith, *Science and Corporate Strategy*, 491–497; and Davis Dyer and David B. Sicilia, *Labors of a Modern Her-*

cules: The Evolution of a Chemical Company (Boston: Harvard Business School Press, 1990), 284–288, 296–303. See also Francis Bello, "The New Breed of Plastics," *Fortune* 56 (November 1957): 172–175, 218, 220, 223–224; Jack R. Ryan, "Du Pont Patents Common Plastic," *New York Times* (December 18, 1957): 55; R. L. Van Boskirk, "The Plastiscope," *MP* 35 (January 1958): 37, 39, 41; and Stephen Stinson, "Discoverers of Polypropylene Share Prize," *Chemical & Engineering News* 65 (March 9, 1987): 30.

[18]   James W. Sullivan, general manager, Union Products, interviewed by the author, October 9, 1986.

[19]   R. L. Van Boskirk, "The Plastiscope," *MP* 26 (February 1949): 168; "Tough New Thermoplastic," *MP* 33 (September 1955): 104–108, 225–228.

[20]   Dolye Smee, "Nylon and Machines," *Wall Street Journal* (November 17, 1953): 1, 5; "'Zytel' Is New Trade-Mark for Du Pont's Nylon Resin," Du Pont Product Information Service release, May 20, 1954, DPA, accession 1410, box 42; and "Making the Wheels Go Round," *New York Times Magazine*, February 15, 1959, clipping, DPA, series II, part 2, box 602.

[21]   Du Pont press release on Delrin, November 19, 1956, DPA, accession 1410, box 42; Herbert Solow, "Delrin: Du Pont's Challenge to Metals," *Fortune* 60 (August 1959): 116–119, 160, 163–164; "Du Pont Makes Substantial Reduction in Price of 'Delrin' Acetal Resin," press release, February 7, 1961, DPA, accession 1410, box 48; and Hounshell and Smith, *Science and Corporate Strategy*, 486–491.

[22]   As quoted by G. M. Miller, "Introducing a New Product: Talk for Educators' Conference," June 22, 1954, typescript, DPA, series II, part 2, box 76. On polyester fiber see also "Market Here Tested for 'Fiber V' Shirts," *New York Times*, November 26, 1950, clipping, DPA, Polyester file; "Pressed and Impressed," *Time*, April 15, 1966, clipping, DPA, series II, part 2, box 604; Isadore Barmash, "Polyester Emerges From Shadow of Nylon," *New York Times* (May 1, 1966): sec. 3, 1, 14; and Allen, *Studies in Innovation in the Steel and Chemical Industries*, 53–95.

[23]   Quotation from Gordon H. Kester, "'Mylar' Is in the Memory Business," *Du Pont Magazine* 53 (April-May 1959): 20. On Mylar see also Du Pont Public Relations Department, "Press Analysis for February 14, 1952," DPA, series II, part 2, box 885; Du Pont Product Information Service, "New Developments in 'Mylar' Polyester Film," June 1954, DPA, accession 1410, box 8; "Can 'Mylar' Do a Job for You?," *Du Pont Magazine* 49 (February-March 1955): 11; Public Relations Department, "Press Analysis for March 6, 1956," DPA, series II, part 2, box 901; Kenneth Smith, "Du Pont's Mylar, Young But Tough, Fights Way Into New Industry Jobs," *Wall Street Journal* (April 17, 1956): 1, 19; "Look What They're Doing with Tape," *Du Pont Magazine* 51 (October-November 1957): 28–29; and Public Relations Department, "Press Analysis for August 24, 1960," DPA, series II, part 2, box 916.

[24]   Gordon M. Kline, "Plastics in Germany, 1939–1945," *MP* 23 (October 1945): 152F-152G; "Polyurethane and Polyester Foams," *MP* 32 (November 1954): 106–108, 214–216; Francis Bello, "The Next Great Synthetic," *Fortune* 51 (March 1955): 110–113, 166, 169; C. T. Ludwig, "Foam on the Rise," *Industrial Design* 20 (May 1973): 32–37; Dan J. Forrestal, *Faith, Hope and $5,000: The Story of Monsanto* (New York: Simon and Schuster, 1977), 159–162; and Kurt C. Frisch, "History of Science and Technology of Polymeric Foams," in Seymour, *History of Polymer Science and Technology*, 28–30.

[25]   On Teflon as a nonstick coating see "Advances Shown in Plastic Items," *New York Times* (March 29, 1950): 51; "Plastics Industry Eliminates Thumping of Catsup Bottles,"

unidentified newspaper clipping, March 12, 1952, DPA, series II, part 2, box 886; *The Cavalcade of America*, no. 756, September 30, 1952, p. 29, Hagley Imprint Collection; and "The Demonstration at Madame Romaine's Place," *The Journal of Teflon* 2 (April 1961): 6. See also Hounshell and Smith, *Science and Corporate Strategy*, 157, 482–486; James J. Bohning, "Live Guinea Pigs and Dead Cylinders: Roy Plunkett and the Discovery of Teflon," *Beckman Center News*, 9, no. 1 (Spring 1992): B3–B6.

[26] "Irradiated Polyethylene," *MP* 31 (April 1954): 100–101, 219; "The Atom—and Plastics," special issue of *MP* 32 (March 1955); and Herbert R. Simonds, *A Concise Guide to Plastics* (New York: Reinhold, 1957), 66–67.

[27] Quotations are from *Washington Daily News* as quoted by Public Relations Department, "Press Analysis for January 29, 1964," DPA, series II, part 2, box 928. See also "'Corfam': A Research to Reality Case History," n.d., typescript, DPA, accession 1410, box 48; *This Is Corfam* (Wilmington, Del.: Du Pont, n.d.), DPA, accession 1410, box 48; Harry E. Davis, "'Corfam': First Focus Is Footwear," *Du Pont Magazine* 57 (November-December 1963): 2–5; Lawrence Lessing, "Synthetics Ride Hell-Bent for Leather," *Fortune* 70 (November 1964): 172–175, 180, 184, 188, 190; and Hounshell and Smith, *Science and Corporate Strategy*, 536–538.

[28] For example see John Peter, "Plastic Age," *Look* 22 (April 15, 1958): 44–49.

[29] Gerald C. Johnson, "Multiple Uses—Wider Markets," *MP* 29 (March 1952): 96. Heralded as "the first issue of any industrial magazine ever to be devoted entirely to the future" (93), the March 1952 issue of *MP* published comments by twelve industrial designers on design and plastic.

[30] "The Plastics Industry and Design: A Review," *Industrial Design* 3 (June 1956): 51.

[31] Jane Fiske Mitarachi, "Plastics and the Question of Quality," *Industrial Design* 3 (June 1956): 64–67; Arthur J. Pulos, "The Future of Fiber-Reinforced Plastics," *Industrial Design* 17 (January-February 1970): 65.

[32] General sources on glass-reinforced polyester include Maurice Lannon, *Polyester and Fiberglas*, 2nd ed. (North Hollywood, Calif.: Maurice Lannon and Gem-O'-Lite Plastics Co., 1954); Ralph H. Sonneborn with Albert G. H. Dietz and Alton S. Heyser, *Fiberglas Reinforced Plastics* (New York: Reinhold, 1954); and Gregory Dunne, Arthur Gregor, and Douglas C. Meldrum, "Reinforced Plastics," *Industrial Design* 5 (October 1958): 35–87.

[33] "Reinforced Plastics by 'Sprayup,'" *MP* 35 (February 1958): 119–121, 124; "The Day of RP Mass Production Has Arrived," *MP* 43 (February 1966): 72–78, 166, 168.

[34] R. L. Van Boskirk, "The Plastiscope," *MP* 25 (September 1947): 198; anonymous molders quoted by Dunne, Gregor, and Meldrum, "Reinforced Plastics," 42, 39; and Hiram McCann, report read at meeting of Policy Committee, SPI Reinforced Plastics Division, White Sulphur Springs, West Virginia, September 5, 1952, typescript, SPIA, microfilm reel 10.

[35] Games [sic] Slater, "Production and Fabricating Possibilities of Glass Fabric Reinforced Plastics," *MP* 24 (December 1946): 142. On Stout see "Automobiles: The Plastics Picture," *Newsweek* 41 (February 9, 1953): 61; William B. Stout, *So Away I Went!* (Indianapolis: Bobbs-Merrill, 1951).

[36] Robert Lee Behme, *Manual of Building Plastic Cars* (Los Angeles: Trend Books, 1954), 11. See also Behme, 5–18; untitled press release, April 9, [1952], typescript, SPIA, microfilm reel 1; "Automobiles: And Now Plastic," *Newsweek* 39 (April 21, 1952):

90; "At the 5th N.P.E.," *MP* 29 (May 1952): 87; and "Reinforced Plastics Automobile Bodies," *MP* 30 (February 1953): 75–82.

[37] As quoted by Dunne, Gregor, and Meldrum, "Reinforced Plastics," 41. Corvette sources include their article, 36, 40, 55, 57; "Here Comes the Corvette!," *MP* 31 (December 1953): 83–91, 201; and Behme, *Manual of Building Plastic Cars*, 19–20.

[38] Behme, *Manual of Building Plastic Cars*, inside front cover.

[39] Lannon, *Polyester and Fiberglas*, 3, 28, 26, 27, 10.

[40] Behme, *Manual of Building Plastic Cars*, 32, 37; *Eero Saarinen on His Work*, ed. Aline B. Saarinen (New Haven, Conn.: Yale University Press, 1962), 68 (the composite paragraph from which these phrases come is dated 1949 to 1960).

[41] The most complete source on both Eliel and Eero Saarinen is *Design in America: The Cranbrook Vision 1925–1950*, ed. Andrea P. A. Belloli (New York: Harry N. Abrams, 1983). See also Allan Temko, *Eero Saarinen* (New York: George Braziller, n.d. [ca. 1962]).

[42] For references to plywood as plastic see "In the Limelight," *MP* 18 (February 1941): 74; "Modeling Postwar Motor Car Bodies," *MP* 20 (February 1943): 50; and "Gordon M. Kline: Transcript of an Interview Conducted by Jeffrey L. Meikle in Lake Worth, Florida, on 15 and 16 May 1987" (Philadelphia: Beckman Center for the History of Chemistry, 1989), 4–5. The most complete source on Eames and (beginning in 1941) his wife Ray Kaiser Eames is John Neuhart, Marilyn Neuhart, and Ray Eames, *Eames Design: The Work of the Office of Charles and Ray Eames* (New York: Harry N. Abrams, 1989). See also Arthur Drexler, *Charles Eames: Furniture from the Design Collection* (New York: Museum of Modern Art, 1973); Ralph Caplan, "Making Connections: The Work of Charles and Ray Eames," in *Connections: The Work of Charles and Ray Eames* (Los Angeles: UCLA Art Council, 1976), 15–54; and Belloli, ed., *Design in America*. On the Saarinen-Eames seating entries, see Eliot F. Noyes, *Organic Design in Home Furnishings* (New York: Museum of Modern Art, 1941), 10–17.

[43] "Contour-Formed Chairs," *MP* 26 (November 1948): 154–155. See also R. Craig Miller, "Interior Design and Furniture," in Belloli, ed., *Design in America*, 120. The Rockwell illustration is reproduced by Thomas Hine, *Populuxe* (New York: Alfred A. Knopf, 1986), 69.

[44] Quotations are from the Eameses' competition submissions as illustrated in *Eames Design*, 96–97.

[45] Edgar Kaufmann, Jr., *What Is Modern Design?* (New York: Museum of Modern Art, 1950), 11.

[46] "Polyester Chair Takes the Prize," *MP* 27 (August 1950): 67, 147.

[47] Saarinen as quoted by Dunne, Gregor, and Meldrum, "Reinforced Plastics," 75.

[48] *Eero Saarinen on His Work*, 68. See also Miller, "Interior Design and Furniture," 120–121.

[49] *Eero Saarinen on His Work*, 11.

[50] Dreyfuss as quoted by Dunne, Gregor, and Meldrum, "Reinforced Plastics," 71. See also "Indestructible Bus Seats," *MP* 35 (April 1958): 94–96; "Bowling Gets New Look for New Business," *MP* 36 (June 1959): 91–93.

[51] Clark as paraphrased in SPI board minutes, June 29–30, 1950, SPIA, microfilm reel 2. Biographical details are from Albert G. H. Dietz interviewed by the author, October 10, 1986. See also a blurb accompanying his article "Physical and Engineering Properties of Plastic," in *Plastics in Building*, ed. Charles R. Koehler (Washington, D.C.:

National Academy of Sciences and National Research Council, 1955), 11; "MIT's Al Dietz: Architect in Plastic," *MP* 39 (June 1962): 48. Clark discussed the MIT fund in his contribution to Formica Foremen's Business Club, "This Is Your Life: Mr. D. J. O'Conor," typescript, June 1, 1955, 65–66 (provided by Formica Corporation).

52   *Plastics in Housing* (Cambridge: MIT Department of Architecture, 1955), 3.

53   On applications then current see Koehler, *Plastics in Building*. On the Smith house see Betty Pepsis, "Plastics Limned in 'Dream Houses,'" *New York Times* (October 29, 1954): 26; idem, "People in Plastic Houses," *New York Times Magazine* (November 28, 1954): 51; and "People Who Live in Plastic Houses," *Nation's Business* 42 (December 1954): 96. On the Noyes model see "Plastic-Dome House To Be Built by 1964," *New York Times* (July 1, 1954): 29; Pepsis, "People in Plastic Houses," 51; "Plastics in Building," *Architectural Forum* 102 (January 1955): 120; and Albert G. H. Dietz, "Better Buildings . . . with Plastics," *MP* 32 (February 1955): 85.

54   "Final Report" of R. P. Whittier and M. F. Gigliotti on "Job #P-514," entitled "Engineering Analysis and Structural Design of the Monsanto House of the Future," submitted to the Engineering Department, Plastics Division, Monsanto Chemical Company, April 24, 1957, typescript, 2–3. I am indebted to Albert G. H. Dietz for a copy of the report.

55   Douglas Haskell, "In Architecture, Will Atomic Processes Create a New 'Plastic' Order?," *Architectural Forum* 101 (September 1954): 100–101.

56   Whittier and Gigliotti, "Engineering Analysis and Structural Design," 1–2.

57   The most comprehensive source is Whittier and Gigliotti's report. A good secondary source is Alan Hess, "Monsanto House of the Future," *Fine Homebuilding*, no. 34 (August-September 1986): 70–75; a shorter version appears in idem, *Googie: Fifties Coffee Shop Architecture* (San Francisco: Chronicle Books, 1985), 49–51. See also "Experimental House in Plastics," *Arts & Architecture* 72 (November 1955): 20–21; "Monsanto-MIT Molded Module," *Progressive Architecture* 36 (December 1955): 71; "New Architectural Concepts in Plastics House," *MP* 33 (December 1955): 188–189; Ernst Behrendt, "Plastic House," *Popular Science Monthly* (April 1956): 144–147, 262; "Big Impact, Bigger Potential," *Newsweek* 47 (June 4, 1956): 80–81; "Plastics—Shaping Tomorrow's Houses," *Architectural Record* 120 (August 1956): 210; Gladwin Hill, "4 Wings Flow From a Central Axis in All-Plastic 'House of Tomorrow,'" *New York Times* (June 12, 1957): 31; Thomas W. Bush, "Push-Button, Pine-Scented Plastic House With 'Floating' Rooms Shown at Disneyland," *Wall Street Journal* (June 13, 1957): 5; "Monsanto Reveals Present and Future of Plastics in Architecture," *Progressive Architecture* 38 (July 1957): 89; Dan MacMasters, "Your Plastic House of the Future?," *Los Angeles Examiner Pictorial Living* (July 28, 1957): cover, 6–10; and "Plastic House of the Future," *Industrial Design* 4 (August 1957): 48–57. I am indebted to Joan Goody for copies of some of these articles.

58   Whittier and Gigliotti, "Engineering Analysis and Structural Design," 6.

59   See C. G. Cullen, "Fabricating the Structural Components of the All-Plastic 'House of the Future,'" *Plastics Technology* 10 (October 1958): 921–927.

60   "San Francisco Golden Gate Exposition 1939," *The Architectural Forum* 70 (June 1939): 464. The comment referred to the New York World's Fair of 1939.

61   Behrendt, "Plastic House," 146.

62   Gerd Wilcke, "Plastics Producers Seek a Here and Now for the House of the Future," *New York Times* (January 11, 1970): sec. 12, 59. See also Paul Sargent Clark,

"Better Living Through Chemistry?," *Industrial Design* 15 (July-August 1968): 32–37; Alastair Best, "GRP: Those in Favour," *Design*, no. 318 (June 1975): 37.

[63] "Whose House Is First?," *Industries*, [1962], clipping provided by Joan Goody.

[64] The survey first appeared in *Architektur und Wohnformen*, November 15, 1969, and was reprinted in "Plastic Buildings," *Architectural Design* 40 (April 1970): 167–168, and in larger format in Robert B. Hartwig, *Fiberglass Buildings* (Los Angeles: n.p., 1970). See also *The Last Whole Earth Catalog: Access to Tools* (New York: Portola Institute and Random House, 1971), 92.

[65] William T. Cruse, "The Potential for Plastics in the Building Construction Market," Plastics in Building Construction Council minutes, September 1, 1964, SPIA, microfilm reel 9; C. L. Condit, "Report to Board of Directors: Progress Report on Activity of the SPI Code Advisory Committee," typescript, November 1959, 4, SPIA, microfilm reel 3; and Albert G. H. Dietz, "Is a Plastics Breakthrough in Building Due in the Sixties?," *Architectural and Engineering News* 2 (July 1960): 4. For typical discussions see David S. Plumb, "The Plastics Building Revolution," *SPE Journal* 20 (February 1964): 121–125; "A Battle for Building," editorial, *MP* 41 (April 1964): 83; *Selected Abstracts on Structural Applications of Plastics* (New York: American Society of Civil Engineers, 1967); and Armand G. Winfield, "A Case Study: The Plastic House," *Progressive Architecture* 51 (October 1970): 80–87.

[66] Cruse's opinion is from "The Potential for Plastics in the Building Construction Market." Most quotations are from "The Formica World's Fair House," *Formica Fair Facts*, ca. 1964. I am indebted to Steven Holt of Formica for providing copies of this and other promotional materials. See also "The World's Fair House," *Good Housekeeping* 158 (May 1964): 100–113, 172, 174; the "wipe-clean" quotation is from 101.

[67] Albert G. H. Dietz, "Plastics: A Decade of Progress," *Progressive Architecture* 51 (October 1970): 65; idem, "Plastics: The Next Decade," *Progressive Architecture* 51 (October 1970): 100–101.

[68] Best, "GRP: Those in Favour," 36, 41.

[69] Armand G. Winfield, "Excursion into a Plastic Future," *AIA Journal* 45 (February 1966): 66; Joel Frados, "A Split Personality," *MP* 44 (February 1967): 81.

[70] Both of these structures gained wide publicity; the plastic industry learned of them through "Geodesic Dome and Underwater Tunnel," *MP* 38 (June 1961): 100–101; R. L. Van Boskirk, "The Plastiscope," *MP* 43 (July 1966): 206; and "Report from Expo 67—No. 2: The U.S. Pavilion," *MP* 44 (May 1967): 111, 190. Albert G. H. Dietz gave the U.S. pavilion a prominent place in his handbook *Plastics for Architects and Builders* (Cambridge: MIT Press, 1969), 9. On Fuller's career see Martin Pawley, *Buckminster Fuller* (New York: Taplinger, 1990).

[71] Buckminster Fuller, *Utopia or Oblivion: The Prospects for Humanity* (New York: Bantam, 1969), 184–185; idem, *Ideas and Integrities: A Spontaneous Autobiographical Disclosure* (1963; rpt., New York: Collier, 1969), 75–76.

[72] "Aerospace Age: Plastics in Orbit," *MP* 39 (February 1962): 74–79, 152, 154, 157, 160, 163–164; quotation from 75. See also Du Pont Public Relations Department, "Du Pont Products in the Space Program," August 1965, typescript, DPA, accession 1410, box 54.

[73] R. L. Van Boskirk, "The Plastiscope," *MP* 43 (April 1966): 45; "Big Things Are Happening in Vinyl Film," *MP* 43 (June 1966): 72–73; and "Go-Go Vinyl," *American Fabrics*, no. 73 (Fall-Winter 1966): 92–94.

74   As in the title of Buckminster Fuller, *Operating Manual for Spaceship Earth* (Carbondale: Southern Illinois University Press, 1969). For a British perspective on pop culture, the space program, and plastic see Nigel Whiteley, "Shaping the Sixties," in *1966 and All That*, ed. Jennifer Harris and Sarah Hyde (London: Trefoil, 1986), esp. 25–29; Nigel Whiteley, *Pop Design: Modernism to Mod* (London: Design Council, 1987), 115–116.

75   Frederick J. McGarry, "Structural Considerations," *Progressive Architecture* 41 (June 1960): 169.

76   Thomas Herzog, *Pneumatic Structures: A Handbook of Inflatable Architecture* (New York: Oxford University Press, 1976), 7. Most illustrations were of European origin—owing to Herzog being German—with spectacular exceptions from Expo 70 at Osaka.

77   R. L. Van Boskirk, "The Plastiscope," *MP* 28 (April 1951): 202. See also "Plastic 'Skyhooks,'" *MP* 25 (May 1948): 100–101.

78   "Air-Supported Dome for Swimming Pools Latest in Pneumatic Buildings," press release, June 5, 1957, DPA, series II, part 2, box 601; "The Radome Goes Civilian," *Du Pont Magazine* 51 (August-September 1957): 18–20; David Allison, "Those Ballooning Air Buildings," *Architectural Forum* 111 (July 1959): 134–139; Harry E. Davis, "Celestial Signal Center," *Du Pont Magazine* 56 (May-June 1962): 18–21; "Fabulous Inflatables," *MP* 40 (October 1962): 99–103, 189; Jack Murphy, "New Air of Structural Importance," *Du Pont Magazine* 61 (May-June 1967): 2–6; and Arthur Quarmby, *The Plastics Architect* (London: Pall Mall Press, 1974), 60–62, 98–114.

79   "Pneumatic Geodesic Dome Pumps Up Like an Air Mattress," *Architectural Forum* 109 (November 1958): 177.

80   Murray Kamrass, a Cornell engineer, as quoted in "Air-Supported Dome for Swimming Pools Latest in Pneumatic Buildings." See also "Builder Cuts Lost Time with Nylon Air Shelter," press release, April 11, 1960, DPA, series II, part 2, box 602; "Two Air-Supported Structures for Athletics," *Architectural Record* 135 (March 1964): 209–210; "Plastic Bubble Evolves from Warehouse to Airhouse," *Architectural Record* 121 (June 1957): 248; and "Airhouse," *Domus*, no. 364 (March 1960): 17–18.

81   Price and Ekstrom as quoted by Evan Jenkins, "Plastic Bubble, New Dimension in U.S. Education," *New York Times* (May 26, 1973): 15. See also "Air Buildings: No Longer Just Castles in the Sky," *MP* 49 (March 1972): 47, which describes the Columbia campus as eventually to enclose seven acres under a single dome. On Price's Fun Palace see Reyner Banham, *Megastructure: Urban Futures of the Recent Past* (New York: Harper & Row, 1976), 84–89; Whiteley, *Pop Design*, 135–138.

82   Reyner Banham, "Triumph of Software," *New Society*, October 31, 1968; reprinted in Banham, *Design by Choice*, ed. Penny Sparke (London: Academy Editions, 1981), 133–136; quotation from 134.

83   Banham, *Megastructure*, 100–101. My comments on megastructures are indebted to his chapter on "Fun and Flexibility," 84–103, and to his essay "Zoom Wave Hits Architecture," *New Society*, March 1966, reprinted in Banham, *Design by Choice*, 64–65. See also Quarmby, *The Plastics Architect*, 143–181; Whiteley, *Pop Design*, 138–150.

84   As quoted by Banham, *Design by Choice*, 65.

85   Banham, *Megastructure*, 97–98.

86   Banham, "A House Is Not a Home," *Art in America*, April 1965, reprinted in Banham, *Design by Choice*, 56–60; quotations from 57, 59.

87   "Libre," *Architectural Design* 42 (December 1971): 727–728; "Inflatable, Portable

House," *Vogue* 158 (August 1, 1971): 117; Winfield, "Excursion into a Plastic Future," 63; and Houston Astros general manager Paul Richards as quoted in "Baseball: Daymares in the Dome," *Time* 85 (April 16, 1965): 97. Actually Richards was not referring to the totality of the Astrodome as an enclosed, artificial world, which Robert Altman captured well in his film *Brewster McCloud* (1970). Richards was criticizing the newly opened stadium for a sunny-day glare through its translucent Lucite roof that prevented outfielders from catching easy fly balls and forced them to wear batting helmets for protection. Du Pont had promised the roof would admit enough light for grass to grow on the field. After the 642-foot-diameter roof was coated with an off-white acrylic paint to keep out the glare, developer Roy Hofheinz bragged that Bermuda grass would still grow, and Du Pont predicted the measure would succeed in "retaining the original concept of indoor baseball with an outdoor flavor." They were both wrong, and Hofheinz was soon paying Monsanto more than half a million dollars to install its artificial ChemGrass, then only a year old and shortly renamed AstroTurf in honor of its first major application. Journalist Red Smith delivered a premature epitaph on baseball when he said of the plastic turf that "under the worn soles of a sportswriter it feels like the carpeting in a funeral home." Quotations are from Du Pont Public Relations Department, press release to sports editors, April 21, 1965, DPA, series II, part 2, box 84; Smith as quoted by Ron Fimrite, "Is It Baseball or Pinball?," *Sports Illustrated* 63 (August 12, 1985): 45. See also William Oscar Johnson, "The Tyranny of Phony Fields," *Sports Illustrated* 63 (August 12, 1985): 34–37, 40, 42; and other materials in DPA, series II, part 2, box 84.

[88] "Air-Supported Structure Hardens to Become Self-Supporting," *MP* 44 (August 1967): 107; "Back to the Cave (Urethane, That Is)," *MP* 48 (March 1971): 57; "New Foam Homes," *American Home* 74 (April 1971): 75–87; and Clinton A. Page, "Foam Home," *Progressive Architecture* 52 (May 1971): 100–103.

[89] "New Foam Homes," 75, 77; as quoted by Page, "Foam Home," 102.

[90] Quotation from Stuart Wood, "When Furniture *Really* Goes to Plastic," *MP* 45 (July 1968): 92.

[91] "First One-Piece Molded PP Chairs Moving Fast," *MP* 39 (September 1961): 94–95, 208.

[92] Quotations are from John Peter, "Good Things from Italy," *Look* 32 (October 1, 1968): 59; Barbara Plumb, "Curves on the Horizon," *New York Times Magazine* (December 28, 1969): 20; Harriet Morrison, "Fantasy Comes Alive," *House Beautiful* 112 (January 1970): 74; "The Furniture of Chemistry," *Time* 98 (October 25, 1971): 72; and Spiros Zakas as quoted in "More Plastic Furniture En Route," *House Beautiful* 115 (September 1973): 12. On the popularity of Italian plastic furniture see also Dorothy Kalins Wise, "One Word of Advice: Plastics," *New York* (November 11, 1968): 45–47; Monica Geran, "Plastics Please," *Interior Design* 42 (June 1971): 114–123; J. Roger Guilfoyle, "The Age of Resin," *Industrial Design* 19 (July-August 1972): 52–59; and "The Plastic Aesthetic," *Industrial Design* 20 (July-August 1973): 42–51. On Kartell see Kate Singleton, "How Plastics Gained an Identity," *Design*, no. 411 (March 1983): 13; Penny McGuire, "Find What's Missing," *Design*, no. 413 (May 1983): 44–45. For a sense of the scope of all-plastic furniture design in Europe see photographs of a hundred chairs exhibited at Milan in 1975, "Cento sedie in materia plastica," *Domus*, no. 555 (February 1976): 41–44. The best survey of plastic in furniture during the 1960s is Thelma R. Newman, *Plastics as Design Form* (Philadelphia: Chilton, 1972).

[93] Victor Papanek, *Design for the Real World: Human Ecology and Social Change* (1972;

rpt., New York: Bantam, 1973), 53; Banham, "Triumph of Software," in *Design by Choice*, 134.

94   Melissa Hattersly, "Blow-Up," *Interiors* 126 (May 1967): 129; Stuart L. Levine as quoted by Stuart Wood, "When Furniture *Really* Goes to Plastics," 91. See also Constantino Corsini, "I mobili gonfiabili pneu," *Domus*, no. 457 (December 1967): 8–21.

95   Joel Frados, "Plastics in the 1980's—A 15-Year Outlook," *MP* 45 (November 1968): 120–138; quotations from 129–130 (ellipses in original).

96   Sandra R. Zimmerman, ed., *Plastic as Plastic* (New York: Museum of Contemporary Crafts, 1968).

97   Barbara Plumb, "Genuine Plastic," *New York Times Magazine* (November 10, 1968): 116–117; "Plastic Moves Uptown," *Esquire* 71 (June 1969): 105; Peg Rumely and Nelda Cordts, "The Plastics Explosion in Home Furnishings," *Better Homes and Gardens* 47 (April 1969): 59; and Hilton Kramer, "'Plastic as Plastic': Divided Loyalties, Paradoxical Ambitions," *New York Times* (December 1, 1968): sec. 2, 39. See also Joel Frados, "Plastics for Art's Sake," *MP* 45 (August 1968): 77.

98   Wolfgang Limmer, "Böse neue Welt," *Der Spiegel* (October 25, 1982): 283; Deeds as quoted in "Plastics: The Future Has Arrived," *Progressive Architecture* 51 (October 1970): 95; Deyan Sudjic, "Plastics Grow Up," *Design*, no. 379 (July 1980): 36; and Mayen as quoted in "Plastics: The Future Has Arrived," 95.

99   "Hearing Planned in Flag Burning," *New York Times* (April 12, 1966): 11. For typical accounts of Exploding Plastic Inevitable see Carter Ratcliff, *Andy Warhol* (New York: Abbeville Press, 1983), 55–56; Tally Brown in *Warhol: Conversations about the Artist*, ed. Patrick S. Smith (Ann Arbor, Mich.: UMI Research Press, 1988), 253–254; and Victor Bockris, *The Life and Death of Andy Warhol* (New York: Bantam, 1989), 186–190. On *Silver Clouds* see Kynaston McShine, ed., *Andy Warhol: A Retrospective* (New York: Museum of Modern Art, 1989), 293, 430–431.

100   "Portraits of the Artists" from the portfolio "Ten from Leo Castelli" is illustrated by McShine, *Andy Warhol*, 303. See also Zimmerman, *Plastic as Plastic*. Warhol made a film called *Vinyl* (1965).

101   This is the first quotation in "Warhol in His Own Words," ed. Neil Printz, in McShine, *Andy Warhol*, 457; it is often cited.

102   On the last point see a color photograph in "People," *Time* 120 (November 15, 1982): 72.

103   Ratcliff, *Andy Warhol*, 23, 37, 46. For numerous parallel opinions see statements gathered in "A Collective Portrait of Andy Warhol," in McShine, *Andy Warhol*, 423–455.

104   Lucy Lippard, "New York Letter," *Art International* 9 (May 1965): 53.

105   Tracy Atkinson, "Introduction," *A Plastic Presence*, ed. Tracy Atkinson and John Lloyd Taylor (Milwaukee: Milwaukee Art Center, 1969), 4; Robert Pincus-Witten, "New York," *Artforum* 8 (January 1970): 69; Grégoire Müller, "A Plastic Presence," *Arts Magazine* 44 (November 1969): 36–37. Several books documented plastic in art and offered practical advice: Nicholas Roukes, *Sculpture in Plastics* (New York: Watson-Guptill, 1968); Thelma R. Newman, *Plastics as an Art Form* (1964; rev. ed., Philadelphia: Chilton, 1969); *The Artists' Plastic Guide to Polyester Resins* (New York: Creative Plastics Institute, ca. 1971); and Thelma R. Newman, *Plastics as Sculpture* (Radnor, Penn.: Chilton, 1974).

106   The most complete account of the competition and commission is Jack Murphy, "Sculpture: The Age of Acrylic," *Du Pont Magazine* 64 (September-October 1970): 11–13; quotation from 12.

107   Palmer D. French, "Plastics West Coast," *Artforum* 6 (January 1968): 48–49; Peter Selz and Carol Lindsley, "Plastics in Art," *Art in America* 56 (May-June 1968): 114–115; Stephen S. Prokopoff, *Plastics and New Art* (Philadelphia: Institute of Contemporary Art, University of Pennsylvania, 1969); and Atkinson and Taylor, *A Plastic Presence*, 26–27.

108   Frank Brevoort as quoted by Murphy, "Sculpture: The Age of Acrylic," 13.

109   As quoted by Ed Hotaling, "The Age of Lucite Dawns in Sacramento," *Art News* 69 (May 1970): 80.

110   The best textual introduction is Martin H. Bush, *Sculptures by Duane Hanson* (Wichita, Kans.: Edwin A. Ulrich Museum of Art, Wichita State University, 1985); the monograph with the best illustrations is Kirk Varnedoe, *Duane Hanson* (New York: Harry N. Abrams, 1985).

111   For a similar comment see Fredric Jameson, *Postmodernism, or, The Cultural Logic of Late Capitalism* (Durham, N.C.: Duke University Press, 1991), 32–34.

112   As quoted by Barbara Rose, *Claes Oldenburg* (New York: Museum of Modern Art, 1970), 9; other quotations from 148, 9.

113   Ibid., 139.

114   As quoted by Martin Friedman, *Oldenburg: Six Themes* (Minneapolis: Walker Art Center, 1975), 45.

115   William James, *The Will to Believe and Other Essays* (1897; rpt., New York: Dover, 1956), 25.

116   As quoted by Friedman, *Oldenburg*, 21.

117   Oldenburg as quoted by Friedman, *Oldenburg*, 45.

118   David Bourdon, "The Plastic Arts' Biggest Bubble: Les Levine Bursts with Elusive Ideas—and Ego," *Life* 67 (August 22, 1969): 62–67; quotation from 62. See also John Perreault, "Plastic Man Strikes," *Art News* 67 (March 1968): 36–37, 72–73; David Bourdon, "Plastic Man Meets Plastic Man," *New York* 2 (February 10, 1969): 44–46; and materials in MOMA, "artist file" on Levine.

119   Levine as quoted by Bourdon, "The Plastic Arts' Biggest Bubble," 65; Les Levine, "For Immediate Release," *Art and Artists*, vol. 4, no. 2 (1969): 47–48; and Perreault, "Plastic Man Strikes," 37.

120   Zimmerman, *Plastic as Plastic*, 4.

121   Perreault, "Plastic Man Strikes," 36–37.

122   Levine, "For Immediate Release," 50; Les Levine, statement of July 1973 in Pierre Restany, *Plastics in Arts* (Paris and New York: Leon Amiel, 1974), 134.

123   Brydon Smith, "An Interview with Les Levine," in *Slipcover* (Toronto: Art Gallery of Ontario, 1966), 2–3. Also see another interview, "Les Levine," in *Schemata* 7, ed. Elayne H. Varian (New York: Finch College Museum of Art, 1967).

124   Varian, *Schemata* 7; Museum of Modern Art press release, April 21, 1967, MOMA artist file; and "Bubble Sculptures Are Problems," *Ardmore* (Oklahoma) *Ardmoreite*, May 23, 1967, clipping, MOMA artist file. See also Rita Reif, "And the Walls Come Tumbling Down," *New York Times*, April 19, 1967, clipping, MOMA artist file.

125   Levine, "For Immediate Release," 49. Other sources include a press release from the City of New York Administration of Parks, Recreation and Cultural Affairs, January 17, 1969; and Louis Calta, "Creator to Dispose of Art Exhibition a Little at a Time," *New York Times*, n.d., clipping; both in MOMA, artist file.

126   As quoted by Bourdon, "The Plastic Arts' Biggest Bubble," 65.

127   Flier for a preview at the Douglas Gallery, Vancouver, March 1, 1968, MOMA, artist file. The health risk was real for artists who worked with plastic. Duane Hanson discussed his lymph cancer in an interview in Varnedoe, *Duane Hanson* (1985), 32–33. Niki de Saint-Phalle discussed lung surgery necessitated by exposure to polyester resin in an interview in Restany, *Plastics in Arts*, 159. In a statement in the same book Levine referred to the danger to liver and kidneys from catalysts used to harden polyester (133).

CHAPTER 8: MATERIAL DOUBTS
1   "How Your World Will Change," *New York Times* (May 26, 1968): sec. 12, 3. On "Plastics 100" events see *SPI Annual Report* for 1967–1968, 26, SPIF.
2   Stuart Siegel, "A Note from the Publisher," *MP* 56 (January 1979): 3.
3   Ruth Carson, "Plastic Age," *Collier's* 120 (July 19, 1947): 49–50.
4   See Spencer R. Weart, *Nuclear Fear: A History of Images* (Cambridge: Harvard University Press, 1988), 183–214; Allan M. Winkler, *Life Under a Cloud: American Anxiety About the Atom* (New York: Oxford University Press, 1993), 84–108. While Weart suggests that people displaced fear of nuclear war onto the more manageable issue of nuclear fallout, I am suggesting that some people also displaced a general nuclear fear onto plastic.
5   Mailer interviewed by Arlene Francis on New York radio station WOR, as quoted by Sidney Gross, "Nuts" (editorial), *MP* 48 (March 1971): 51; Paul H. Weaver, "On the Horns of the Vinyl Chloride Dilemma," *Fortune* 90 (October 1974): 150.
6   Norman Mailer, "The Big Bite," *Esquire*, April, May 1963; reprinted in Mailer, *The Presidential Papers* (New York: Berkley Medallion, 1970), 159, 178–179.
7   Norman Mailer, "In the Red Light: A History of the Republican Convention in 1964," reprinted in Mailer, *The Idol and the Octopus* (New York: Dell, 1968), 177; idem, *The Armies of the Night: History as a Novel/The Novel as History* (1968; rpt., New York: Signet, 1968), 135; and idem, *St. George and the Godfather* (1972), reprinted in idem, *Some Honorable Men: Political Conventions 1960–1972* (Boston: Little, Brown, 1976), 349.
8   Norman Mailer, *Of a Fire on the Moon* (Boston: Little, Brown, 1970), 130, 186.
9   Ibid., 141–142, 316.
10   John Kerouac, *The Town and the City* (New York: Harcourt, Brace, 1950), 369–373; Allen Ginsberg, *Howl and Other Poems* (San Francisco: City Lights Books, 1956), 9–10; and Gary Snyder, "LMFBR," *Turtle Island* (New York: New Directions, 1974), 67.
11   Mastroianni as quoted by Robert Jay Lifton, "Protean Man," in *The Psychoanalytic Interpretation of History*, ed. Benjamin B. Wolman (New York: Basic Books, 1971), 40; the essay first appeared in *Partisan Review* 35 (Winter 1968): 13–27.
12   Philip Smith, "She's Allergic to Modern Living," *National Enquirer* (February 7, 1984): 3; G. S. Wiberg, "Consumer Hazards of Plastic," *Environmental Health Perspectives* 17 (October 1976): 221, 224–225. See also Sidney Gross, "VCM—What Now," *MP* 51 (November 1974): 49; "Answering the Critics," *MP* 57 (January 1980): 52; and "Nationwide Ban for Urea-Formaldehyde," *MP* 59 (April 1982): 24.
13   Quotations are from "Fluorocarbons Move into Consumer Goods—In Volume," *MP* 42 (February 1965): 88–89, 142, 144; Gilbert M. Miller, "Talk for Footwear Marketing Conference," June 19, 1968, typescript, DPA, series II, part 2, box 76. See also "The Demonstration at Madame Romaine's Place," *The Journal of Teflon* 2 (April 1961):

6; "Out of the Fat, Onto the Fryer," *Du Pont Magazine* 55 (November-December 1961): 24–25; R. L. Van Boskirk, "The Plastiscope," *MP* 42 (May 1965): 220; Philip C. Hauck, "Du Pont's Teflon," *Wall Street Journal* (May 15, 1967): 1; R. W. Vaughn, "Footwear Marketing Conference," June 19, 1968, typescript, DPA, series II, part 2, box 76; "'SilverStone': A Demonstrated Success," *Du Pont Magazine* 71 (November-December 1977): 1; and David A. Hounshell and John Kenly Smith, Jr., *Science and Corporate Strategy: Du Pont R&D, 1902–1980* (Cambridge: Cambridge University Press, 1988), 485–486.

[14]   "Plastics: Into the Fire?," *Newsweek* 59 (April 16, 1962): 80.

[15]   John A. Zapp, Jr., *The Anatomy of a Rumor* (Wilmington, Del.: E. I. du Pont de Nemours & Co., 1962). See also "Out of the Fat, Onto the Fryer," 24; "The Anatomy of a Rumor," *Du Pont Magazine* 56 (May-June 1962): 32; and Hounshell and Smith, *Science and Corporate Strategy*, 566–569.

[16]   On reactions to strontium-90 in milk see Weart, *Nuclear Fear*, 213–214.

[17]   Paul Sargent Clark, "The Coming (Any Minute) Revolution for Plastics," *Industrial Design* 14 (October 1967): 65; Walter McQuade, "Encasement Lies in Wait for All of Us," *Architectural Forum* 127 (November 1967): 92; and "S.F. Baby Killed By Plastic Bag," *San Francisco News*, May 20, 1959, clipping scrapbook, SPIF.

[18]   "This Bag Spells Business," *Du Pont Magazine* 50 (February-March 1956): 24.

[19]   Sales figures are from an Associated Press story, "Ad Campaign to Warn of Plastic Bag Danger," appearing among other places in the *Wilkes-Barre Times-Leader-News*, June 18, 1959, clipping scrapbook, SPIF.

[20]   "This Bag Spells Business," 25.

[21]   Paul B. Jarrett as quoted by the AMA in a passage quoted in a brief history of the "Polyethylene Bag Problem" inserted in SPI board minutes, July 23–24, 1959, 28, SPIA, microfilm reel 3.

[22]   Mildred Murphy, "Plastic Industry to Warn on Bags," *New York Times* (June 18, 1959): 33; "79 Infant[s] Have Died in Plastic Bags," *New Kensington* (Penn.) *Dispatch*, July 24, 1959, clipping scrapbook, SPIF (this UPI story appeared in dozens of newspapers).

[23]   Du Pont spokesman as paraphrased by Morris Kaplan, "Industry Warns on Plastic Bags," *New York Times* (June 4, 1959): 33; Hiram McCann, "Hazards in Film Misuse Must Be Taught Parents," *MP* 36 (June 1959): 262; "This Bag Spells Business," 25; and "Dry Cleaning: Big New Market for Film," *MP* 35 (April 1958): 106–107.

[24]   "Warning: 28 Died Like This," *Toronto Telegram* (May 25, 1959): 41, SPIA, dry cleaning bag scrapbook no. 3; "S.F. Baby Killed By Plastic Bag"; "Plastic Sheeting Claims 2 Lives," *New York Journal*, June 12, 1959, clipping scrapbook, SPIF; Cruse as quoted by William M. Freeman, "Producers Shape Plastic-Bag Code," *New York Times* (July 29, 1959): 60; "Of Plastic Bags . . . And Our Future," *Redwood City Tribune*, July 6, 1959, SPIA, dry cleaning bag scrapbook no. 3; and "How Many Must Die?," *San Francisco News*, May 20, 1959, clipping scrapbook, SPIF. Only a few months earlier, in March 1959, the public reacted with anxiety but considerably less hysteria to a *Consumer Reports* article on strontium-90 in milk. See Norman Isaac Silber, *Test and Protest: The Influence of Consumers Union* (New York: Holmes & Meier, 1983), 108–112; Winkler, *Life Under a Cloud*, 102–103.

[25]   "Thin Bag of Death," *Life* 46 (June 8, 1959): 117. For other photographs see

"Wave of Publicity Warns Parents of Plastic Bag Danger," *Santa Ana* (Calif.) *Register*, June 7, 1959, SPIA, dry cleaning bag scrapbook no. 3; and "Warning: 28 Died Like This."

26   SPI's general counsel Jerome H. Heckman as quoted by Robert Leaversuch, "The SPI at 50: Looking into an Exciting, Uncertain Future," *MP* 64 (September 1987): 50–51.

27   SPI press release, May 7, 1959, SPIA, microfilm reel 1; "Polyethylene Film Death Toll Mounts—Education Apparent Solution," *SPI Bulletin*, no. 710 (May 14, 1959): 1–2, SPIA.

28   SPI press release, May 29, 1959, SPIA, microfilm reel 1. The first quotation is a direct statement from W. G. Johnson of the National Safety Council. See also *SPI Bulletin*, no. 713 (June 1, 1959): 1–2, SPIA; SPI press release, June 8, 1959, SPIA, microfilm reel 1; and *SPI Annual Report 1960*, SPIF.

29   "BBDO Preparing Plastics Campaign," *Printers Ink* 267 (June 19, 1959): 7; "Society Sponsored Educational Program Launched . . . ," *SPI Bulletin*, no. 719 (June 24, 1959): 1–3, SPIA; Hiram McCann, "A Lesson in Public Relations," *MP* 36 (July 1959): 218; "Progress Report on Plastic Bag Education Program . . . ," *SPI Bulletin*, no. 723 (July 14, 1959): 1–3; and SPI board minutes, November 17–18, 1960, SPIA, microfilm reel 3.

30   Cruse as quoted in "Ad Campaign to Warn of Plastic Bag Danger," *Wilkes-Barre Times-Leader-News*, June 18, 1959, clipping scrapbook, SPIF. See also Mildred Murphy, "Plastic Bag Industry to Warn on Bags," *New York Times* (June 18, 1959): 33.

31   "An important message to parents about PLASTIC BAGS," *New York Times* (June 18, 1959): C21; McCann, "A Lesson in Public Relations," 218.

32   "Final Report on Plastics Bags Legislation," *SPI Bulletin*, no. 776, September 13, 1960, SPIA; quotations from 1. I am indebted to Jerome Heckman for a telephone interview on July 20, 1994; also see Leaversuch, "The SPI at 50," 50–51. See also "Educational Program Outlined at Meeting . . . ," *SPI Bulletin*, no. 726 (July 31, 1959): 3–4, SPIA.

33   Pinsky as quoted in "Death Toll Rises: Campaign Fails to Break Plastic Bag Menace," *New York World-Telegram and Sun*, January 11, 1960, SPIA, dry cleaning bag scrapbook no. 3; the article's pessimistic thrust was mistaken. See also "Standard for Polyethylene Film Expected Soon," *SPI Bulletin*, no. 726 (July 31, 1959): 1, SPIA.

34   Robert Martino, "Promote the Promotable," *MP* 62 (March 1985): 41; "Why Not Call It Plastics?," *MP* 41 (September 1963): 85; and "Let's Use the Word 'Plastics' with Pride!," *MP* 28 (February 1951): 5.

35   James Beck et al., *Linear Polyethylene and Polypropylene: Problems and Opportunities: A Report for Businessmen* (Kansas City: Polymer Associates, 1958), 109.

36   McQuade, "Encasement Lies in Wait for All of Us," 92.

37   In general see Miles Orvell, *The Real Thing: Imitation and Authenticity in American Culture, 1880–1940* (Chapel Hill: University of North Carolina Press, 1989).

38   "Vinyl Plants Are True to Life," *MP* 28 (June 1951): 94; actress and Margaret Mead as quoted by Sherwood Kohn, "The Flowering of Fake Flowers," *New York Times Magazine* (August 23, 1964): 54, 60. See also "Polyethylene Blossoms," *MP* 36 (November 1958): 139; "Plastic Plants and Flowers," *MP* 39 (April 1962): 94–97, 205–206.

39   "Pacific Coast Plastics Products," *MP* 34 (March 1957): 158, 281, 283.

40   "PVC Christmas Trees Ring Up Record Season," *MP* 46 (December 1969): 67;

shopper quoted by Michael C. Jensen, "Plastic Vies With Pines in Christmas-Tree Sales," *New York Times* (December 20, 1972): 63.

41   On metallicized trees see "Plastics in the Product Revolution: Christmas Decorations," *MP* 38 (December 1960): 89. For a wonderful tirade on the disappearance of "genuine" lead Christmas-tree icicles and their third-rate plastic substitutes see Russell Baker, "The Right Trim," *New York Times Magazine* (December 23, 1984): 10.

42   "Wall Covering," *MP* 32 (July 1955): 179; "'Stone' Facing," *MP* 35 (August 1958): 107; "Business Bulletin," *Wall Street Journal* (April 7, 1966): 1; "Plastics Marble: Cutting the Cost of Beauty," *MP* 41 (July 1964): 92–95, 154; A. Stuart Wood, "Plastics Marble: A Little-Known Market Takes Off," *MP* 47 (June 1970): 76–79; "Embossing Puts Woodgrain into Vinyl Siding," *MP* 48 (January 1971): 34, 36; and Robert Martino, "Vinyl Dresses Up for New Era in Siding," *MP* 53 (June 1976): 34–37.

43   Quoted from answers by designers Dave Chapman and Herbert V. Gosweiler, Jr., to a questionnaire circulated by Egmont Arens and included in his mimeographed "Permanence and Change: Inter-Society Color Council Report from the American Society of Industrial Designers," *ASID Newsletter Supplement*, no. 6 (August 1964): 3, 10. I am indebted to Harper Landell for making this available.

44   C. T. Ludwig, "Foam on the Rise," *Industrial Design* 20 (May 1973): 32–37 (quotation from 33); "Furniture, An Industry Custom-Made for Custom Molders, Finds Exciting Uses for Plastics," *MP* 43 (May 1966): 174–179, 294 (quotation from 175). See also "Woodgraining Brings New Dimension to Plastics Design," *MP* 44 (November 1966): 84–88, 155, 157, 160; "Plastics Are in Furniture to Stay," *MP* 45 (September 1967): 118–119, 246; "Now It's Vinyl Veneer for Furniture," *MP* 45 (December 1968): 81–83; Rita Reif, "Plastic That Looks Like Anything But," *New York Times* (October 22, 1969): 54; Stuart Wood, "Furniture 1970: Plastics Transforms a Tradition-Bound Industry," *MP* 46 (December 1969): 62–66; and "Furniture Is Coming Back—And Maybe Bigger Than Ever," *MP* 52 (August 1975): 52–53.

45   On furniture shows see "A Plastic Trend in Furniture's Future," *Business Week* (September 26, 1970): 112.

46   "Plastics: The Future Has Arrived," *Progressive Architecture* 51 (October 1970): 92.

47   "The New Excitement in Furniture," *MP* 45 (January 1968): 89.

48   Quotations are from Peter H. Prugh, "Against the Grain," *Wall Street Journal* (July 21, 1966): 1; "A Plastic Trend in Furniture's Future," 112.

49   "History You Can Hang on a Wall," *Du Pont Magazine* 64 (September-October 1970): 18–21.

50   Tom Wolfe, *The Electric Kool-Aid Acid Test* (1968; rpt., New York: Bantam, 1969), 28.

51   Marshall McLuhan, *Understanding Media: The Extensions of Man* (1964; rpt., New York: Signet, 1966), 33.

52   John Cheever, "The Brigadier and the Golf Widow," in *The Stories of John Cheever* (1978; rpt., New York: Ballantine, 1980), 599; John Updike, *Rabbit, Run* (1960; rpt., New York: Fawcett Crest, 1969), 10, 16, 83, 87, 188.

53   Detroit Artists Workshop, "White Panther Manifesto" (1968); Yippie broadside, New York City (ca. 1967); and "Acid-Armed Consciousness," broadside, New York City (ca. 1968); all reprinted in *Bamn: Outlaw Manifestos and Ephemera 1965–70*, ed. Peter Stansill and David Zane Mairowitz (Baltimore: Penguin, 1971), 176–177, 107, 161.

54   Leonard Wolf with Deborah Wolf, *Voices from the Love Generation* (Boston: Little,

Brown, 1968), 216; Lewis Yablonsky, *The Hippie Trip* (1968; rpt., Baltimore: Penguin, 1973), 97, 272. The word straight in 1968 commonly referred to a member of mainstream society; that is, to a non-hippie.

[55] Charles A. Reich, *The Greening of America* (1970; rpt., New York: Bantam, 1971), 6–7, 171–172, 193, 203, 408–409, 252, 430.

[56] Theodore Roszak, *Where the Wasteland Ends: Politics and Transcendence in Postindustrial Society* (Garden City, N.Y.: Doubleday, 1972), 14, 18–19, 63–64. Roszak admitted being influenced in this passage by Jean-Luc Godard's film *Alphaville* (1965); his vision of a crumbling plastic utopia also seems eerily prophetic of Terry Gilliam's film *Brazil* (1985) and of the disintegration of the Soviet Union (1991).

[57] Wolfe, *The Electric Kool-Aid Acid Test*, 261; idem, "What If He Is Right," in *The Pump House Gang* (1968; rpt., New York: Bantam, 1969), 107; idem, "Introduction" and "The Put-Together Girl," in *The Pump House Gang*, 1, 74; idem, *The Electric Kool-Aid Acid Test*, 182; and idem, *The Kandy-Kolored Tangerine-Flake Streamline Baby* (1965; rpt., New York: Pocket Books, 1970), 87.

[58] Ross Macdonald, *The Zebra-Striped Hearse* (1962; rpt., New York: Bantam, 1984), 38; Robert B. Parker, *God Save the Child* (Boston: Houghton Mifflin, 1974), 8; James Crumley, *The Wrong Case* (1975; rpt., New York: Vintage Contemporaries, 1986), 137, 15; Dick Francis, *Rat Race* (1971; rpt., New York: Pocket Books, 1978), 29; Simon Brett, *An Amateur Corpse* (1978; rpt., New York: Berkley, 1980), 152; and Robert Ludlum, *The Bourne Supremacy* (1986; rpt., New York: Bantam, 1987), 307.

[59] Robert Stone, *A Hall of Mirrors* (1967; rpt., Boston: Houghton Mifflin, 1981), 352–353; Ishmael Reed, *Mumbo Jumbo* (1972; rpt., New York: Avon, 1978), 70–71; Stanley Elkin, *The Franchiser* (1976; rpt., Boston: Nonpareil Books, 1980), 227; and Gilbert Sorrentino, *Mulligan Stew* (New York: Grove Press, 1979), 22.

[60] Swissair advertisement as quoted in "Answering the Critics," *MP* 57 (May 1980): 34; KitchenAid advertisement from *Good Housekeeping*, ca. December 1986; interviewee as quoted by Mihaly Csikszentmihalyi and Eugene Rochberg-Halton, *The Meaning of Things: Domestic Symbols and the Self* (Cambridge: Cambridge University Press, 1981), 198; "Elementary Students Fed Up with New Plastic Silverware," AP story in *The Daily Texan* (January 20, 1988): 3; L. E. Sissman, "Innocent Bystander: Plastic English," *The Atlantic Monthly* 230 (October 1972): 32; Raymond Loewy, *Industrial Design* (New York: Overlook Press, 1979), 36; and Sidney Gross interviewed by the author, December 9, 1986.

[61] Barry Commoner, *The Closing Circle: Nature, Man, and Technology* (New York: Alfred A. Knopf, 1971), 12, 15, 127, 162–164, 185, 12.

[62] Barry Commoner, *The Poverty of Power: Energy and the Economic Crisis* (New York: Alfred A. Knopf, 1976), esp. 198–210; quotations from 206–207.

[63] Joel Frados, "There's Something in the Air," *MP* 44 (October 1966): 89.

[64] On 1950s packaging and disposables see "One Portion at a Time," *MP* 29 (May 1952): 84–85; "Flip to Close," *MP* 33 (April 1956): 119; "Plastics Applications in the Years Ahead," *MP* 33 (June 1956): 171; R. L. Van Boskirk, "The Plastiscope," *MP* 34 (October 1956): 304; and idem, "The Plastiscope," *MP* 36 (September 1958): 39–41.

[65] On 1960s packaging see R. L. Van Boskirk, "The Plastiscope," *MP* 38 (May 1961): 43; "New Day for Thin-Wall Containers," *MP* 41 (May 1964): 84–88, 160, 162; "Coming Up—Multi-Million Lb. Market," *MP* 41 (December 1963): 93; "Packaging's Versatile Vessels," *Du Pont Magazine* 58 (January-February 1964): 12–13; "Designing Closures

for a Dual Purpose," *MP* 42 (September 1964): 116–119, 192, 194; R. L. Van Boskirk, "The Plastiscope," *MP* 43 (April 1966): 45, 47; "What's in Store for Polystyrene Meat Trays?," *MP* 44 (November 1966): 89–91; "Business Bulletin," *Wall Street Journal* (May 25, 1967): 1; "$65,000,000 Bet on Plastics-Paper Vs. Glass," *MP* 39 (March 1962): 87; "$65 Million Throw-Away!," *MP* 39 (June 1962): 88–91; and "Coming Market for Polyethylene: Milk Bottles," *MP* 42 (December 1964): 84–88, 162–166.

66 Wyeth's impatience with his lack of recognition as a chemist was reported by Jon Eklund, National Museum of American History, Smithsonian Institution, Washington, D.C.

67 Dominick V. Rosato, William K. Fallon, and Donald V. Rosato, *Markets for Plastics* (New York: Van Nostrand Reinhold, 1969), 6.

68 Both quotations, including direct quotation from Lloyd Stouffer, are from "Plastics in Disposables and Expendables," *MP* 34 (April 1957): 93.

69 Sidney Gross, "Garbage (2)," *MP* 47 (January 1970): 63; idem, "Garbage," *MP* 46 (April 1969): 81; and idem, "Garbage (4)," *MP* 48 (August 1971): 37.

70 "Nonreturnables Face Legislative Ban in Madison, Wisconsin," *Plastics and the Environment*, no. 1 (April 3, 1970): 2–4, SPIA. See also Julian Kestler, "Localities May Tax Plastics Packaging," *MP* 48 (May 1971): 14, 16.

71 Copy of letter from Ralph L. Harding, Jr., SPI executive vice president, to Executive Committee of SPI Environment Policy Committee, June 10, 1971, SPIA, looseleaf notebook entitled "Executive VP—Chronological or Reading File, 1971–77" (hereafter referred to as "VP Reading File"); "Plastics Return the Ecologists' Fire," *Business Week* (July 10, 1971): 25; Julian Kestler, "New York's Plastic Container Tax Poses Grave Threat to Plastics Packaging," *MP* 48 (August 1971): 10; and idem, "What Are the Implications of the Overturning of New York City's Tax Law?," *MP* 48 (December 1971): 18.

72 "Plastics in Canada," *MP* 44 (April 1967): 106; SPI Polyethylene Refuse Bag Committee minutes, October 31, November 17, 1967, February 2, 1968, March 28, 1969, SPIA, microfilm reel 9; "Plastics Refuse Bags Get Wide Test," *MP* 46 (September 1969): 228; "Refuse Bags: How Big, How Fast, How Profitable?," *MP* 47 (February 1970): 10, 12; "New York City Goes For Plastic Refuse Bags," *Plastics and the Environment*, no. 3 (June 29, 1970): 5–6, SPIA; and a copy of Ralph L. Harding, Jr., to Charles Luce, chairman, Consolidated Edison, February 11, 1971, SPIA, VP Reading File.

73 For typical discussions see A. Stuart Wood, "Plastics' Challenge in Packaging: Disposability," *MP* 47 (March 1970): 50–54; "A Plastic for Ecologists," *Time* 95 (May 11, 1970): 86; "Plastics—Mostly PVC—Under Attack in Press," *Plastics and the Environment*, no. 3 (June 29, 1970): 1–3, SPIA; "SPI Position Paper Outlines Industry's Stand on Coping with the Garbage Crisis," *MP* 47 (October 1970): 184; and F. Rodriguez, "Prospects for Biodegradable Plastics," *MP* 48 (September 1971): 92, 94. For technical discussions with similar conclusions see J.J.P. Staudinger, ed., *Plastics and the Environment* (London: Hutchinson, 1974).

74 Robert H. Wehrenberg II, "Plastics Recycling: Is It Now Commercially Feasible?," *Materials Engineering* 89 (March 1979): 34. For the lower estimate and a discussion of the difficulty of measuring volumes of garbage see William Rathje and Cullen Murphy, *Rubbish!: The Archaeology of Garbage* (New York: HarperCollins, 1992), 99–102. See also Robert D. Leaversuch, "Industry Begins to Face Up to the Crisis of Recycling," *MP* 64 (March 1987): 44–47; Cass Peterson, "Recycling: Making Cents of Trash," *Washington Post* Service, in *Austin American-Statesman* (June 14, 1987): D1, D13; Guy Darst, "Plastic

Bottles to Carry Code for Recycling" (AP), *Austin American-Statesman* (April 9, 1988): B11; Myra Klockenbrink, "Plastics Industry, Under Pressure, Begins to Invest in Recycling," *New York Times* (August 30, 1988): 19; "Recycling: Don't Trash That Foam," *Time* 133 (January 9, 1989): 48; Barbara Rudolph, "Second Life for Styrofoam," *Time* 133 (May 22, 1989): 84; SPI Council for Solid Waste Solutions, "The Urgent Need to Recycle" (advertising section), *Time* 134 (July 17, 1989): 17–28; and *Plastics Recycling: Problems and Possibilities*, Serial No. 102–63, 102nd Congress, Second Session, House Committee on Small Business, Subcommittee on Environment and Employment (Washington, D.C.: Government Printing Office, 1992).

[75] Boyce Rensberger, "Plastic Is Found in Sargasso Sea," *New York Times* (March 19, 1972): 68; John B. Colton, Jr., Frederick D. Knapp, and Bruce R. Burns, "Plastic Particles in Surface Waters of the Northwestern Atlantic," *Science* 185 (August 9, 1974): 491–497.

[76] Quotations are from Carpenter to Harding, April 20, 1972; Harding to A. W. Andrews of Monsanto Company (not necessarily the culprit), April 10, 1973. See also Harding to George Ingle of Monsanto Company, April 20, 1972; memorandum from Harding to U.S. polystyrene producers, "Polystyrene Contamination in Long Island Sound," June 2, 1972. All documents SPIA, VP Reading File.

[77] Christine Duerr, "Plastic Is Forever: Our Nondegradable Treasures," *Oceans* 13 (November 1980): 59–60; Robert D. Leaversuch, "Ocean Litter Becomes Public, Legislative Issues," *MP* 63 (October 1986): 10–11; and James Pinkerton, "International Ocean Pollution Treaty Approved," *Austin American-Statesman* (January 1, 1988): A17.

[78] Harding, memorandum to PAC Steering Committee on "PAC Strategy Ahead," August 16, 1972, SPIA, VP Reading File.

[79] For background on flammability see "Appliances—New Front for Attack on Plastics," *MP* 36 (April 1959): 246; "What About Flame Resistance in Plastics?," *MP* 37 (September 1959): 81–84, 176, 178; Joel Frados, "The Flammability Gap," *MP* 45 (May 1968): 77; and Stuart Wood, "Facing the Problems of Fire," *MP* 45 (June 1968): 82–86.

[80] Richard O. Simpson, Department of Commerce, to Ralph L. Harding, Jr., undated, received December 1, 1972; Harding to Simpson, December 18, 1972; and memorandum from Cleveland Lane of B. F. Goodrich Chemical Company to members of SPI Ad Hoc Public Affairs Committee on Flammability, December 21, 1972; all documents SPIA, VP Reading File. See also "F.T.C. Accuses Plastic Industry of Deception on Fire Hazards," *New York Times* (May 31, 1973): 1; Sidney Gross, "Why Now, FTC?," *MP* 50 (July 1973): 37.

[81] J. H. Petajan et al., "Extreme Toxicity from Combustion Products of a Fire-Retarded Polyurethane Foam," *Science* 187 (February 28, 1975): 742–744. Herman F. Mark discussed his work on flammability during a presentation at the University of Texas at Austin on June 19, 1989.

[82] U.S. Representative Richard L. Ottinger of New York as quoted by Frados, "The Flammability Gap," 77; Martin M. Brown of American International Group Companies and William J. Kelley of Frankel & Co. as quoted by C. Otis Port, "Fire!," *MP* 50 (August 1973): 37, 38.

[83] Sheldon Samuels as quoted by Janet H. Weinberg, "Toxic Surprises from the Plastics Industry," *Science News* 106 (September 7, 1974): 155; Sidney Gross interviewed by the author, December 9, 1986. See also Roland R. MacBride, "OSHA and Materials Tox-

icity: The Issue Grows, and the Impact Begins," *MP* 50 (August 1973): 16; Paul H. Weaver, "On the Horns of the Vinyl Chloride Dilemma," *Fortune* 90 (October 1974): 150–153, 200, 202–204; and "Of Mice and Men: Alarm Over Plastics," *Time* 104 (October 14, 1974): 64.

[84]   Heckman as quoted by David Burnham, "Ban Is Asked on Vinyl Chloride in Food Packages," *New York Times* (July 2, 1975): 18; Sidney Gross, "Nader's Nadir," *MP* 52 (August 1975): 37; and Caryl Roberts, "Plastics Foul Up in Outer Space," *Science Digest* 75 (June 1974): 33. On FDA and migration of vinyl chloride see also Roland R. Mac-Bride, "A Closer FDA Eye on Extraction," *MP* 51 (June 1974): 22; Sidney Gross, "VCM—What Now," *MP* 51 (November 1974): 49; and papers on *Safety and Health with Plastics* (n.p.: Society of Plastics Engineers, 1977). The phasing-out of Du Pont's slogan is mentioned by Graham D. Taylor and Patricia E. Sudnik, *Du Pont and the International Chemical Industry* (Boston: Twayne, 1984), xx. With the exception of the final phrase— "Through Chemistry"—the slogan was later reinstated.

[85]   As quoted by Stephen J. Sansweet, "Plastics Firms Face Materials Shortages Believed Likely to Last at Least 2 Years," *Wall Street Journal* (September 11, 1973): 38; Kline paraphrased by Roland R. MacBride, "More Feedstocks?," *MP* 51 (June 1974): 20; and Sidney Gross, ".500 for the Hill," *MP* 51 (July 1974): 39. For other views of the immediate crisis see Julian Kestler, "Plastics and the Energy Crisis: Big Changes Ahead in Price and Availability," *MP* 49 (January 1972): 14, 16; "Plastics Producers Sort Out the Oil Shortage," *Business Week* (January 19, 1974): 84, 86. Kline retired from his position as technical editor in 1991.

[86]   Mike Kenward, "Better Than Burning It," *Design*, no. 318 (June 1975): 34. See also Sidney Gross, "Challenges of the Eighties," *MP* 57 (January 1980): 55; idem, "Critical Buzzwords," *MP* 57 (October 1980): 41.

[87]   Robert C. Lohnes, "Back to Wood, Back to Paper," *New York Times* (June 6, 1979): 27.

[88]   Sidney Gross interviewed by the author, December 9, 1986.

[89]   Quoted by John J. Keville, "Third Annual Salute to the Plastics Industry, January 12–February 20, 1981," radio script for WLMS, Leominster, Massachusetts, 47, provided by John J. Keville.

[90]   Christopher Flavin, *The Future of Synthetic Materials: The Petroleum Connection*, Worldwatch Paper 36 (Washington: Worldwatch Institute, April 1980), 41, 7, 37–38, 46.

[91]   Richard D. Lyons, "Study Finds Synthetics a Benefit To Man Despite Petroleum Use," *New York Times* (April 27, 1980): 20. See also Flavin's own summary, "Synthetic Materials' Future," *New York Times* (May 12, 1980): 19.

[92]   Sidney Gross, "What Next for the Plastics Industry," *MP* 50 (October 1973): 61; idem, "Milestone," *MP* 54 (October 1977): 37.

[93]   "New Chicago Municipal Incinerator Handles Plastics Wastes Without Problems," *MP* 48 (February 1971): 92; Julian Kestler, "Results of SPI Incineration Study," *MP* 48 (August 1971): 10, 12, 14.

[94]   Harding to Hugh H. Connolly, Bureau of Solid Waste Management, Environmental Control Administration, HEW, April 12, 1971, SPIA, VP Reading File.

[95]   Harding, "Personal and Confidential" memorandum, June 15, 1971, SPIA, VP Reading File.

[96]   Harding, "Challenges Facing the Plastics Industry," typescript of talk, a copy of

which was enclosed with a letter from Laura E. Jones to Alex D. Kischitz, Hooker Chemical Corporation, December 8, 1971, SPIA, VP Reading File.

97   "Plastics Council Meeting Summary," November 4, 1971; see also memorandum from Harding to Plastics/Packaging Council, July 20, 1971; both documents from SPIA, VP Reading File.

98   Quotations from E. S. Nuspliger of SPI to Harding, memorandum summarizing PAC's activities, April 11, 1973, SPIA, VP Reading File. See also "Financial Statements and Supplementary Information, Year Ended May 31, 1972," SPIA, carton labeled "Financial Statements 1938–1972"; and "Summary: Planning Committee Meeting," March 29, 1973; "Proposed SPI Public Affairs Organization," April 11, 1973; both documents SPIA, VP Reading File. I am indebted to Jerome H. Heckman for a telephone interview, July 20, 1994.

99   Ralph L. Harding, Jr., "Positive Perspective," typescript, August 7, 1972, SPIA, VP Reading File. Harding based his comments on two polls conducted for SPI by Research Strategies Corporation of New York and Opinion Research Corporation of Princeton (in collaboration) in 1971 and 1972. A larger, more complicated poll in 1975 came to similar conclusions regarding the impact on the general public of the antiplastic attitudes of "community leaders." See Walter G. Barlow and Joseph R. Goeke, "Plastics and the Environment: An Interrelated Group of Pilot Surveys Conducted for the Public Relations Committee of the Plastics Council," Fall 1971; Walter G. Barlow and Joseph R. Goeke, "The General Public Appraises Plastics and the Plastics Industry: A Nationwide Survey Conducted for the Public Affairs Council of the Society of the Plastics Industry," 1972; and "Members of Special Public Groups Evaluate Plastics and Their Impact on the Environment: A Nationwide Study Conducted for the SPI, Inc.," ORC Study No. 62046, December 1975; all documents SPIA, miscellaneous carton.

100   Sidney Gross, "Image, Please," *MP* 46 (August 1969): 43; "What *Is* Plastics' Image, Anyway," *MP* 47 (July 1970): 66.

101   Monsanto used the word "chemophobia" when beginning a $4 million public relations campaign in 1978. See Nick Fountas, "Tired of the Flak the Industry Talks Back," *Plastics World* 36 (December 1978): 39.

102   I am indebted to John J. Beer, University of Delaware, for providing his notes for the meeting, December 26, 1973, and a copy of a memorandum from Melvin Kranzberg, president of the Society for the History of Technology, to members of the editorial board of *Technology and Culture*, November 29, 1973, by which the meeting was organized.

103   Bob Martino, "Contempt Is Not a Strategy," *MP* 62 (November 1985): 43; idem, "Industry? What Industry?," *MP* 64 (September 1987): 45.

104   Sidney Gross interviewed by the author, December 9, 1986. The quotation is from Sidney Gross, "Plastics Age Arrives Early," *MP* 57 (February 1980): 45.

105   Kit Pedler and Gerry Davis, *Mutant 59: The Plastic-Eaters* (New York: Viking, 1972), 109.

CHAPTER 9: BEYOND PLASTIC

1   Umberto Eco, *Travels in Hyperreality*, trans. William Weaver (1975; rpt., New York: Harcourt Brace Jovanovich, 1986), 57; Jean Baudrillard, *Simulations*, trans. Paul Foss, Paul Patton, and Philip Beitchman (New York: Semiotext[e], 1983), 91–92; idem, *America*, trans. Chris Turner (1986; rpt., London: Verso, 1989), 29–30; and Michael

Talbot, "The Universe As Hologram," *The Village Voice* 32 (September 22, 1987): 33, 34. I observed the Epcot motto in May 1987.

2   Daniel J. Boorstin, *The Image: A Guide to Pseudo-Events in America* (1962; rpt., New York: Atheneum, 1971; originally published as *The Image or What Happened to the American Dream*), viii, title page, 57, 14, 5, 253, 36, 256, 181–182.

3   Charles W. Moore, "You Have To Pay for the Public Life," *Perspecta 9/10* (1965): 65. Disney's supposed preservation was familiar as urban folklore before I found it noted in Baudrillard, *Simulations*, 24.

4   Moore, 65; William Irwin Thompson, *At the Edge of History: Speculations on the Transformation of Culture* (1971; rpt., New York: Harper & Row, 1972), 22, 12. See also Nathanael West, *The Day of the Locust* (1939; rpt., New York: New Directions, 1969), 130–135.

5   Joseph Morgenstern, "What Hath Disney Wrought?," *Newsweek* 78 (October 18, 1971): 38. On "purified" used to imply diminishment and reduction, see Kimberly Dovey, "The Quest for Authenticity and the Replication of Environmental Meaning," in *Dwelling, Place and Environment: Towards a Phenomenology of Person and World*, ed. David Seamon and Robert Mugerauer (Dordrecht: Martinus Nijhoff, 1985), 41.

6   Tom Shales, "The ReDecade," *Esquire* 105 (March 1986): 68, 70; Michael Sorkin, "See You in Disneyland," in *Variations on a Theme Park: The New American City and the End of Public Space*, ed. Michael Sorkin (New York: Hill and Wang, 1992), 208. On pastiche see Fredric Jameson, *Postmodernism, or, The Cultural Logic of Late Capitalism* (Durham, N.C.: Duke University Press, 1991), 16–25.

7   Philip Langdon, *Orange Roofs, Golden Arches: The Architecture of American Chain Restaurants* (New York: Alfred A. Knopf, 1986), 159, 157.

8   David Lowenthal, "The American Scene," *The Geographical Review* 58 (January 1968): 78, 76. On American culture's endless attempts to preserve, reconstruct, or fabricate the past see idem, *The Past Is a Foreign Country* (Cambridge: Cambridge University Press, 1985).

9   Robert B. Riley, "Speculations on the New American Landscapes," *Landscape* 24, no. 3 (1980): 6. See also John Pastier, "The Architecture of Escapism," *AIA Journal* 67 (December 1978): 26–37.

10   William Severini Kowinski, *The Malling of America: An Inside Look at the Great Consumer Paradise* (New York: William Morrow, 1985), 25, 49, 270, 145–146, 339, 63.

11   Shopper as quoted by Kowinski, ibid., 151; Anna McLean Meikle, the author's grandmother, in casual conversation, May 1987; and Roland R. MacBride, "In Fire Safety, the Future Is Now," *MP* 50 (November 1973): 65. On the Disney parks as "plastic" see Charles A. Reich, *The Greening of America* (1970; rpt., New York: Bantam, 1971), 184; Margaret King, "The New American Muse: Notes on the Amusement/Theme Park," *Journal of Popular Culture* 15 (Summer 1981): 59.

12   Richard V. Francaviglia, "Main Street U.S.A.: A Comparison/Contrast of Streetscapes in Disneyland and Walt Disney World," *Journal of Popular Culture* 15 (Summer 1981): 152; Eco, *Travels in Hyperreality*, 46. Information on construction techniques is from "Construction of Rock Formations and Building Facades," press release no. 0487, provided in May 1987 by EPCOT Outreach Resource Service, CommuniCore West, EPCOT Center.

13   Edward W. Soja, "Inside Exopolis: Scenes from Orange County," in Sorkin, *Varia-*

tions on a Theme Park, 101. My use of "foregrounded" is borrowed from Gerhard Hoff-
mann, "The Foregrounded Situation: New Narrative Strategies in Postmodern American
Fiction," in *The American Identity: Fusion and Fragmentation*, ed. Rob Kroes (Amsterdam:
Amerika Instituut, Universiteit van Amsterdam, 1980), 289–344.

[14]   I am indebted to a letter and accompanying materials from Max Imgrueth, president of SWATCH Watch, U.S.A., Inc., June 25, 1985.

[15]   Suzanne Slesin, "Fake Stone Finishes on a Realistic Budget," *New York Times*
(July 16, 1981): C1, C6. See also Robert Jensen and Patricia Conway, *Ornamentalism:
The New Decorativeness in Architecture & Design* (New York: Clarkson N. Potter, 1982).

[16]   Patricia Cornelison, class essay, University of Texas at Austin, May 1984.

[17]   Edward Relph, *Place and Placelessness* (London: Pion, 1976), 93, 133–135.

[18]   Ron Loewinsohn, "Looking for Love After Marriage," *New York Times Book Review*
(September 30, 1984): 1, 43.

[19]   Reich, *The Greening of America*, 219–220.

[20]   David Riesman, Nathan Glazer, and Reuel Denney, *The Lonely Crowd: A Study in
the Changing American Character* (1950; rpt., Garden City, N.Y.: Doubleday, 1955), 23,
31, 42, 65, 349.

[21]   John Barth, *The End of the Road* (1958; rpt., New York: Bantam, 1969), 120, 89,
88; Jerzy Kosinski, *Being There* (1971; rpt., New York: Bantam, 1972), 33, 5.

[22]   Garry Wills, *Nixon Agonistes: The Crisis of the Self-Made Man* (Boston: Houghton
Mifflin, 1970), 9, 162–163, 406; Irving Howe, "Melting Down the Plastic Man," *Dissent*
21 (Summer 1974): 365–366.

[23]   Wills, *Nixon Agonistes*, 406.

[24]   Gail Sheehy, *Passages: Predictable Crises of Adult Life* (New York: E. P. Dutton,
1976), 20; Tom Wolfe, "The Me Decade and the Third Great Awakening," *New York*,
August 23, 1976, reprinted in Wolfe, *Mauve Gloves & Madmen, Clutter & Vine* (New
York: Farrar, Straus and Giroux, 1976), 143.

[25]   Robert Jay Lifton, "Protean Man," in *The Psychoanalytic Interpretation of History*,
ed. Benjamin B. Wolman (New York: Basic Books, 1971), 33, 42, 36–37; the essay first
appeared in *Partisan Review* 35 (Winter 1968): 13–27. Lifton updated his insight and
extended its interpretation, relying on individual case studies, on literary analysis, and
on a sensitive reading of recent history and contemporary events, in *The Protean Self:
Human Resilience in an Age of Fragmentation* (New York: Basic Books, 1993). See also
Reich, *The Greening of America*, 406; Thompson, *At the Edge of History*, 19.

[26]   Walter Benjamin, "The Work of Art in the Age of Mechanical Reproduction"
(1936), reprinted in Benjamin, *Illuminations* (New York: Harcourt, Brace & World,
1968), 223, 225; Jean-François Lyotard, *The Postmodern Condition: A Report on Knowledge* (1979; trans., Minneapolis: University of Minnesota Press, 1984), 51.

[27]   Charles Jencks, *The Language of Post-Modern Architecture*, rev. ed. (New York: Rizzoli, 1977), 7, 95, 127.

[28]   Christopher Lasch, *The Culture of Narcissism: American Life in an Age of Diminishing Expectations* (1978; rpt., New York: Warner, 1980), 91.

[29]   The phrase, which defines technology in general, is from Philip Slater, *The Pursuit
of Loneliness: American Culture at the Breaking Point*, rev. ed. (1970; rpt., Boston: Beacon
Press, 1976), 58.

[30]   Edward Mendelson, "Gravity's Encyclopedia," in *Mindful Pleasures: Essays on*

Thomas Pynchon, ed. George Levine and David Leverenz (Boston: Little, Brown, 1976), 161–165.

[31]   Roland Barthes as quoted by Gore Vidal, "American Plastic: The Matter of Fiction," *New York Review of Books* (July 15, 1976): 38; Roland Barthes, "Plastic," reprinted in Barthes, *Mythologies*, trans. Annette Lavers (London: Paladin, 1980), 97–98.

[32]   Rather than following current practice of assuming the text as an independent entity, I retain the convention of the author as creator of the text. This decision renders more obvious my (perhaps mistaken) projections of authorial intent and yields a less contorted prose. Page references in parentheses are to Thomas Pynchon, *Gravity's Rainbow* (New York: Viking, 1973). An excellent page-by-page guide is Steven Weisenburger, *A Gravity's Rainbow Companion: Sources and Contexts for Pynchon's Novel* (Athens: University of Georgia Press, 1988). I am developing ideas first explored in my essay "The Culture of Plasticity: Observations on Contemporary Cultural Transformation," *Amerikastudien* 28 (1983): 205–218. Since then I have found the following helpful on Pynchon and plastic: Joel D. Black, "Probing a Post-Romantic Paleontology: Thomas Pynchon's *Gravity's Rainbow*," *Boundary 2* 8 (Winter 1980): 229–254; Michael Vannoy Adams, "The Benzene Uroboros: Plastic and Catastrophe in *Gravity's Rainbow*," in *Spring 1981* (Dallas: Spring Publications, 1981), 149–161; Frederick R. Karl, *American Fictions 1940–1980* (New York: Harper & Row, 1983), 451–453; Stephen de Paul, "Plastic and the Parabola: The New Vocabulary of Attention in Pynchon's *Gravity's Rainbow*," *The CEA Critic* 49 (Winter 1986–Summer 1987): 180–184; and Robert L. McLaughlin, "IG Farben and the War Against Nature in *Gravity's Rainbow*," in *Germany and German Thought in American Literature and Cultural Criticism*, ed. Peter Freese (Essen: Die Blaue Eule, 1990), 319–336.

[33]   The name also indulges a pun: Imipolex = *imitation pole*.

[34]   Ruth Carson, "Plastic Age," *Collier's* 120 (July 19, 1947): 50.

[35]   Ellipsis in original.

[36]   Pynchon, *The Crying of Lot 49* (1966; rpt., New York: Bantam, 1967), 134.

[37]   On decentering see Jacques Derrida, "Structure, Sign and Play in the Discourse of the Human Sciences," in Derrida, *Writing and Difference* (Chicago: University of Chicago Press, 1978), 278–293; on paradigm shift see Thomas S. Kuhn, *The Structure of Scientific Revolutions* (Chicago: University of Chicago Press, 1962).

[38]   Karl Marx, *The Communist Manifesto* (1848; rpt., Baltimore: Penguin, 1967), 83. See the interpretation by Marshall Berman, *All That Is Solid Melts Into Air: The Experience of Modernity* (New York: Simon and Schuster, 1982), esp. 99.

[39]   Ezio Manzini, *The Material of Invention*, trans. Antony Shugaar (Milan: Arcadia, 1986), 25, 31, 37, 29, 45; Marco Diani, "Immateriality Takes Command," *Design Issues* 4, no. 1/2 (1988): 6–11.

[40]   "Computers Manufacture Humans?," Associated Press story, *The Daily Texan*, undated clipping, ca. autumn 1985; Philip Elmer-Dewitt, "Through the 3-D Looking Glass," *Time* 133 (May 1, 1989): 66. On virtual reality see Michael Benedikt, ed., *Cyberspace: First Steps* (Cambridge: MIT Press, 1991); Howard Rheingold, *Virtual Reality* (New York: Summit, 1991); and Benjamin Woolley, *Virtual Worlds: A Journey in Hype and Hyperreality* (Oxford: Blackwell, 1992).

[41]   William Gibson, *Neuromancer* (New York: Ace, 1984), 51; concluding ellipsis in original.

42   Gibson, *Neuromancer*, 6, 9, 4, 91, 5; Pynchon, *Gravity's Rainbow*, 699.

43   Michael Benedikt, "Cyberspace: Some Proposals," in Benedikt, *Cyberspace*, 124, 161, 160. James referred to "the total push and pressure of the cosmos" and described reality as "something resisting, yet malleable." See William James, *Pragmatism* (New York: Longmans, Green, 1907), 4, 258.

INDEX

Note: Page numbers in italics indicate illustrations.

About the Author

Jeffrey L. Meikle is a professor of American studies and art history at the University of Texas at Austin. He is the author of *Twentieth Century Limited: Industrial Design in America, 1925–1939*.